MICROBIAL BIOTECHNOLOGY IN FOOD PROCESSING AND HEALTH

Advances, Challenges, and Potential

MICROBIAL BIOTECHNOLOGY IN FOOD PROCESSING AND HEALTH

Advances, Challenges, and Potential

Edited by

Deepak Kumar Verma

Ami R. Patel, PhD

Sudhanshu Billoria, PhD

Geetanjali Kaushik, PhD

Maninder Kaur, PhD

AAP | APPLE ACADEMIC PRESS

First edition published 2023

Apple Academic Press Inc.
1265 Goldenrod Circle, NE,
Palm Bay, FL 32905 USA

760 Laurentian Drive, Unit 19,
Burlington, ON L7N 0A4, CANADA

CRC Press
6000 Broken Sound Parkway NW,
Suite 300, Boca Raton, FL 33487-2742 USA

4 Park Square, Milton Park,
Abingdon, Oxon, OX14 4RN UK

Library and Archives Canada Cataloguing in Publication

Title: Microbial biotechnology in food processing and health : advances, challenges, and potential / edited by Deepak Kumar Verma, Ami R. Patel, PhD, Sudhanshu Billoria, PhD, Geetanjali Kaushik, PhD, Maninder Kaur, PhD.
Names: Verma, Deepak Kumar, 1986- editor. | Patel, Ami R., editor. | Billoria, Sudhanshu, editor. | Kaushik, Geetanjali, editor. | Kaur, Maninder, editor.
Description: First edition. | Includes bibliographical references and index.
Identifiers: Canadiana (print) 20220204993 | Canadiana (ebook) 20220205051 | ISBN 9781774637289 (hardcover) | ISBN 9781774637432 (softcover) | ISBN 9781003277415 (ebook)
Subjects: LCSH: Probiotics. | LCSH: Microbial metabolites. | LCSH: Functional foods. | LCSH: Microbial biotechnology. | LCSH: Food industry and trade.
Classification: LCC RM666.P835 M53 2023 | DDC 615.3/29—dc23

Library of Congress Cataloging-in-Publication Data

CIP data on file with US Library of Congress

ISBN: 978-1-77463-728-9 (hbk)
ISBN: 978-1-77463-743-2 (pbk)
ISBN: 978-1-00327-741-5 (ebk)

Dedication

This book is dedicated to the eminent Professor Hari Niwas Mishra, of the AgFE Department, IIT Kharagpur, West Bengal, India.

Dr. Mishra has made wonderful and remarkable contributions to the growth of knowledge and research in the field of food science and technology with his vast experience in the administration and management of industrial-scale collaborative research projects ranging from new processes, product development, and also creation of pilot-scale facilities. He focuses on developing, enhancing, and commercializing food technologies by collaborating with industry to promote product diversification and enhancement, increase market growth, and aid competitiveness.

About the Editors

Deepak Kumar Verma, PhD
Department of Agricultural and Food Engineering, Indian Institute of Technology Kharagpur, West Bengal, India

Deepak Kumar Verma, PhD, is an agricultural science professional and has received PhD degree with specialization in Food Processing Engineering from Agricultural and Food Engineering Department, Indian Institute of Technology Kharagpur (West Bengal), India. In 2012, he received a DST-INSPIRE Fellowship for PhD study by the Department of Science and Technology (DST), Ministry of Science and Technology, Government of India. Mr. Verma is currently assigned for research on "Aroma Volatile and Flavoring Compounds from Indian Rice Cultivars" whereas during his master's degree, his research was on "Physicochemical and Cooking Characteristics of Azad Basmati (CSAR 839-3): A Newly Evolved Variety of Basmati Rice (*Oryza sativa* L.)." He earned his BSc degree in agricultural science in 2009 from the Faculty of Agriculture, Gorakhpur University, Gorakhpur, and MSc (Agriculture) in Agricultural Biochemistry in 2011 with the first rank and also received a department topper award from the Department of Agricultural Biochemistry, Chandra Shekhar Azad University of Agricultural and Technology, Kanpur, India. Apart from his area of specialization in plant biochemistry, he has also built-up a sound background in plant physiology, microbiology, plant pathology, genetics, and plant breeding, plant biotechnology, and genetic engineering, seed science and technology, food science and technology, etc. In addition, he is a member of various professional bodies, and his activities and accomplishments include conferences, seminars, workshops, training, and also the publication of research articles, books, and book chapters.

Ami R. Patel, PhD
Assistant Professor, Division of Dairy and Food Microbiology, Mansinhbhai Institute of Dairy & Food Technology, Dudhsagar Dairy Campus, Gujarat, India

Ami R. Patel, PhD, is an Assistant Professor in the Division of Dairy and Food Microbiology, Mansinhbhai Institute of Dairy & Food Technology-MIDFT, Dudhsagar Dairy Campus, Gujarat, India. He has expertise in specialized areas that involve isolation, screening, and characterization of exopolysaccharides from potential probiotic cultures and employing them for food and health applications. In addition, she is engaged with teaching undergraduate, postgraduate, and research students in food microbiology, microbial biotechnology, food biotechnology, food science and technology, clinical microbiology and immunology, etc. She has authored peer-reviewed papers and technical articles in international and national journals as well as book chapters, books, proceedings, and technical bulletins. She has received awards and honors, including the Vice Chancellor Gold Medal for her PhD work, selection as a BiovisinNxt11 fellow to attend an international conference at Lyon, France, and an Erasmus Mundus scholarship from the European Union for three years to visit Lund University, Sweden, as a guest researcher. She is also serving as an expert reviewer for several scientific journals. She is a member of several academic and professional organizations, including the Indian Dairy Association and the Swedish South Asian Network for Fermented Foods.

Dr. Patel earned her BSc (Microbiology) and her MSc (Microbiology) degree from Sardar Patel University, Gujarat, and received her doctorate degree in Dairy Microbiology from the Dairy Department of SMC College of Dairy Science, Anand Agricultural University, Gujarat, India.

Sudhanshu Billoria, PhD
Assistant Professor of Production (Food Processing), Center for Research, Consultancy and Publication, Vaikunth Mehta National Institute of Cooperative Management (VAMNICOM), Pune, Maharashtra, India

Sudhanshu Billoria, PhD, is an Assistant Professor of Production (Food Processing) at the Center for Research, Consultancy and Publication, Vaikunth Mehta National Institute of Cooperative Management (VAMNICOM), Pune, Maharashtra, India, where she teaches various postgraduate-level courses and guides research projects. Her research interests include postharvest management of fresh produce, mainly fruits and vegetables, by extending shelf life. She utilizes her expertise for addressing value-chain and supply chain management and industrial cooperatives such as dairy, sugar, and fruit and vegetable cooperatives. She has published many research papers in peer-reviewed international journals and conference proceedings along with many book chapters. She is a member of various professional bodies and has attended, presented papers, and delivered invited lectures at workshops, conferences, and training programs.

Dr. Billoria earned an MSc degree in Food Science and Technology from the University of Jammu, Jammu and Kashmir, India, and her PhD in Food Process Engineering from the Agricultural and Food Engineering Department, Indian Institute of Technology Kharagpur, wherein her research work was "Shelf-Life Extension of Light Red Color Tomatoes (Lycopersicon esculentum cv. Vaishali) with the Help of Active Packaging and Edible Coating Technologies." Her master's project work was entitled "Standardization and Preparation of Ghee based Butter."

Geetanjali Kaushik, PhD
Associate Professor, Department of Civil Engineering and Head of R&D, Hi Tech Institute of Technology, Waluj, Maharashtra, India

Geetanjali Kaushik, PhD, is an Associate Professor in Environment at Hi Tech Institute of Technology Waluj Maharashtra. Dr. Geetanjali Kaushik completed her PhD in Pesticide residues in Food, from the prestigious Indian Institute of Technology Delhi with more than 75 International papers and chapters in Journals and Books of repute. She has participated in International Conferences held in Germany (won first prize) and USA (received American Society of Nutrition travel grant). Subsequent to her PhD she has also worked with University of California, Berkley on an air quality monitoring project for over 40 locations across Delhi. She has also edited 6 International books. Her taught assignments include stints as Assistant Professor with the University of Petroleum and Energy Studies, Dehradun, and Amity University, Noida. In 2012, Dr. Kaushik received full scholarship from School of Management, Swansea University UK to undertake MBA. She topped the MBA class and has worked in UK and Turkey in various positions. She returned to India in 2015 and has worked as an Associate Professor with MGM's Jawaharlal Nehru Engineering College, Aurangabad. Currently she is working as an Associate Professor in Dept of Civil Engineering and Head R&D at Hi Tech Institute of Technology Waluj Maharashtra. She has prepared Air Action Plans for cities of Aurangabad, Ulhasnagar, Jalna, Latur, Nashik, Thane, Jalgaon, and Badlapur which have been approved for funding and further implementation by CPCB under NCAP (National Clean Air Program). She has also prepared and submitted the Environment Status Report 2017–2018, 2018–2019 for Aurangabad city and 2018–2019 for Ulhasnagar city. Dr. Kaushik is also undertaking Maha YUVA (Maharashtra Youth Understanding Value of Air) Campaigns and Awards 2019–2020 in cities of Aurangabad, Jalna, and Latur for school and college students. This project is sponsored by MPCB. Dr. Geetanjali also organized 11 workshops in Aurangabad city in collaboration with EPIC (Energy Policy Institute at University of Chicago). Dr. Kaushik is also the nodal person for implementation of Unnat Bharat Abhiyan for Hi-Tech Institute of Technology. This is a flagship program of MHRD being implemented by IIT Delhi.

Maninder Kaur, PhD
Associate Professor, Department of Food Science and Technology, Guru Nanak Dev University, Amritsar, Punjab, India

Maninder Kaur, PhD, is working as an Associate Professor in the Department of Food Science and Technology, Guru Nanak Dev University, Amritsar, Punjab, India. In 1997, She earned her BSc (Home Science); in 1999, MSc (Food Technology) degree and received her doctorate degree in Food Technology from the Department of Food Science and Technology, Guru Nanak Dev University, Amritsar, Punjab, India in the year 2005. Dr. Kaur has her expertise in a specialized area which involves Characterization of flours, starches, and proteins from different botanical sources. In addition, she is engaged with teaching UG, PG, and PhD students at Guru Nanak Dev University, Amritsar, India. She has authored a number of peers reviewed papers, technical articles, book chapters, in international and national journals, books, proceedings, and technical bulletins. Prof. Kaur has received a number of awards and honors, such as: awarded *Post-Doctoral Research Fellowship* by Korean Ministry of Education, South Korea and selected for *Young Scientist Award-2010* by Association of Food Scientists and Technologists (India). She is serving as an expert reviewer for several scientific journals. She is a member of academic and professional organizations like Association of Food Scientists and Technologists (AFSTs), Mysore, India, and Punjab Academy of Sciences (PASs).

Contents

Contributors

Adekemi T. Adesulu-Dahunsi
Department of Microbiology, Landmark University, Omu-Aran, Kwara State, Nigeria,
E-mail: adesuluchemmy@yahoo.com

Afshin Babazadeh
Department of Biomedical Sciences, Faculty of Medicine and Health Sciences, Macquarie University, Sydney, NSW – 2109, Australia, E-mail: babazadeh.afshin@gmail.com

Kolawole Banwo
Department of Microbiology, University of Ibadan, Oyo State, Nigeria, E-mail: kolabanwo@yahoo.com

Carlos Eduardo Barão
Federal Institute of Paraná (IFPR), Paranavaí – 87703-536, Paraná, Brazil,
E-mail: carlos.barao@ifpr.edu.br

Tânia Aparecida Becker-Algeri
Dra. Research in Food Science and Technology, Quality Management Department, Cooperative Agro-Industrial Lar, Avenida 24 de Outubro, n 59, Industrial Area, Medianeira, Paraná, CEP – 85884000, Brazil, E-mail: taniabecker86@yahoo.com.br

Sudhanshu Billoria
Center for Research, Consultancy, and Publication, Vaikunth Mehta National Institute of Cooperative Management, Pune, Maharashtra – 411007, India, E-mail: sudharihant@gmail.com

Chiranjit Chowdhury
Amity Institute of Molecular Medicine and Stem Cell Research, Amity University Uttar Pradesh, Sector-125, Noida – 201313, Uttar Pradesh, India,
E-mails: cchowdhury@amity.edu; cchowdhury82@gmail.com

Eliane Colla
Dra. Research in Food Science and Technology Post Graduate Program in Food Technology (PPGTA) Department, Federal Technological University of Paraná (UTFPR), Medianeira, Paraná, Brazil,
E-mail: ecolla@utfpr.edu.br

Adriano Gomes da Cruz
Rio de Janeiro Federal Institute of Education, Science, and Technology (IFRJ), Professional Master's in Food Science and Technology of (PGCTA), Maracanã – 20270-021, Rio de Janeiro, Brazil, E-mail: food@globo.com

H. W. Deshpande
Department of Food and Industrial Microbiology, College of Food Technology, Vasantrao Naik Marathwada Krishi Vidyapeeth, Parbhani – 431402, Maharashtra, India,
E-mail: hemantd22@gmail.com

Kajal Dhawan
Research Scholar, Department of Food Technology and Nutrition, Lovely Professional University, Phagwara – 144411, Punjab, India, E-mail: kajaldhawan42@gmail.com

Tejpal Dhewa
Department of Nutrition Biology, Central University of Haryana, Mahendergarh – 123031, Haryana, India, E-mails: tejpal_dhewa07@rediffmail.com; tejpaldhewa@cuh.ac.in

Deisy Alessandra Drunkler
Dra. Research in Food Science and Technology, Post Graduate Program in Food Technology (PPGTA) Department, Federal Technological University of Paraná (UTFPR), Medianeira, Paraná, Brazil, E-mail: deisydrunkler@utfpr.edu.br

Sushma Gurumayum
Assistant Professor, Department of Basic Engineering and Applied Sciences, College of Agricultural Engineering and Post-Harvest Technology, Central Agricultural University, Ranipool – 737135, Sikkim, India, E-mail: sushmagurumayum@gmail.com

Bibiana Juan
Center for Innovation, Research, and Transfer in Food Technology (CIRTTA), XaRTA, TECNIO, MALTA Consolider, Department of Animal Science and Food Sciences, Faculty of Veterinary Medicine, Universitat Autònoma de Barcelona, Bellaterra – 08193, Spain, Tel.: +34-935811447, Fax: +34-935812006, E-mail: bibiana.juan@uab.es

Damanpreet Kaur
Research Scholar, Department of Food Technology and Nutrition, Lovely Professional University, Phagwara – 144411, Punjab, India, E-mail: preet.daman9624@gmail.com

Navneet Kaur
School of Engineering Technology and Applied Science, Centennial College, Toronto, Canada, E-mail: nkaur253@my.centennialcollege.ca

Sawinder Kaur
Department of Food Technology and Nutrition, Lovely Professional University, Phagwara – 144411, Punjab, India, E-mail: sawi_raman@yahoo.co.in

Geetanjali Kaushik
Department of Civil Engineering, HiTech Institute of Technology, Waluj Aurangabad – 431136, Maharashtra, India, E-mail: geetanjaliac@gmail.com

Suellen Jensen Klososki
Federal Institute of Paraná (IFPR), Paranavaí – 87703-536, Paraná, Brazil, E-mail: suellen.jensen@ifpr.edu.br

Anu Kumar
Department of Biotechnology, Chandigarh University, Gharuan, Mohali – 140301, Punjab, India, E-mails: anubiotech88cu@gmail.com; anubiotech88@gmail.com

Ashwani Kumar
Assistant Professor, Department of Food Technology and Nutrition, Lovely Professional University, Phagwara – 144411, Punjab, India, E-mail: ashwanichandel480@gmail.com

Vanessa Aparecida Marcolino
Federal Institute of Paraná (IFPR), Paranavaí – 87703-536, Paraná, Brazil, E-mail: vanessa.marcolino@ifpr.edu.br

Majid Nooshkam
Department of Food Science and Technology, Faculty of Agriculture, Ferdowsi University of Mashhad (FUM), Mashhad, Iran, Tel.: +98-51-38795620, Mobile: +98-9169448412, Fax: +98-51-38787430, E-mail: Nooshkamma@gmail.com

Omotade R. Ogunremi
Department of Microbiology, Technical University, Ibadan, Oyo State, Nigeria, E-mail: tadeogunremi@yahoo.com

Vanessa Guimarães Alves Olher
Federal Institute of Paraná (IFPR), Paranavaí – 87703-536, Paraná, Brazil,
E-mail: vanessa.olher@ifpr.edu.br

Ami R. Patel
Assistant Professor, Mansinhbhai Institute of Dairy and Food Technology (MIDFT),
Dudhsagar Dairy Campus, Mehsana – 384002, Gujarat, India,
E-mails: amimicro@gmail.com; ami@midft.com

Tatiana Colombo Pimentel
Federal Institute of Paraná (IFPR), Paranavaí – 87703-536, Paraná, Brazil,
E-mail: tatiana.pimentel@ifpr.edu.br

Prasad Rasane
Assistant Professor, Department of Food Technology and Nutrition,
Lovely Professional University, Phagwara – 144411, Punjab, India,
E-mails: rasaneprasad@gmail.com; prasadrasane4u@gmail.com

Shagun Rastogi
Amity Institute of Biotechnology, Amity University Uttar Pradesh, Sector-125, Noida – 201313,
Uttar Pradesh, India, E-mail: shagun177@gmail.com

Michele Rosset
Federal Institute of Paraná (IFPR), Colombo – 83403-515, Paraná, Brazil,
E-mail: michele.rosset@ifpr.edu.br

Jyoti Singh
Assistant Professor, Department of Food Technology and Nutrition, Lovely Professional University,
Phagwara – 144411, Punjab, India, E-mail: jyotisingh9377@gmail.com

Mahnaz Tabibiazar
Nutrition Research Center and Department of Food Science and Technology,
Faculty of Nutrition and Food Sciences, Tabriz University of Medical Sciences, Tabriz, Iran,
E-mail: mahnaz_tabibiazar@yahoo.com

Mamta Thakur
Department of Food Engineering and Technology, Sant Longowal Institute of Engineering and
Technology, Longowal – 148106, Punjab, India,
E-mails: thakurmamtafoodtech@gmail.com; mamta.ft@gmail.com

Antonio José Trujillo
Center for Innovation, Research, and Transfer in Food Technology (CIRTTA), XaRTA, TECNIO,
MALTA Consolider, Department of Animal Science and Food Sciences,
Faculty of Veterinary Medicine, Universitat Autònoma de Barcelona, Bellaterra – 08193, Spain,
Tel.: +34.935813292, Fax: +34-935812006, E-mail: Toni.Trujillo@uab.es

Deepak Kumar Verma
Agricultural and Food Engineering Department, Indian Institute of Technology, Kharagpur – 721302,
West Bengal, India, Tel.: +91-3222281673, Mobile: +91-7407170259, Fax: +91-3222282224,
E-mails: deepak.verma@agfe.iitkgp.ernet.in; rajadkv@rediffmail.com

Katia Francine Wochner
Ma. Research in Food Science and Technology, Post Graduate Program in Food Technology (PPGTA)
Department, Federal Technological University of Paraná (UTFPR), Avenida Brasil, 4232,
Bairro Parque Independência, Medianeira, Paraná, CEP – 85884-000,
Brazil, E-mail: katia_kfw29@hotmail.com

Zahra Zareie
Department of Food Science and Technology, Gorgan University of Agricultural Sciences and Natural Resources, Gorgan, Iran, E-mail: Zareie91@gmail.com

Abbreviations

ACE	angiotensin converting enzyme
AFB_1	aflatoxin B_1
AFLP	amplified-fragment length polymorphism
AFM_1	aflatoxin M_1
ARDRA	amplified ribosomal DNA restriction analysis
BEN	Balkan endemic nephropathy
BLIS	bacteriocin like inhibitory substances
BMV	bacterial microcompartment vertex
BOD	biological oxygen demands
BPs	bioactive peptides
CAGR	compound annual growth rate
CC	codex committees
CCPs	casein calcium peptides
CD	Crohn's disease
CE-MS	capillary electrophoresis mass spectrometry
CF	curd firmness
Cfu	colony-forming unit
COD	chemical oxygen demands
CODES	colon targeted delivery system
CPC	consumer protection commission
CPP	casein phosphor-peptides
EC	European Commission
EFSA	European Food Safety Authority
EPS	exopolysaccharide
EU	European Union
Eut	ethanolamine utilization
EutM	Eut MCP
FAO	Food and Agriculture Organization
FB_1	fumonisin B_1
FDA	Food and Drug Administration
FOS	fructo-oligosaccharides
FSSAI	Food Safety and Standards Authority of India
FT-IR	Fourier transform infrared
GC-MS	gas chromatography-mass spectrometry

GFP	green fluorescent protein
GI	gastrointestinal
GIT	gastrointestinal tract
GMPs	good manufacturing practices
GOS	galactooligosaccharides
GRAS	generally recognized as safe
GST	glutathione S-transferase
HACCP	hazard analysis and critical control point
HHP	high hydrostatic pressure
HPH	high pressure homogenization
HPMC	hydroxyl propyl methyl cellulose
HPP	high-pressure processing
HTS	high-throughput sequence
HTST	high temperature-short time
IAT	impinging aerosol technology
IBD	inflammable bowel disease
IBS	irritable bowel syndrome
Igs	immunoglobulins
IMO	isomaltooligosaccharide
LAB	lactic acid bacteria
LC-MS	liquid chromatography-mass spectrometry
LTLT	low-temperature long-time
MAP	modified atmosphere packaging
MBP	maltose-binding protein
MC	moisture content
mCOLD-PCR	modified CO-amplification at low denaturation temperature PCR
MCP	microcompartment
MEA	malt extract agar
MLSA	multilocus sequence analysis
MLST	multilocus sequence typing
MRPs	Maillard reaction products
MRS	Man Rogosa Sharpe
MS	mass-spectrometry
NAFDAC	National Agency for Food and Drug Administration and Control
NFSAP	National Food Safety Action Plan
NGS	next generation sequencing
NMR	nuclear magnetic resonance

O/W	oil-in-water
OTA	ochratoxin
OTU	operational taxonomic unit
PCA-BCP	plate count agar with bromocresol purple
PCR-DGGE	polymerase chain reaction-denaturing gradient gel electrophoresis
Pdu	propanediol utilization
PEFs	pulsed electric fields
PMA	propidium monoazide dye
PR	pressure-treated
PVOH	polymer of polyvinyl alcohol
qPCR	quantitative PCR
RAPD	random amplification of polymorphic DNA
RCF	rate of curd firming
RCT	rennet clotting time
RFLP	restriction fragment length polymorphism
SAB	Slanetz and Bartley medium
SDF	soluble dietary fiber
SERB	Science and Engineering Research Board
SHR	spontaneously hypertensive rats
SMEs	small and medium-scale enterprises
SPI	soy protein isolate
SSCP	single stranded confirmation polymorphisms
TCA	tricarboxylic acid cycle
UC	ulcerative colitis
UHPH	ultra-high-pressure homogenization
UHT	ultra-high temperature
UNL	University of Nebraska-Lincoln
USA	United States of America
UTI	urinary tract infection
VBNC	viable but non-culturable
W/O	water-in-oil
W/O/W	water-in-oil-in-water
WGPH	wheat germ protein hydrolysates
WHO	World Health Organization
WPC	whey protein concentrate
WPI	whey protein isolate
WTO	World Trade Organization
YGC	yeast glucose chloramphenicol

YPD	yeast extract potato dextrose agar
ZEA	zearalenone
α-La	α-lactalbumin
α-ZOL	α-zearalenol
β-Lg	β-lactoglobulin

Symbols

%	percentage
°C	degree Celsius
μm	micrometer
h	hour
Kg	kilogram
mg	milligram
ml	milliliter
pH	power of hydrogen
Tg	glass transition temperature
w/v	weight/volume

Preface

The use of foods to improve health and the state of well-being is progressively more accepted by both science and society apart from the primary role of food to provide nutrients for performing normal physiological functions. Probiotics (health beneficiary bacteria) or their metabolites such as bacteriocins, bioactive peptides (BPs), etc., and prebiotics (usually functional non-digestible oligosaccharides) are considered as a key biotechnological field with tremendous potential for innovation. Additionally, knowledge of microbial communities is essential for shaping the final characteristics of diverse food products. Consequently, this book describes the understanding and dissemination of knowledge surrounding probiotics and/or their metabolites, prebiotics in various food products, and human health, along with recent trends in microbial fermentation and innovative technologies in academic researches.

This book *Microbial Biotechnology in Food Processing and Health: Advances, Challenges, and Potential*, is divided into three main parts: Part I: Probiotics and Their Metabolites: Food Industry Perspective; Part II: Probiotics and Potential Health Benefits; and Part III: Industrial Aspects of Microbial Food Technologies. Probiotic cultures are employed for the manufacturing of dairy as well as non-dairy-based fermented or non-fermented food products. Recently, screening and designing functional starters with technological aid advantages are fascinating areas of investigation in the food sector. In this context, chapters compiled in Part I discuss the types and roles of beneficial microbes and/or their metabolites in food products such as enhancing food safety by decontaminating or neutralizing toxic components like mycotoxins associated with foods or improving their stability in functional foods. The health-beneficial microorganisms known as probiotics have been proven for many immunomodulatory and immunostimulatory effects in a number of chronic infections and diseases.

Hence, in Part II of the book, recent breakthroughs in the development of novel probiotics-incorporated dairy and non-dairy (fruits and vegetables) based food products, challenges associated with their commercialization, and studied health benefits have been covered. Innovative approaches are employed in food science for various means; hence in Part III, emerging technologies dealing with assessing microbial diversities or management of microbiological hazards in food products are discussed.

The book represents a concise state of the art addressing the current status as well as various challenges for the development of new probiotics and prebiotic-based foods and the niches for future research. All of the chapters of the book have been contributed by a group of international contributors who are well recognized and excellent in the frontier field of functional foods, food microbiology, and microbial biotechnology. Together we have produced an outstanding reference book that is expected to be a valuable resource for researchers; teachers; students; food, nutrition, and health practitioners; and all those working in the dairy, food, and nutraceutical industry. We extend our sincere thanks to all the contributing authors whose co-operation has made our task as editors a pleasure. We hope that this book will result in enlightening and inspiring action for readers.

—Editors

Part I

Probiotics and Their Metabolites: A Food Industry Perspective

CHAPTER 1

Action of Probiotic Microorganisms on Mycotoxin Decontamination in Milk and Milk Products

KATIA FRANCINE WOCHNER,* TÂNIA APARECIDA BECKER-ALGERI, ELIANE COLLA, and DEISY ALESSANDRA DRUNKLER

**Corresponding author. E-mail: katia_kfw29@hotmail.com*

ABSTRACT

Milk is a foremost source of vital nutrients required for the development plus human health maintenance. Nevertheless, it can also be a vehicle for toxic agents, causing serious hazards in individuals who consume it, especially children. Mycotoxins are among the main chemical contaminants in milk, which are important secondary fungal metabolites in human and animal health. When ingested by animals, they can be excreted in the milk, and once present in milk, these mycotoxins can withstand most treatments (pasteurization and ultra-high-temperature processing) to obtain dairy products. Therefore, mycotoxins may be present in cheeses and yogurts. It is imperative that milk and milk products should be free from toxic components with regard to the high levels of consumption; therefore, the study of efficient decontamination protocols that are safe environmentally and precise is an important and current research topic. Amongst the category of biological methods pertinent for the chemical contaminants removal, many of the studies are mainly focused on employing probiotic cultures, with promising outcomes. In this context, the current chapter aims to review the literature on mycotoxins and their decontamination processes from milk and milk products by means of probiotics.

1.1 INTRODUCTION

There has been increasing concern related to the milk and dairy products consumption because of contamination by mycotoxins, since, in recent years, countless studies have demonstrated their incidence in fluid milk and dairy products (Scaglioni et al., 2014; Bahrami et al., 2016; Becker-Algeri et al., 2016; Michling et al., 2016; Al-Hilphy et al., 2016; Verma et al., 2017; Kangethe et al., 2017; Zheng et al., 2017; Mao et al., 2018). Dairy products consumption accounts for more than 80% of children's dietary habits and intake worldwide (Campagnollo et al., 2016). Mycotoxins are synthesized as secondary metabolites by the molds of the genera *Aspergillus*, *Fusarium*, and *Penicillium*, divided into six main classes: aflatoxins, ochratoxins (OTAs), fumonisins, zearalenone (ZEA), trichothecenes, and ergot alkaloids (Ahlberg et al., 2015). The structure of this secondary metabolites is depicted in Figure 1.1.

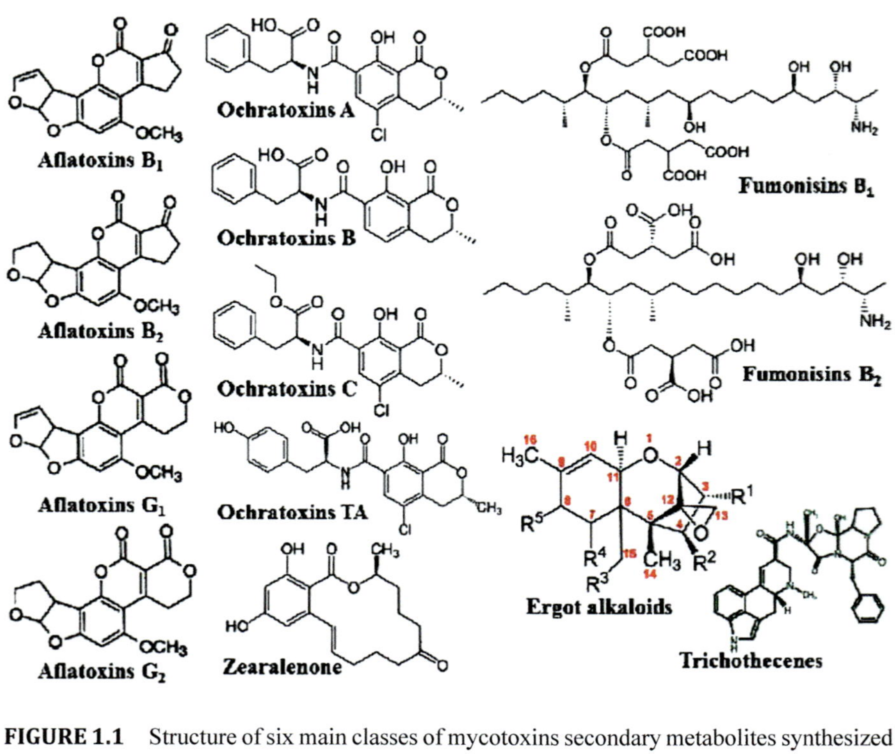

FIGURE 1.1 Structure of six main classes of mycotoxins secondary metabolites synthesized by the genera *Aspergillus*, *Fusarium*, and *Penicillium*.

The Food and Agriculture Organization (FAO) and the World Health Organization (WHO) have emphasized mycotoxins in food because of the great toxicological and carcinogenic potential that some of them pose to animal and human health (International Agency for Research on Cancer (IARC), 2012). Taking into account the high consumption of milk and dairy products by the people and the risks to human health linked with the intake of the most varied mycotoxins frequently present in them (Table 1.1), some nations have established tolerance borders (Codex Alimentarius Commission, 2001; Mercosul, 2002; European Commission, 2006). In addition, socio-economic issues should be taken into account, since global legislation is not compatible, as each country has its own regulations, which makes it difficult to monitor, besides the lack of legislation for all milk-based products manufactured and consumed around the world (Campagnollo et al., 2016).

TABLE 1.1 Mycotoxins of Milk and Dairy Products and it is Consumption Effects on Human Health

Mycotoxins	Chemical Nature	Milk and Dairy Products	Effects on Human Health
Aflatoxin B_1 (AFB_1)	Polyketide-coumarin (difurocoumaro-cyclopentenone)	Cheeses (Manchego, Prato, Parmesan)	Acute toxicity, immunosuppressive, teratogenic, mutagenic, carcinogenic
Aflatoxin M_1 (AFM_1)	Polyketide-coumarin (difurocoumaro-cyclopentenone)	Goat milk and Cow milk (raw, pasteurized, ultra-high temperature UHT), cream, butter, ice cream, yogurt, cheeses (white cheese, pickled cheese, Tilsit, Brick, Parmesan, Feta, Camembert, Mozzarella, Manchego, Kashar, Tulum, goat hard cheese)	Immunotoxic, teratogenic, mutagenic, hepatotoxic, carcinogenic
Andrastin A-D	Meroterpenoids	21 brands of Blue-veined cheeses (Danablue, Castello, Klosterkrone, St. Clemens, Bleu D'auvergne, Castello, Fourme d'Ambert, Gorgonzola, Saint Agur Bresse Bleu, Stilton, and Magor)	Impaired milk production and reproduction, inappetence, immunosuppression

TABLE 1.1 *(Continued)*

Mycotoxins	Chemical Nature	Milk and Dairy Products	Effects on Human Health
Citrinin	Polyketide (quinonemethine)	Smear cheeses, surface mold semi-soft cheeses, Prato, Parmesan, goat soft cheese, and Saint Marcellin TM, (SoignonTM), Kashar	Neurotoxicity, hepatotoxicity (occasionally), nephrotoxicity (acute and fatal)
Cyclopiazonic acid	Indole-tetramic acid	White cheese, Camembert, Taleggio Italian, Kashar	Immunotoxicity, cytotoxicity, kodua poisoning (acute intoxication causing nerve problems)
Isofumigacla-vines A and B	Alkaloids	Blue-veined cheeses	Neurotoxic
Mycophenolic acid	Phthalide	German blue cheese, Tilsit, Stilton, Roquefort, Bleu des Causses, Bleu d'Auvergne, Fourme d'Ambert, Manchego	Immunosuppressive (A selective inhibitor of lymphocyte proliferation)
Ochratoxin A	Polyketide (dihydroisocoumarin coupled with L-phenylalanine)	Cow milk-based infant formulae, and Cheeses (different types of semi-hard raw-milk cheese, Gorgonzola, Roquefort).	Nephrotoxic, teratogenic, and immunotoxic, decreasing human growth in malnourished children. Endemic toxicity-nephropathy (Balkan region); urothelial urinary tract tumors (Bulgaria) and chronic interstitial nephropathy (North Africa)
Patulin	Polyketide lactone	Raw milk semi-hard cheese, Tulum, Blue-veined, Manchego	Acute toxicity including neurotoxicity (paralysis of motor nerves, brain damage, convulsion), nephrotoxicity, and hepatotoxicity, teratogenicity

TABLE 1.1 *(Continued)*

Mycotoxins	Chemical Nature	Milk and Dairy Products	Effects on Human Health
Penicillic acid	Polyketide butyrolactone	Cheddar, Tulum, Roquefort, various Swiss-type cheeses	Cytotoxic, hepatotoxic, nephrotoxic, genotoxic, carcinogenesis
PR-toxin	Eremophilane sesquiterpene	Various Blue-veined cheeses	Carcinogenic, mutagenic, acute toxicity (increase of capillary permeability and a direct damage to the heart, kidneys, lungs, liver)
Roquefortine C	Indole alkaloid (diketopiperazine)	Tulum, blue-white mold cheese (bovine) Manchego, Blue-veined cheese (ewe, goat, bovine, and mixture), ripened blue, Roquefort (ovine), Gorgonzola, Stilton, Cabrales, dressing of blue cheese	Neurotoxin (acute intoxication)

Source: Reprint from: Benkerroum (2016); with permission. © Elsevier.

There are different control measures for the exclusion of mycotoxins from milk-based products; nonetheless, these processes do not provide sufficient protection or are very costly to implement (El-Nezami et al., 1998; Shcherbakova et al., 2015). In turn, the biological degradation of chemical contaminants employing microbial cells or their metabolic products is a specific alternative to reduce levels or eliminate these contaminants in food while maintaining the food's safety, since several foods contain microflora added during their development (Ji et al., 2016; Sarlak et al., 2017). Several lactic acid bacteria (LAB) and yeasts, because of their GRAS (generally recognized as safe) standing as well as application as probiotics, are the most studied to lessen the mycotoxins' bioaccessibility in milk and its derivatives (Bovo et al., 2013; Elsanhoty et al., 2014; Sarlak et al., 2017; Wochner et al., 2018; Patel et al., 2018).

The aim of the current chapter is to examine the major mycotoxins associated with milk and milk products and discusses the biological decontamination methods with the use of probiotic microorganisms.

1.2 MYCOTOXINS: AN OVERVIEW

Mycotoxins are among the most serious public health contaminants due to their presence in food and the adverse effects they may cause in both animals and humans (Benkerroum, 2016). They are a group of extremely toxic secondary metabolites, which usually develop in places with extreme conditions, such as low water availability, abrupt temperature variations, and long periods of rain (Ahlberg et al., 2015). Contamination by mycotoxins occurs widely in foods of plant origin, all through their pre-and post-harvest or mainly during storage, especially in cereals, feed, fodder, and other agricultural foodstuffs intended for animal or human consumption. As said by the United Nations' FAO, more than 25% of the food crops of the world are appreciably polluted by means of mycotoxins (Campagnollo et al., 2016). When ingested through contaminated food, mycotoxins are usually transmitted to animal products including milk, egg, or meat, after it get metabolized and biotransformed; most are chemically and thermally stable in processing, including cooking, boiling, frying, and pasteurizing, posing a potential risk, even in processed foods (Bruerton, 2001; Creppy et al., 2002; Murphy et al., 2006).

Currently, more than 300 mycotoxins have been identified, and this number continues to increase. Scientific attention is focused mainly on mycotoxins that are carcinogenic and/or toxigenic, among which are aflatoxins, OTAs, fumonisins, trichothecenes, and ZEA (Becker-Algeri et al., 2016). It is possible to correlate many human diseases with an ingestion of mycotoxins, especially the long-term consumption, the core lethal effects being carcinogenicity, nephrotoxicity, hepatotoxicity, genotoxicity, dermal irritation, immunosuppression, and reproductive disorders (Bovo et al., 2013; Lee and Ryu, 2015). According to IARC (2012), aflatoxins such as aflatoxin B_1 (AFB_1) plus aflatoxin M_1 (AFM_1) are basically classified as group 1 (carcinogenic to humans). Fumonisin B_1 (FB_1) and OTA belong to group 2B (most likely human carcinogen). Although other toxins with ZEA and its derivatives (alpha-zeralanol, beta-zeralanol, alpha-zearalenol, beta-zearalenol) are considered non-carcinogenic, they cause other adverse effects, in particular, estrogenic effects, which may affect reproduction in mammals (Huang et al., 2014).

Considering all mycotoxins present in foods, aflatoxin is the widely distributed toxin in milk and food products and is expressed as the most toxic among all known toxins (IARC, 2012). A total of 18 comparable compounds are explained as aflatoxins, however the ones found naturally in foods are six: B_1 (AFB_1), G_1 (AFG_1), M_1 (AFM_1), B_2 (AFB_2), G_2 (AFG_2), and M_2 (AFM_2) (Campagnollo et al., 2016; Patel et al., 2018). However, the foremost prevalent type of form and the highest potent of among all mycotoxins are AFB_1 (Murphy et al., 2006; Corassin et al., 2013). These toxins are usually soluble in organic solvents like benzene, chloroform, and methanol, have a low molecular weight, are considered lipophilic molecules and are extremely stable above 100°C temperatures (Murphy et al., 2006). It is important to note that milk or dairy products can get contaminated through aflatoxin in two ways: when lactating milk animals eat AFB_1 polluted feed which will be biotransformed in AFM_1 and excreted in milk (indirect contamination), or by direct contamination on milk-derived product as deliberate additives or unintentional accidental contamination (Campagnollo et al., 2016).

The symptoms of aflatoxicosis (poisoning that results from ingesting aflatoxins) depend on the aflatoxin nature; the quantity ingested and the exposure period; the sex, age, and health of the exposed person (Elsanhoty et al., 2014). The main target organ is the liver due to the hydrophobic nature of aflatoxins, once is a predominately lipophilic organ, and it is in hepatocytes where each and every component that are transported through the blood stream are accumulated and hence get concentrated (Campagnollo et al., 2016). Acute aflatoxicosis where there is exposure to high-level of aflatoxins may result in nausea and vomiting until serious liver damage such as cirrhosis or liver failure, and death. Already the chronic aflatoxicosis can results in cancer, especially liver cancer, immunosuppression, and interferes with metabolism of protein as well as micronutrients that are vital to health (Bennet and Klich, 2003; Williams et al., 2004). Aflatoxins are of major concern in tropical and subtropical climates, especially in developing countries where safe food storage is not guaranteed (Aldars-García et al., 2016). Recently, the Centers for Disease Control and Prevention reported cases of acute aflatoxicosis in Kenya causing 125 deaths among 317 cases of poisoning caused by high levels of AFB_1 found in homegrown maize (CDC, 2004). Therefore, there is also a great concern regarding the health of children major consumers of milk and dairy products and whose sensitivity is remarkable and significantly higher as compared to adults (Scaglioni et al., 2014). So, because of the regular utilization of milk or milk products at higher level, it is extremely essential that they are free of poisonous ingredients (Flores-Flores et al., 2015).

1.2.1 MYCOTOXINS IN MILK AND MILK PRODUCTS

The incidence of mycotoxins in milk or dairy products has been accounted for in different countries, demonstrating the global concern about these toxins (Table 1.2).

TABLE 1.2 Incidence of Mycotoxins in Milk and Milk Products in Different Countries

Country	Product	Mycotoxin Type	Number of Samples	Degree of Contamination (μg L^{-1}/μg kg^{-1})	References
Argentina	Raw milk	AFM_1	160	0.003–0.293	Michling et al. (2016)
Brazil	Cheese	AFM_1	58	0.01–0.3	Iha et al. (2011)
	Yogurt		53	0.01–0.53	
	Dairy drink		12	0.01–0.05	
China	Raw milk	α-ZOL	30	0.024–0.073	Huang et al. (2014)
	Pasteurized milk		12	0.036–0.045	
	Powdered milk		8	0.043–0.064	
Iran	Traditional cheese	AFM_1	360	0.05–0.30	Shahbazi et al. (2017)
Italy	UHT[a] and raw milk	FB_1	10	0.26–0.43	Gazzoti et al. (2009)
Mexico	Pasteurized and UHT[a] milk	AFB_1	290	0.05–0.42	Carvajal et al. (2003)
Serbia	UHT[a] milk	AFM_1	438	0.025–1.00	Tomasevic et al. (2015)
	Raw milk		678	0.025–>1.00	
	White cheese		47	0.025–1.00	
	Yogurt		56	0.026–0.50	

Note: [a]UHT: Ultra-high temperature.

The mainly studied mycotoxin found in fluid milk and its derivatives is aflatoxin AFM_1, characterized as a hydroxylated derivative of AFB_1 and originating from the precursor's ingestion by milch cattle by the consumption of feed, hay, and, mainly, contaminated silage. Hepatic biotransformation occurs through enzymes of the cytochrome P450 complex partially hydroxylated to AFM_1 (Figure 1.2). According to Creppy (2002), in the milk, about 0.3% to 6.2% of the entirety AFB_1 ingested by the cattle are generally converted into AFM_1. The solubility of AFM_1 in water facilitates its excretion in the biological fluids of animals, such as milk and urine, accumulating in them (Oatley et al., 2000; Murphy et al., 2006).

Before some researchers investigated and confirmed AFB_1 occurrence in milk (Carvajal et al., 2003; Scaglioni et al., 2014), it was assumed that it was fully biotransformed in other compounds, including its hydroxylated AFM_1. Due to the scarcity of these studies, only AFM_1 is regulated and has its maximum limits stipulated in milk-based food matrices. The European Community and the *Codex Alimentarius* recommend a limit of 0.05 μg per L in milk, milk powder, as well as other dairy products (Codex Alimentarius Commission, 2001; European Commission, 2006). Other nations like the United States of America (USA) and China set a boundary of 0.5 μg per L for milk and dairy products. Brazil pursues the same limits established by the Common Market of the South (Mercosul), who maximally permitted AFM_1 concentrations in fluid milk, milk powder, and cheeses are 0.5 μg L^{-1}, 5.0 μg per kg, and 2.5 μg per kg, respectively (Mercosul, 2002; Brasil, 2011). On the other hand, there is still a lack of regulation of the definition of the highest permitted levels of AFB_1 in fluid milk and dairy products.

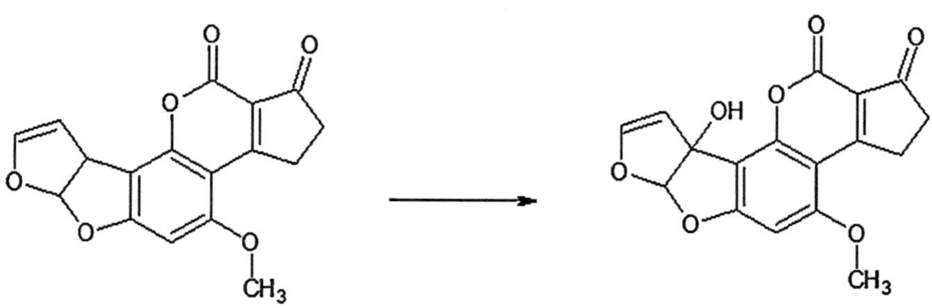

FIGURE 1.2 Hydroxylation of aflatoxin B_1 in M_1 by the P-450 system.

In addition to the problems related to the natural AFM_1 contamination of milk, further there is a great apprehension regarding the stability of contaminant in foods (Iqbal et al., 2015), which includes thermal resistant applied on the raw material during the manufacturing of dairy products, such as pasteurization and sterilization, allowing the risk of this contamination to be present also in products such as cheeses and yogurts (Tomasevic et al., 2015; Li et al., 2017; Shahbazi et al., 2017). Furthermore, in some cases, dairy products such as cheeses may present a higher concentration (from three to five times greater) of AFM_1 than in natural milk, because AFM_1 binds predominantly to the milk protein fraction and preferably in hydrophobic regions of casein (Iha et al., 2011; Chavarría et al., 2017). In milk-based products such as yogurt, the effect of fermentation reduces the original concentrations of aflatoxin initially present in milk (Serrano-Niño et al., 2013; Elsanhoty et al., 2014; Sarlak et al., 2017).

Few research articles show the incidence of other mycotoxins in milk and derivatives, and unlike AFM_1 there is no legislation for them. Gazzoti et al. (2009); and Coffey et al. (2009) analyzed the mycotoxin FB_1 presence in milk, finding a mean level of 0.26 and 0.36 μg per L, respectively, which, even at low concentrations, is worrisome, since the coexistence of fumonisin and aflatoxin-mycotoxins could have toxicological effects and synergistic carcinogens (Gelderblom et al., 2002). In Italy, Pattono et al. (2011) found the incidence of mycotoxin OTA in samples of fluid milk produced in organic farms, ranging between 0.07 and 0.110 μg L^{-1}. Pattono et al. (2013) analyzed the OTA mycotoxin in traditional cheeses produced with raw milk, where six samples were contaminated with a concentration of between 18.4 and 146.0 μg L^{-1} on the inside and between 1.0 and 262.2 μg L^{-1} on the outside. OTA is mainly linked with Balkan endemic nephropathy (BEN) which is a fatal human renal syndrome (Bennet and Klich, 2003). Moreover, when associated to AFM_1, it presents a high level of toxicity in human cells (Tavares et al., 2013). Huang et al. (2014) detected mycotoxins ZEA, α-zearalenol (α-ZOL), and OTA, with their respective mean concentrations 0.014, 0.024, and 0.056 μg per L in raw milk; 0.025, 0.036, and 0.026 μg per L, respectively, in pasteurized milk; and 0.011, 0.043, and 0.027 μg per L, respectively, in milk powder samples collected from various milk farms and supermarkets in Beijing (China) reinforcing the problem of the occurrence of other mycotoxins in milk and dairy products.

1.3 BIOLOGICAL DECONTAMINATION OF MYCOTOXINS IN MILK AND DAIRY PRODUCTS: USE OF PROBIOTIC MICROORGANISMS

Since mycotoxins are present in milk like food products, their removal can be a difficult step because of the several mycotoxins' resistance to tremendous ecological conditions (pH, temperature, radiation), in addition to specific biological, chemical, and physical processes for their inhibition

(Benkerroum, 2016). There are different control measures for the removal of mycotoxins in fluid milk plus related products; conversely, these methods endow with an inadequate degree of protection or are difficult to implement (El-Nezami et al., 1998; Shcherbakova et al., 2015). Mycotoxins' biological removal by making use of microorganisms (fungi, yeasts, and bacteria) and/or enzymes is an alternative to reduce levels or eliminate mycotoxins in foods, maintaining the safety and quality of mycotoxins (Sarlak et al., 2017; Wochner et al., 2018).

Currently, attention has been paid to probiotics due to their ability to reduce and/or inhibit the action of toxic compounds, including mycotoxins (Bilandžić et al., 2011; Skrbic et al., 2014; Shaker and Elsharkawy, 2015; Bahrami et al., 2016). Several species of LAB, bifidobacteria, and yeast *Saccharomyces cerevisiae var. boulardii*, are considered as probiotics, "living microorganisms which, when administered in suitable amounts, have a beneficial effect on host health" (FAO/WHO, 2001). Although, the positive actions linked with the regular probiotics consumption include improved intestinal function by regulation of the endogenous beneficial microbes, immunostimulation, increased bioaccessibility of nutrients (minerals and vitamins), reduced symptoms of lactose intolerance, in addition to anticancer and antimutagenic activity, acting in the colon cancer prevention and other intestinal illnesses (Mallebrera et al., 2013: Shori, 2015).

Biological removal approach involves microbial cells (live or dead) to eliminate mycotoxins, by cell wall absorbance, thus reducing their bioaccessibility (Kabak and Var, 2008; Serrano-Niño et al., 2013; Elsanhoty et al., 2014). Bioaccessibility may be defined as 'the amount of a compound released from eaten food matrix into the gastrointestinal (GI) tract and made accessible for consequent absorption by the intestine' (Serrano-Niño et al., 2013). The structure and strength of the microorganism-mycotoxin complex are affected by various factors such as concentration, type, and specificity of the microorganism, incubation temperature, pH, inactivation treatment, nutrient addition, and food matrix (Bovo et al., 2013). The strains of microorganisms most studied in *in-vitro* decontamination of mycotoxins involve the genera of LAB such as *Lactobacillus, Lactococcus*, *Streptococcus,* and *Propionibacterium* genera (Dalié et al., 2010; El-Khoury et al., 2011; Patel et al., 2018). Besides bacteria, the yeast *Saccharomyces cerevisiae* has also been successful in the detoxification of mycotoxins (Shetty and Jespersen, 2006; Armando et al., 2011; Karazhiyan et al., 2016).

Even though the exact mechanism of action for these microorganisms on mycotoxins has not yet been known, the most relevant hypothesis is a non-covalent physical adsorption of these microorganisms with the cell wall components of bacterial cell or yeast cell, especially in specific peptidoglycans and polysaccharides (Shetty and Jespersen, 2006). The peptidoglycan layer contains mainly C-O, OH, and NH as functional groups and has been reported as the main component involved in the biological adsorption of mycotoxins (Zoghi et al., 2014). On the other hand, Serrano-Niño et al. (2015) affirmed the significance of the teichoic acid during such interaction, where they suggested that the mode of interaction happened between glucose and glycerol of the acid of the Lactobacillus strains examined with the hydroxyl groups and with the carbonyl oxygen of AFB_1. In addition, it is believed that some treatments of inactivation, be they chemical, physical, or enzymatic, are competent to enhance the aptitude of the microorganisms to attach to the mycotoxin of the specific medium, since such disorders increase the hydrophobic nature of the micro surface-organism, exposing the plasma membrane components that were occupied when it was intact (Harkard et al., 2001; Sarlak et al., 2017). The use of non-viable microbial cells is an advantage when fermentation is undesirable and the alteration of the product is not desired. In addition, the different strains of LAB and *S. cerevisiae* yeasts are not equivalent in terms of removal of toxin; in contrast, mycotoxin removal aptitude is a feature of specific strains only, with their efficacy varying distinctly. Such kinds of variations between strains propose dissimilarities in cell wall or/and cell envelope structure as well as the existence of diverse binding positions (Bovo, 2013; Karazhiyan et al., 2016).

El Khoury et al. (2011) evaluated the efficiency of AFM_1 removal during the production of yogurt (6 h) by cultures of *Lactobacillus bulgaricus* and *Streptococcus thermophilus*, in isolation and together, obtaining a removal of 58.5%, 37.7%, and 46.7% of mycotoxin, respectively. Using an *in vitro* digestibility model, Serrano-Niño et al. (2013) evaluated the capability of five probiotics (*L. acidophilus, L. rhamnosus, L. johnsonii and Bifidobacterium bifidum*) to lower AFM_1 aflatoxin bioaccessibility in milk that was artificially spiked with it. The AFM_1 bioaccessibility was decreased between 22.72 and 45.17%, depending on the used probiotic strain. Elsanhoty et al. (2014) evaluated the prospective of five LAB strains (*L. plantarum*, *L. acidophilus*, *L. bulgaricus* and *S. thermophilus*) and one strain of bifidobacterium (*Bif. angulatum*) to eliminate AFM_1 from fermented milk, yogurt. The highest reduction found (87%) was the fermented test with 50% *L. plantarum* and 50% yogurt (*S. thermophilus* and *L. bulgaricus*). Sarlak et al. (2017) analyzed the effects of probiotic bacteria (*L. casei, L. acidophilus, L. rhamnosus, B. lactis*) on the reduction of AFM_1 in an established Iranian fermented drinkable milk (Doogh), obtaining a result of 60.2% aflatoxin reduction in 28 days of storage.

Taheur et al. (2017) investigated the biodegradation and/or adsorption capacity of bacteria and yeasts isolated from kefir of Tunisia. The results showed an adsorption of 82%, 94%, and 100% for the mycotoxins AFB_1, OTA, and ZEA, respectively; when kefir was grown in milk, with emphasis on the *L. kefiri* strain KFLM3. These results propose that kefir consumption may help to lessen mycotoxin toxic effects by reducing the GI absorption of these. In addition to studies performed in a dairy matrix, there are studies that demonstrate the potential for LAB decontamination in buffer solutions. Sangsila et al. (2016) studied ZEA removal capability of 5 *L. pentosus* strains from a sodium acetate buffer solution, also obtained good removal capacities, ranging from 29.74 to 83.17%, indicating promise to also decrease the contamination of mycotoxins from food.

The use of yeasts combined with LAB or even in isolation has also been studied as a possible biological decontamination technique. Corassin et al. (2013) assessed the aptitude of a strain of *S. cerevisiae* and a set of 3 bacteria (*Lactobacillus delbrueckii subsp. bulgaricus*, *L. rhamnosus*, and *B. lactis*), isolated or in mixture, to remove AFM_1 from skimmed UHT milk samples. The mean percentage of removal was 90% using isolated yeast, 11% for LAB, and 100% at AFM_1 levels when *S. cerevisiae* yeast was used in conjunction with LAB. Karazhiyan et al. (2016) reported in their study a mean removal of 75% of AFM_1 in yogurt from the *S. cerevisiae* strain (PTCC 5177). For comparison, viable and non-viable cells (acid, heat, and ultrasound) were used. It was shown that the non-viable cells had a higher aptitude to bind to aflatoxin and, consequently, to reduce it. For both Corassin et al. (2013); and Karazhiyan et al. (2016), the aflatoxin removal method employing *S. cerevisiae* alone or in conjunction with LAB, mainly strains used in several food products, has prospective application to reduce AFM_1 levels in milk and other fermented foods such as yogurt.

1.4 SUMMARY AND CONCLUSION

Mycotoxin contamination in milk and dairy products plays a significant role in food safety, due to ingestion in isolation or in combination, mainly in a chronic way, related to serious health problems. The use of microorganisms, such as LAB and some yeasts, is a promising approach for the safe, specific, and effective decontamination of these foods. Several probiotic microorganisms have been studied and are of inquisitiveness for the expansion of fermented dairy products intended for the consumption of human being and currently have a novel probiotic attribute, the mycotoxin adsorption and hence ultimately decontamination from food products.

KEYWORDS

- **acid lactic bacteria**
- **aflatoxin**
- **bacteria cell wall**
- ***Bifidobacterium* spp.**
- **biological degradation**
- **contaminated milk**
- **dairy products**
- **decontamination methods**
- **fermentation**
- **food safety**
- **human health**
- **immunosuppression**
- ***Lactobacillus* spp.**

REFERENCES

Ahlberg, S. H., Joutsjoki, V., & Korhonen, H. J., (2016). Potential of lactic acid bacteria in aflatoxin risk mitigation. *International Journal Food Microbiology, 207*, 87–102.

Aldars-García, L., Ramos, A. J., Sanchis, V., & Marin, S., (2016). Modeling postharvest mycotoxins in foods: Recent research. *Current Opinion Food Science*, *11*, 46–50.

Al-Hilphy, A. R. S., Verma, D. K., Niamah, A. K., Billoria, S., & Srivastav, P. P., (2016). Principles of ultrasonic technology for treatment of milk and milk products. In: Meghwal, M., & Goyal, M. R., (eds.), *Food Process Engineering: Emerging Trends in Research and Their Applications* (Vol. 5, pp. 177–200). As part of book series on innovations in agricultural and biological engineering, Apple Academic Press, USA.

Armando, M. R., Dogi, C. A., Pizzolitto, R. P., Escobar, F., Peirano, M. S., Salvano, M. A., Sabini, L. I., et al., (2011). *Saccharomyces cerevisiae* strains from animal environment with*in vitro* aflatoxin B_1 binding ability and anti-pathogenic bacterial influence. *World Mycotoxin Journal, 4*, 59–68.

Bahrami, R., Shahbazi, Y., & Nikousefat, Z., (2016). Aflatoxin M_1 in milk and traditional dairy products from west part of Iran: Occurrence and seasonal variation with an emphasis on risk assessment of human exposure. *Food Control, 62*, 250–256.

Becker-Algeri, T., Castagnaro, D., Bortoli, K., Souza, C., Drunkler, D. A., & Badiale-Furlong, E., (2016). Mycotoxins in bovine milk and dairy products: *A review. Journal of Food Science, 81*, 544–552.

Benkerroum, N., (2016). Mycotoxins in dairy products: A review. *International Dairy Journal, 62*, 63–75.

Bennett, J. W., & Klich, M., (2003). Mycotoxins. *Clinical Microbiology Revision, 16*, 497–516.

Bilandžić, N., Đokić, M., Sedak, M., Solomun, B., Varenina, I., Knežević, Z., & Benić, M., (2011). Trace element levels in raw milk from northern and southern regions of Croatia. *Food Chemistry, 127*, 63–66.

Bovo, F., Corassin, C. H., Rosim, R. E., & Oliveira, C. A. F., (2013). Efficiency of lactic acid bacteria strains for decontamination of aflatoxin M_1 in phosphate buffer saline solution and in skimmed milk. *Food Bioprocess Technology, 6,* 2230–2234.

Brasil, Agência Nacional de Vigilância Sanitária (ANVISA). Brazil, National Health Surveillance Agency (ANVISA). Collegiate Board Resolution No. 7, of February 18, 2011. Maximum Tolerated Limits for Mycotoxins in Food. Daily Official of the Federative Republic of Brazil.

Bruerton, K., (2001). Finding practical solutions to mycotoxins in commercial production: A nutritionist's perspective. In*: Proceedings of the Alltech's 17th Annual Symposium* (pp. 161–168). Queensland, Australia.

Campagnollo, B. F., Ganev, K. C., Khaneghan, A. M., Portela, J. B., Cruz, A. G., Granato, D., Corassin, C. H., et al., (2016). The occurrence and effect of unit operations for dairy products processing on the fate of aflatoxin M1: A review. *Food Control, 68*, 310–329.

Carvajal, M., Rojo, F., Méndez, I., & Bolaños, A., (2003). Aflatoxin B_1 and its interconverting metabolite aflatoxicol in milk: The situation in Mexico. *Food Additives & Contaminants, 20*, 1077–1086.

Cavallarin, L., Antoniazzi, S., Giaccone, D., Tabacco, E., & Borreani, G., (2014). Transfer of aflatoxin M_1 from milk to ripened cheese in three Italian traditional production methods. *Food Control, 38*, 174–177.

Centers for Disease Control & Prevention (CDC), (2004). Outbreak of aflatoxins poisoning-eastern and central provinces, Kenya. *Morbidity and Mortality Weekly Report, 3*, 790–793.

Chavarría, G., Molina, A., Leiva, A., Méndezm, G., Wong-González, E., Cortés-Muñoz, M., Rodríguez, C., & Granados-Chinchilla, F., (2017). Distribution, stability, and protein interactions of aflatoxin M_1 in fresh cheese. *Food Control, 73*, 581–586.

Codex Alimentarius Commission, (2001). Comments submitted on the draft maximum level for aflatoxin M_1 in milk. *Codex Committee on Food Additives and Contaminants 33rd Session*. Hague, The Netherlands.

Coffey, R., Cummins, E., & Ward, S., (2009). Exposure assessment of mycotoxins in dairy milk. *Food Control, 20*, 239–249.

Corassin, C. H., Bovo, F., Rosim, R. E., & Oliveira, C. A. F., (2013). Efficiency of Saccharomyces cerevisiae and lactic acid bacteria strains to bind aflatoxin M_1 in UHT skim milk. *Food Control, 31*, 80–83.

Creppy, E. E., (2002). Update of survey, regulation and toxic effects of mycotoxins in Europe. *Toxicology Letters, 27*, 19–28.

Dalié, D. K. D., Deschamps, A. M., & Richard-Forget, F., (2010). Lactic acid bacteria – potential for control of mould growth and mycotoxins: A review. *Food Control, 21*, 370–380.

El-Khoury, A., Atoui, A., & Yaghi, J., (2011). Analysis of aflatoxin M_1 in milk and yogurt and AFM_1 reduction by lactic acid bacteria used in Lebanese industry. *Food Control, 22*, 1695–1699.

El-Nezami, H., Kankaanpaa, P., Salminen, S., & Ahokas, J., (1998). Ability of dairy strains of lactic acid bacteria to bind a common food carcinogen, aflatoxin B_1. *Food and Chemical Toxicology, 36*, 321–326.

Elsanhoty, R. M., Salam, S. A., Ramadan, M. F., & Badr, F. H., (2014). Detoxification of aflatoxin M_1 in yogurt using probiotics and lactic acid bacteria. *Food Control, 43*, 129–134.

European Commission, (2006). European Commission, regulation (EC) no. 1881 (2006): 19 of December 2006, setting maximum levels for certain contaminants in foodstuffs. *Official Journal, L364*, 5–24.

Flores-Flores, M. E., Lizarraga, E., Cerain, A. L., & González-Peñas, E., (2015). Presence of mycotoxins in animal milk: A review. *Food Control, 53*, 163–176.

Gazzotti, T., Lugoboni, B., Zironi, E., Barbarossa, A., Serraino, A., & Pagliuca, G., (2009). Determination of fumonisin B_1 in bovine milk by LCeMS/MS. *Food Control, 20*, 1171–1174.

Gelderblom, W. C. A., Marasas, W. F. O., Lebepe-Mazur, S., Swanevelder, S., Vessey, C. J., De-La, M., & HALL, P., (2002). Interaction of fumonisin B_1 and aflatoxin B_1 in a short-term carcinogenesis model in rat liver. *Toxicology, 171*, 161–173.

Haskard, C. A., El-Nezami, H. S., Kankaanpaa, P. E., Salminen, S., & Ahokas, J. T., (2001). Surface binding of aflatoxin B_1 by lactic acid bacteria. *Applied and Environmental Microbiology, 67*, 3086–3091.

Huang, L. C., Zheng, N., Zheng, B. Q., Wen, F., Cheng, J. B., Han, R. W., Xu, X. M., Li, S. L., & Wang, J. Q., (2014). Simultaneous determination of aflatoxin M_1, ochratoxin A, zearalenone and α-zearalenol in milk by UHPLC–MS/MS. *Food Chemistry, 146*(9), 242–249.

Iha, M. H., Barbosa, C. B., Okada, I. A., & Truksess, M. W., (2011). Occurrence of aflatoxin M_1 in dairy products in Brazil. *Food Control, 22*, 971–1974.

International Agency of Research on Cancer- IARC, (2002). *Monographs on the Evaluation of Carcinogenic Risks to Humans* (Vol. 82). WHO, Lyon.

Iqbal, S. Z., Jinap, S., Pirouz, A. A., & Faizal, A. A. R., (2015). Aflatoxin M_1 in milk and dairy products, occurrence and recent challenges: A review. *Trends in Food Science & Technology, 46*, 110–119.

Ji, C., Fan, Y., & Zhao, L., (2016). Review on biological degradation of mycotoxins. *Animal Nutrition, 2*, 127–133.

Kabak, B., & Var, I., (2008). Factors affecting the removal of aflatoxin M_1 from food model by *Lactobacillus* and *Bifidobacterium* strains. *Journal of Environmental Science and Health Part B, 43*, 617–624.

Kang'ethe, E. K., Sirma, A. J., Murithi, G., Mburugu-Mosoti, C. K., Ouko, E. O., Korhonen, H. J., Nduhiu, G. J., et al., (2017). Occurrence of mycotoxins in food, feed, and milk in two counties from different agroecological zones and with historical outbreak of aflatoxins and fumonisins poisonings in Kenya. *Food Quality and Safety,* 1–9.

Karazhiyan, H., Sangatash, M., Karazhyan, R., Mehrzad, A., & Haghighi, E., (2016). Ability of different treatments of saccharomyces cerevisiae to surface bind aflatoxin M_1 in yoghurt. *Journal Agriculture Science Technology, 18*, 1489–1498.

Lee, H. J., & Ryu, D., (2015). Advances in mycotoxin research: Public health perspectives. *Journal of Food Science, 80*, 2970–2983.

Li, S., Min, L., Wang, P., Zhang, Y., Zheng, N., & Wang, J., (2017). Occurrence of aflatoxin M_1 in pasteurized and UHT milks in China in 2014–2015. *Food Control, 78*, 94–99.

Mallebrera, B., Meca, G., Manyes, L., Mañes, J., & Font, G., (2013). Influence of pro- and prebiotics on gastric, duodenal and colonic bioaccessibility of the mycotoxin beauvericin. *Journal of Food Composition and Analysis, 32*, 141–149.

Mao, J., Zheng, N., Wen, F., Guo, L., Fu, C., Ouyang, H., Zhong, L., Wang, J., & Lei, S., (2018). Multi-mycotoxins analysis in raw milk by ultra high performance liquid chromatography coupled to quadrupole orbitrap mass spectrometry. *Food Control, 84*, 305–311.

Mercosul, GMC, (2002). Resolution 25/02. *Mercosul Technical Regulation about Maximum Tolerable Limits for Aflatoxin in Milk, Peanuts, and Corn.*

Michling, N., Signorini, M., Gaggiotti, M., Chiericatti, C., Basílico, J. C., Repetti, M. R., & Beldomenico, H. R., (2016). Risk factors associated with the presence of aflatoxin M_1 in raw bulk milk from Argentina. *Food Control, 64*, 151–156.

Murphy, P. A., Hendrich, S., Landgren, C., & Bryant, C. M., (2006). Food mycotoxins: An update. *Journal Food Science, 71*, 51–65.

Oatley, J. T., Rarick, M. D., Ji, G. E., & Linz, J. E., (2000). Binding of aflatoxin B1 to bifidobacteria *in vitro*. *Journal Food Protection, 63*, 1133–1136.

Patel, A., Thakkar, P., Shah, N., Verma, D. K., & Srivastav, P., (2018). Chapter 7: Significance of food-grade microorganisms in biodegradation and biotransformation of harmful toxic compounds. In: *Bioprocess Technology in Food and Health: Potential Applications and Emerging Scope* (pp. 227–250). Publisher: Apple Academic Press, Inc., [AAP]- CRC Press.

Pattono, D., Gallo, P. F., & Civera, T., (2011). Detection and quantification of ochratoxin A in milk produced in organic farms. *Food Chemistry, 127*, 374–377.

Pattono, D., Grosso, A., Stocco, P. P., Pazzi, M., & Zeppa, G., (2013). Survey of the presence of patulin and ochratoxin A in traditional semi-hard cheeses. *Food Control, 33*, 54–57.

Sangsila, A., Faucet-Marquis, V., Pfohl-Leszkowicz, P., & Itsaranuwat, P., (2016). Detoxification of zearalenone by *Lactobacillus pentosus* strains. *Food Control, 62*, 187–192.

Sarlak, Z., Rouhi, M., Mohammadi, R., Khaksar, R., Mortazavian, A. M., Sohrabvandi, S., & Garavand, F., (2017). Probiotic biological strategies to decontaminate aflatoxin M_1 in a traditional Iranian fermented milk drink (Doogh). *Food Control, 71*, 152–159.

Scaglioni, P., Becker-Algeri, T., Drunkler, D., & Badiale-Furlong, E., (2014). Aflatoxin B_1 and M_1 in milk. *Analytica Chimica Acta,* 829, 68–74.

Serrano-Niño, J. C., Cavazos-Garduño, A., Cantú-Cornelio, F., González-Córdova, A. F., Vallejo-Córdoba, B., Hernández-Mendoza, A., & García, H. S., (2015). *In vitro* reduced availability of aflatoxin B_1 and Acrylamide by bonding interactions with teichoic acids from *Lactobacillus* strains. *LWT – Food Science and Technology, 64*, 1334–1341.

Serrano-Niño, J. C., Cavazos-Garduño, A., Hernandez-Mendoza, A., Applegate, B., Ferruzzi, M. G., Martin-González, M. F. S., García, H. S., et al., (2013). Assessment of probiotic strains ability to reduce the bio-accessibility of aflatoxin M_1 in artificially contaminated milk using an *in vitro* digestive model. *Food Control, 31*, 202–207.

Shahbazi, Y., Nikousefat, Z., & Karami, N., (2017). Occurrence, seasonal variation and risk assessment of exposure to aflatoxin M1 in Iranian traditional cheeses. *Food Control, 79*, 356–362.

Shaker, E. M., & Elsharkawy, E. E., (2015). Organochlorine and organophosphorus pesticide residues in raw buffalo milk from agro-industrial areas in Assiut, Egypt. *Environmental Toxicology and Pharmacology, 39*, 433–440.

Shcherbakova, L., Statsyuk, N., Mikityuk, O., Nazarova, T., & Dzhavakhiya, V., (2015). Aflatoxin B_1 degradation by metabolites of phomaglomerata PG41 isolated from natural substrate colonized by aflatoxigenic *Aspergillus flavus*. *Jundishapur Journal Microbiology, 8*(1), e24324.

Shetty, P. H., & Jespersen, L., (2006). *Saccharomyces cerevisiae* and lactic acid bacteria as potential mycotoxin decontaminating agents. *Trends in Food Science and Technology, 17*, 48–55.

Shori, A. B., (2015). The potential applications of probiotics on dairy and non-dairy foods and their viability during storage. *Biocatalysis and Agricultural Biotechnology, 4*, 423–431.

Skrbic, B., Zivancev, J., Antic, I., & Godula, M., (2014). Levels of aflatoxin M_1 in different types of milk collected in Serbia: Assessment of human and animal exposure. *Food Control, 40*, 113–119.

Taheur, F. B., Fedhila, K., Chaieb, K., Kouidhi, B., Bakhrouf, A., & Abrunhosa, L., (2017). Adsorption of aflatoxin B_1, zearalenone and ochratoxin A by microorganisms isolated from Kefir grains. *International Journal of Food Microbiology, 25*, 1–7.

Tavares, A. M., Alvito, P., Loureiro, S., Louro, H., & Silva, M. J., (2013). Multi-mycotoxin determination and *in vitro* combined cytotoxic effects of aflatoxin M_1 and ochratoxin. *World Mycotoxin Journal, 6*, 375–388.

Tomasevic, I., Petrovic, J., Jovetic, M., Raicevic, S., Milojevic, M., & Miocinovic, J., (2015). Two year survey on the occurrence and seasonal variation of aflatoxin M_1 in milk and milk products in Serbia. *Food Control, 56*, 64–70.

Verma, D. K., Mahato, D. K., Billoria, S., Kapri, M., Prabhakar, P. K., Ajesh Kumar, V., & Srivastav, P. P., (2017). Microbial spoilage in milk products, potential solution, food safety and health issues. In: Verma, D. K., & Srivastav, P. P., (eds.), *Microorganisms in Sustainable Agriculture, Food and the Environment* (Vol. 1, pp. 171–196). As part of book series on innovations in agricultural microbiology, Apple Academic Press, USA.

Williams, J. H., Phillips, T. D., Jolly, P. E., Stiles, J. K., Jolly, C. M., & Aggarwal, D., (2004). Human aflatoxicosis in developing countries: A review of toxicology, exposure, potential health consequences, and interventions. *The American Journal of Clinical Nutrition, 5*, 1106–1122.

Wochner, K. F., Becker-Algeri, T. A., Colla, E., & Drunkler, D. A., (2018). The action of probiotic microorganisms on chemical contaminants in milk. *Critical Reviews in Microbiology, 44*, 112–123.

Zheng, N., Li, S. L., Zhang, H., Min, L., Gao, N., & Wang, J. Q., (2017). A survey of aflatoxin M_1 of raw cow milk in China during the four seasons from 2013 to 2015. *Food Control, 78*, 176–182.

Zoghi, A., Khosravi-Darani, K., & Sohrabvandi, S., (2014). Surface binding of toxins and heavy metals by probiotics. *Mini-Reviews in Medicinal Chemistry, 14*, 84–98.

CHAPTER 2

New Technological Trends in Probiotics Encapsulation for Their Stability Improvement in Functional Foods and Gastrointestinal Tract

MAJID NOOSHKAM* and ZAHRA ZAREIE

*Corresponding author. E-mail: Nooshkamma@gmail.com

ABSTRACT

The ability of probiotics to alter gut community and render health-promoting effect, induce food producers to use them as bioactive agents. However, they are sensitive to adverse conditions that may encounter through production and in the human stomach. As a matter of fact, researchers tried to increase their bioavailability through various methods. Encapsulation is known as a promising technique that has been frequently applied to protect and deliver probiotic bacteria (i.e., *Lactobacillus* and *Bifidobacterium* species). The present chapter discusses the encapsulation of probiotics in many delivery systems such as emulsion, hydrogels, microgels, microcapsules, colloidosomes, and the Maillard reaction products (MRPs). It then highlights the viability and bioavailability of the encapsulated probiotics in food products and gastrointestinal (GI) tract.

2.1 INTRODUCTION

In recent years, in the light of significant demands for health-enhancing foods, the concept of 'functional foods' has been emerged in the food industry to fortify food products with bioactive ingredients having health benefits. In this way, the incorporation of probiotic microorganisms into food is an outstanding approach to produce functional foods. Probiotic strains have many documented health-related properties and it is believed that the living LAB with probiotic effect play an important and beneficial role in the intestinal tract of the host (Heidebach et al., 2012). However, the loss of the number of living probiotic cells (i.e., bioactivity loss) during production, storage, and gastrointestinal (GI) transition is a key issue, and therefore several attempts have been made to minimize the bioactivity loss of probiotics (Mattila-Sandholm et al., 2002; Siuta-Cruce and Goulet, 2001; Shah, 2000).

In this area, the most prominent technique is encapsulation with potential to significantly protect living probiotic cells or other microorganisms from adverse conditions (Nualkaekul et al., 2013). Encapsulation has received special attention in the food science to add functional ingredients, enzymes, colorants, antioxidants, and microbial products into foods; in the medical industry to produce drugs and vaccines; and in the tissue-engineered medical products as scaffolds (Borges et al., 2012). Encapsulation is known as a technology of packing compounds with solid, liquid, or gas nature in sealed capsules to release them through a controlled mode in the right place or time. The packed compounds or materials are commonly called fill, internal phase, payload, actives, and core material; whilst, the coating material, wall material, carrier, shell, capsule, and membrane are the terms used for the packaging materials. It is worth to pointing out that polysaccharides (natural or modified), proteins, lipids, synthetic polymers, gums, and sugars can be employed as wall materials (Fang and Bhandari, 2010).

In general, the encapsulation technique is used in the pharmaceutical and food sectors for different reasons; it is applied to: (i) decrease the reactivity of core material to its environment and, in turn, lower its degradation rate; (ii) lower the transfer/evaporation rate of core material to its surrounding environment; (iii) provide an easier handling by modifying the physical properties of the core material; (iv) separate the mixture components for lowering their reactivity; (v) dilute the encapsulated compound when its low level is required; (vi) cover or mask the unpleasant taste or flavor of the encapsulant; (vii) ameliorate the food safety *via* suppressing the microbial proliferation; (viii) boost water solubility of bioactive ingredients; and (ix) tailor the releasing rate of the encapsulant for providing a slower release and in a controlled mode (Cohen et al., 2011; Desai and Park, 2005; Wang et al., 2015; Zhu, 2017).

In probiotic encapsulation, the microbial cells are randomly embedded and immobilized in a continuous matrix as core materials (Desai and Park, 2005), and therefore the terms "encapsulation," "immobilization," and "entrapment" are commonly used as synonyms in this regard (Krasaekoopt et al., 2003; Anal and Singh, 2007). Different microorganisms have been encapsulated in biocompatible and semipermeable materials capable of modulating the delivery of probiotic cells (Gbassi and Vandamme, 2012). The aim of probiotic encapsulation is to decrease the loss of living cell numbers in foods and finally in the human gut. In addition, the encapsulated probiotic should be released completely from the capsule in the intestinal tract. It is noteworthy that the size of the capsule should be remarkably small to prevent adverse sensorial properties on the functional foods containing probiotics (Naidu et al., 1999).

This chapter reviews the importance of probiotic microorganisms in food systems, factor influencing their bioactivity, and different approaches of probiotic encapsulation. In addition, the stability of the encapsulated probiotics in food products and gastrointestinal tract (GIT) has been discussed.

2.2 THE IMPORTANCE OF PROBIOTICS IN FOOD SYSTEMS

In the last decades, there has been a great deal of attention in the beneficial role of probiotic microorganisms in human health. Probiotics render their positive health effects when they reach to the action sites alive and proliferate in certain numbers. It has been recommended by the International Dairy Federation that the probiotic bacteria should be present at least 10^7 CFU g^{-1} in active form in the product to exert their health-promoting effects (Sultana et al., 2000).

Probiotics are recognized as live microbial supplements with the ability to improve the microbial balance of the host intestine. They have many claimed benefits such as, controlling cholesterol levels, inhibition of the growth of food pathogenic and poisoning bacteria in the digestive tract, anticarcinogenic effect, acid production, bacteriocin formation, the improvement of immune system, and the production of β-galactosidase with lactose hydrolyzing activity, suitable for people suffering from lactose intolerance (Krasaekoopt et al., 2003).

The positive effects are dependent on the survivability and proliferation of the probiotic strains in the GI of the host. Furthermore, most products containing probiotics on the market have a very short shelf-life, do not usually contain the required viable cell numbers to possess a prebiotic effect, and viable cells content decreases significantly after consumption due to bile secretions and acidic environment in the GI (Todorov et al., 2012). The survivability of probiotic cells in food products is influenced by many factors include storage temperature, pH, H_2O_2 production, and post-acidification during the fermentation process. Hence, the protection of probiotic bacteria within a physical barrier against harsh circumstances is a technique currently gaining remarkable research and industrial attentions (Gbassi and Vandamme, 2012). In this context, encapsulation approaches have been employed to ameliorate the survivability and delivery of probiotic microorganisms at controlled rates. The encapsulated or entrapped bacteria indicate prolonged viability, manifest heat stability, and improved survivability under *in vitro* digestion conditions (Santillo et al., 2012).

2.3 PROBIOTIC ENCAPSULATION

The encapsulation of probiotic in a matrix with a physical barrier in order to protect them towards harsh environmental conditions has been trailed by using different materials and approaches. Many demands have therefore been demonstrated to be associated with a successful probiotic encapsulant includes: (i) the materials used to encapsulate probiotics must be inexpensive, food-grade, and compatible with food; (ii) the encapsulation approach must be inexpensive, simple, and must not lower the viability of the probiotic microorganisms; (iii) the efficiency of the encapsulation process must be close to 100%; (iv) the obtained capsules must contain a significantly high loading of the living cells; (v) the texture and taste of the food and beverages must not be adversely impacted by the capsules (small capsules with <30 mm sizes); (vi) the capsules must provide superb conditions to protect the probiotic bacteria against environmental stresses such as moisture and acidic conditions during production and storage period; and (vii) the capsules must protect the living cells during GI transit and eventually release them in the gut at the right time or action sites (Lee and Salminen, 2009).

The encapsulation processes are currently performed by several approaches such as, extrusion, emulsification, spray drying, hydrogels, electrospinning, coacervation, colloidosomes, and impinging aerosol technology (IAT). The capsules obtained by the encapsulation have nano (<1.0 μm), micro (1.0–5,000 μm), or macro (>5,000 μm) sizes (Esfanjani et al., 2018).

2.3.1 SPRAY DRYING

Spray drying is known as “the atomization of an oily or aqueous suspension, containing probiotics and encapsulating materials, into hot gas in a vessel to evaporate water rapidly.” It is commonly used in the food sector as a relatively cheap and effective method for probiotic encapsulation. However, high temperatures applied in this method may lead to a decrease in a viable number of probiotic strains. In this regard, some protectants such as, soluble fiber, non-fat milk solids, trehalose, and granular starch should be added to the suspension before the drying process to minimize mechanical, oxidative, and osmotic stresses subjected to probiotic cells during drying at high temperatures (Manojlović et al., 2010). Furthermore, an additional layer can be used on the surface of spray-dried beads to provide further protection ability for probiotics (Semyonov et al., 2010). Various polymers have been applied to encapsulate probiotics through the spray drying method (Table 2.1).

TABLE 2.1 Examples of Encapsulated Probiotics by Spray Drying Technique

Carrier Materials	Probiotic Microorganisms	References
Modified starch	*Bifidobacterium* PL1	O'Riordan et al. (2001)
Reconstituted skim milk and polydextrose	*Lactobacillus rhamnosus GG*	Corcoran et al. (2004)
Gum Arabic	*Bifidobacterium bifidum*	Rodríguez-Huezo et al. (2007)
Whey protein and resistant starch	*L. rhamnosus* GG	Ying et al. (2013)
Whey solution	*Lactobacillus reuteri*	Jantzen et al. (2013)
Whey protein isolate	*Lactobacillus acidophilus* Ki	Rodrigues et al. (2011)
	Lactobacillus paracasei L26	
	Bifidobacterium animalis BB-12	
Maltodextrin + gum Arabic	*L. acidophilus* BCRC 14079	Su et al. (2007)
	Bifidobacterium longum BCRC 14605	
β-cyclodextrin + gum Arabic	*L. acidophilus*	Zhao et al. (2008)
Gelatin	*B. longum B6*	Lian et al. (2003)
	Bifidobacterium infantis CCRC 14633	
Gelatin, whey protein concentrate, modified starch, maltodextrin, pea protein isolate, gum Arabic	*Saccharomyces cerevisiae* var. *boulardii*	Arslan et al. (2015)
Maltodextrin, inulin, oligofructose	*Bifidobacterium animalis* spp. Lactis	Paim et al. (2016)
Inulin	*L. acidophilus* La-5	dos Santos et al. (2019)

2.3.2 EXTRUSION TECHNIQUE

Extrusion is a physical immobilization or microencapsulation method to encapsulate probiotic microorganisms because it can be a cheap and simple process in conjugation with operation at low temperatures (Favaro-Trindade et al., 2011). In this approach, a probiotic-rich solution is pressingly projected through a nozzle (Figure 2.1). The technique is sometimes known as prilling when the droplet formation is occurred in a controlled environment *via* vibration or pulsation of the jet nozzle. In addition, the droplets can be formed through other common technique by applying an electrostatic field or coaxial flow (Burgain et al., 2011).

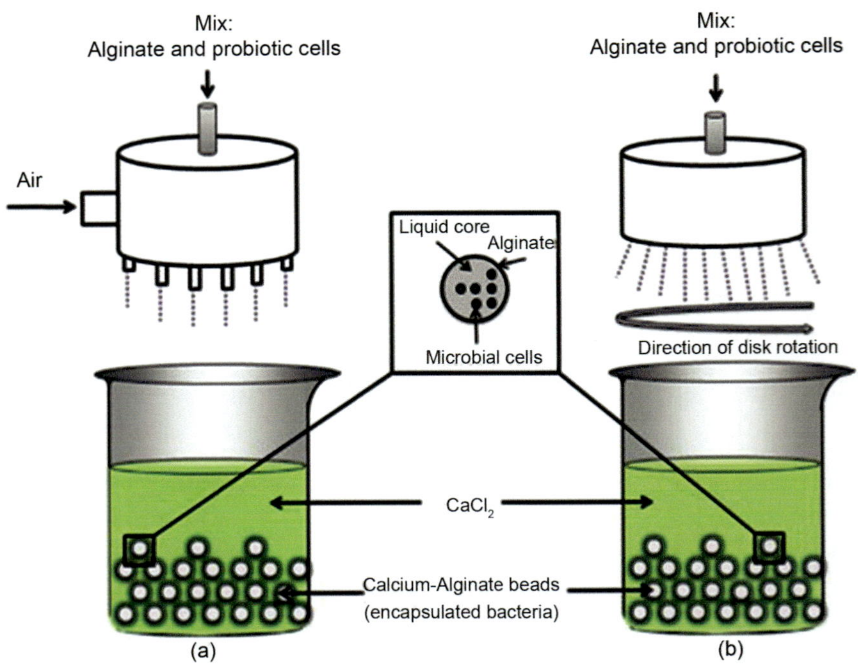

FIGURE 2.1 Extrusion techniques: A schematic representation of air-driven (a); and pinning disk device (b) based needle droplet-generator. The hydrocolloidally probiotic-rich solution is dripped through a nozzle spray machine or a syringe needle in droplet forms that are dropped freely into a hardening solution like $CaCl_2$ (Burgain et al., 2011).

Different polymers can be applied to produce capsules by the extrusion method (Table 2.2). However, whey proteins, k-carrageenan, and alginate are the most used agents for this purpose, and the latter has been frequently used as a superb hydrocolloid in the method. The obtained capsules by the extrusion procedure are commonly named as beads with different sizes ranged between micrometers and millimeters (Favaro-Trindade et al., 2011).

Milk-alginate microspheres were used to encapsulate *Lactobacillus bulgaricus* through the extrusion method to improve its survivability against harsh conditions, i.e., low pHs (2.0 and 2.5), high bile salt concentration (1.0 and 2.0%), and long storage time (30 days). The encapsulation did not change the probiotic viability after 2.0 h incubation in the simulated gastric digestion with pH 2.5, and more than 8 log CFU g^{-1} of the encapsulated probiotic was kept in gastric fluid pH 2.0 after incubation time of 2.0 h. After 1.0 and 2.0 h exposure in 1% bile salt solution, the encapsulated probiotic underwent a viability decrease from 9.98 log CFU g^{-1} to 9.24 and 8.48 log CFU g^{-1}, respectively. For 2% bile salt solution, approximately 1.3 and 2.1 log CFU g^{-1} of the encapsulated probiotic was reduced after 1 and 2 h incubation. Interestingly, the extrusion procedure preserved the full survivability of encapsulated probiotic after a long storage period of 30 days at 4°C. In addition, the entrapped probiotic was completely released in alginate-milk microspheres during 1 h digestion (Shi et al., 2013).

Whey proteins have also been used in the extrusion method to encapsulate probiotic bacteria. In this context, protein hydrolysis by digestive enzymes could liberate bioactive peptides (BPs) with multi-physiological activities. Doherty et al. (2011) reported that probiotics can be protected in the particles during 3 h digestion in simulated gastric conditions when whey protein was employed as an encapsulation material. The mixture of hydrocolloids has been applied for probiotic encapsulations. In a study conducted by Gebara et al. (2013), pectin microparticles containing *L. acidophilus* were coated with whey protein. This resulted in a greater protective effect to the probiotic in comparison to the non-encapsulated cells. However, the microencapsulation system did not confer extra protective role to *L. acidophilus* under simulated GI conditions compared to un-coated pectin microparticles. In contrast to these results, the improved survivability of probiotics in whey protein-coated pectin particles compared to uncoated counterparts was found by Gerez et al. (2012).

Generally, in extrusion technique, the flowability and dispersibility of capsules in food products is achieved when the small capsules are formed. It is worth mentioning that the capsules with larger sizes will result in more harmful effects on the food texture. The extrusion process is simply performed in laboratory scales, but its low productivity is a main problem to use it in large-scale productions. As well, the disintegration of particles in media containing phosphate and citrate is another drawback of the technique (Favaro-Trindade et al., 2011).

TABLE 2.2 Different Carrier Materials Employed to Improve the Survivability of Probiotic Strains in Extrusion Technology

Encapsulation Materials	Probiotic Strains	References
Alginate	*B. bifidum, B. lactis, B. longum, L. acidophilus, L. lactis*	Özer et al. (2008, 2009); Prevost and Divies (1992); Lee and Heo (2000)
Gellan/xanthan gum	*B. infantis*	Sun and Griffiths (2000)
Raftilose and raftiline	*L. acidophilus*	Anjani et al. (2004)
Alginate/chitosan	*B. bifidum, B. lactis, L. acidophilus, L. casei, L. gasseri*	Krasaekoopt et al. (2006); Urbanska et al. (2007); Kailasapathy et al. (2008); Mortazavian et al. (2008); Chávarri et al. (2010)
Whey protein	*L. rhamnosus*	Reid et al. (2007)
Gellan/xanthan/pullulan/jamilan	*L. plantarum, L. rhamnosus*	Jiménez-Pranteda et al. (2012)
Whey protein/pectin	*L. rhamnosus*	Gebara et al. (2013)
Alginate /pectin	*B. longum, L. casei, L. plantarum*	Sandoval-Castilla et al. (2010); Nualkaekul et al. (2013)
Alginate/milk	*L. bulgaricus*	Shi et al. (2013)
Algniate/human-like collagen	*B. longum*	Su et al. (2011)
κ-carrageenan	*L. acidophilus*	TSEN et al. (2008)
Sodium caseinate/cow milk, goat milk, and casein hydrolysate	*B. longum*	Prasanna and Charalampopoulos (2018)

2.3.3 EMULSION TECHNIQUE

Probiotics have been successfully encapsulated by emulsion technique because it can be scaled up easily, flexibility adjust the capsule size, and provide a high probiotic survival rate. In this technique, emulsifier addition, energy input during emulsion, and the viscosity ratio between continuous and dispersed phases, are the main parameters that should be considered regarding the control of microbead sizes. It is noteworthy that a second polymer can be used to coat microbeads in order to improve the survivability of probiotic bacteria further (Dong et al., 2013). Although spray drying and extrusion are the most common methods for probiotic encapsulation, reduction in bacterial viability due to high air temperature applied in spray-drying along with the unpleasant sensory perception in food containing large-sized beads (2–5 mm) obtained by the extrusion technique have led to the use of water-in-oil (W/O) emulsions as a promising method to add health-enhancing bacteria into food matrices because of the lack of severe processing conditions, leading to the preservation of cell viability (Picone et al., 2017). Oil-in-water (O/W), W/O, and water-in-oil-in-water (W/O/W) emulsions have been used for probiotics encapsulation (Table 2.3).

TABLE 2.3 Examples of Encapsulated Probiotics by Emulsion-based Techniques

Emulsion Type and Carrier Materials	Probiotic Microorganisms	References
O/W Emulsions		
Corn oil micro-droplets emulsified with peptides	*Pediococcus acidilactici*	Papagianni and Anastasiadou (2009)
W/O Emulsions		
Alginate	*L. rhamnosus*, *L. plantarum*, *L. paracasei*, *L. acidophilus*, *L. salivarius*, *B. longum*, and *B. lactis*	Ding and Shah (2007)
Alginate	*L. acidophilus*	Sabikhi et al. (2010)
Double Emulsions		
Decaglycerol monolaurate and hexaglyceryl condensed ricinoleate	*L. acidophilus*	Shima et al. (2006)
Decaglycerol monolaurate and hexaglyceryl condensed ricinoleate	*L. acidophilus*	Shima et al. (2009)
Concentrated sweet whey	*L. rhamnosus*	Pimentel-González et al. (2009)
Milk	*L. paracasei*	El Kadri et al. (2018)

By encapsulating the probiotics (i.e., *Lactobacillus rhamnosus*, *L. plantarum*, *L. paracasei*, *L. acidophilus*, *L. salivarius*, *Bifidobacterium longum*, *B. lactis* B1-04, and *B. lactis* Bi-07) in an alginate matrix *via* a W/O emulsion approach and exposing them to ox gall (bile salt) and 65°C, Ding and Shah (2007) demonstrated that the viability of probiotics was decreased by 6.51 and 3.36 log CFU mL^{-1} in free and encapsulated probiotic bacteria, respectively, in the presence of bile salt. The heat treatment for 30 min led to a more decrease in non-encapsulated bacteria (6.74 log CFU mL^{-1}) in comparison to encapsulated ones (4.17 log CFU mL^{-1}). However, a similar loss was observed in both free and encapsulated strains upon heat treatment for 1.0 h at 65°C. In another research, *L. acidophilus* was successfully encapsulated in sodium alginate through W/O method (Sabikhi et al., 2010). The encapsulation resulted in a higher probiotic survivability at high temperatures (72, 85, and 90°C) and high salt levels (1.0, 1.5, and 2.0%); 4.14 log reduction and a complete destruction were observed in encapsulated and free cells, respectively, after heat treatment at 90°C. In addition, the incubation of probiotic at 2.0% bile salt for 3.0 h decreased the number of viable cells by 2.16 log reduction in the encapsulated cells and 5.47 log reductions in free probiotic cells.

O/W emulsion system containing corn oil micro-droplets emulsified with peptides was developed to encapsulate *Pediococcus acidilactici* cells. About 85% viability rates of the encapsulated cells was preserved by the encapsulation. As well, it was declared that the encapsulation process could release approximately 92% of the entrapped cells (Papagianni and Anastasiadou, 2009). W/O/W was also designed for the encapsulation of *L. acidophilus*. The emulsion presented 49% relative viability of the probiotic after 120 min digestion in the simulated gastric juice; whilst, the survivability of the non-encapsulated probiotic in the juice decreased to 1.3% at 40.2 min contacting time (Shima et al., 2006), suggesting that the emulsion technology is an efficacious technique to protect probiotics towards harsh processing circumstances and simulated digestive juices.

2.3.4 EMULSION GEL

Emulsion gel is recognized as a complex colloidal system that consists of both gel and emulsion phases. Emulsion-filled protein gel and protein-stabilized emulsion gel are two types of emulsion gels (Figure 2.2). In emulsion-filled protein gel, emulsion droplets are embedded in a protein gel matrix, whereas protein-stabilized emulsion gel is known as a particulate gel type, which its rheological properties is mainly determined by the network of aggregated emulsion droplets in the system (Dickinson, 2012).

Emulsion gels are able to simultaneously deliver both hydrophilic and hydrophobic nutraceuticals within a single formulation. In this way, probiotic cells and lipophilic drugs can be loaded in the water and oil phases of O/W emulsion gels, respectively. However, the susceptibility of oils, particularly vegetable ones towards oxidation reaction is a major issue faced with the development and storage of many oil-based emulsion gels. This is very important in probiotic encapsulation and delivery due to the fact that many of probiotics are anaerobic and sensitive to oxygen (Pandey et al., 2016). In this context, a modified emulsion gel was developed by using water as polar phase (contains *L. plantarum* and natural gums guar and xanthan), sunflower oil as non-polar phase (contains antimicrobial metronidazole), and sorbitan monopalmitate as a non-ionic emulsifier. The emulsion gel had higher probiotic viability at different storage conditions of 4°C, –20°C, and –196°C compared to the control sample. In addition, the emulgel provided a sustained delivery of metronidazole, which is a hydrophobic therapeutic compound usually used with *L. plantarum* (Pandey et al., 2016).

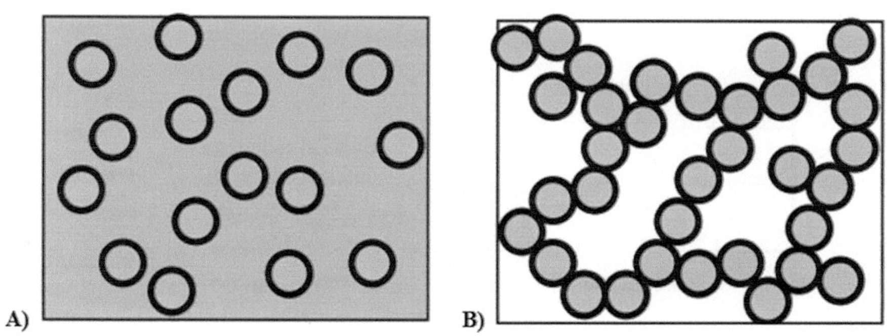

FIGURE 2.2 (A) emulsion-filled protein gel; and (B) protein-stabilized emulsion gel.
Source: Reprinted with permission from Dickinson, 2012. © Elsevier

In a study conducted by Picone et al. (2017), a novel delivery system was fabricated by the gelation of a W/O emulsion containing gellan gum as a carrier for *L. rhamnosus*. The obtained gels (microgels) were able to resist oral and gastric digestion, whilst the enteric fluid led the microgel destabilization. The gelled emulsions provided more than 77% probiotic viability under *in vitro* digestion, whereas non-gelled pairs showed a 66% decrease in the viability of *L. rhamnosus*.

2.3.5 HYDROGELS

Hydrogels are tridimensional polymer networks, which due to their inherent hydrophilicity have potential to entrap a large amount of solvent without being dissolved. They can form various physical structures such as coatings and microparticles or nanoparticles, with a unique structural tunability and high deformable properties. Therefore, hydrogels are suitable vehicles with versatile applications like drug delivery, biomedicine, tissue engineering, and encapsulation techniques (Falco et al., 2017).

Heidebach et al. (2009) gelled casein suspensions containing *B. lactis* Bb12 and *L. paracasei ssp. paracasei* F19 through transglutaminase-catalyzed gelation method (Figure 2.3). The encapsulation yields of water insoluble-spherical capsules were 70% for *L. paracasei* and 93% for *B. animalis*. The microencapsulation process provided a protective effect to the prebiotics under simulated gastric juice without pepsin (pHs 2.5 and 3.6 at 37°C for 90 min), mainly because of the covalent network obtained by the transglutaminase, which was resistant against dissolution at acidic pHs.

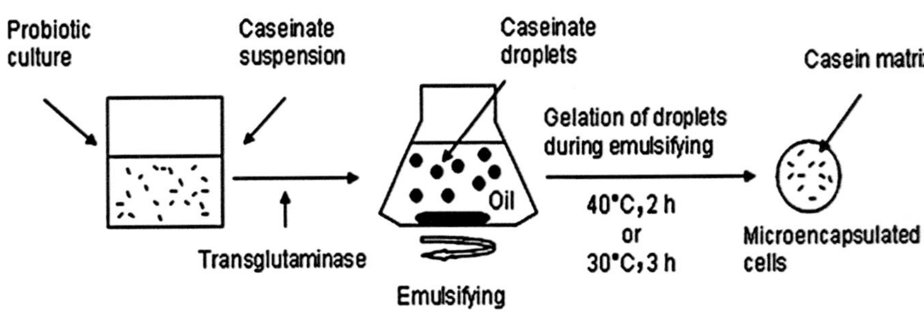

FIGURE 2.3 Schematic illustration of transglutaminase-induced gelation of casein suspension containing probiotic cells.

Source: Reprinted with permission from Heidebach et al., 2009. © Elsevier.

In another research, *L. plantarum* was successfully encapsulated in starch-pectin binary hydrogels. The viability of the encapsulated strain in adverse conditions of bile salt solution and simulated gastric juice was remarkably higher compared to the non-encapsulated form; free cells were almost completely destroyed under 2.0 h exposure to simulated gastric and intestinal fluids (Dafe et al., 2017).

Chitosan coated-alginate microgels enhanced the survivability of *B. longum* during passing through a mimicked GI (Yeung et al., 2016). Nonetheless, only a few studies demonstrated the successfully preserved viability of probiotic microorganisms in colloidal delivery vehicles after 2.0 h gastric digestion (Lee and Heo, 2000; Yeung et al., 2016). The poor gastric stability of probiotics embedded in hydrogels is mainly due to the fact that hydrogen ions could be easily diffused through the gel network of the hydrogels and inactivate the entrapped probiotics. In this context, antacid agents such as $Mg(OH)_2$ were added into alginate microgels to neutralize hydrogen ions and subsequently maintain the neutral internal pH during incubation of probiotic-loaded microgels in strong acidic gastric fluids (Zhang et al., 2017). Therefore, these antacid microgels are efficient carriers for improving the survivability of probiotic cells for oral delivery. Gu et al. (2019) encapsulated *B. pseudocatenulactum* G7 in alginate microgels containing $Mg(OH)_2$ and $CaCO_3$ as antacids in order to control the internal pH value of the microgel within the stomach. Their results showed that no live cells were quantified in antacid-free microgels after *in vitro* gastric digestion; whilst, only 1.5 log CFU of the probiotic cells decreased in antacid-loaded ones under the same conditions. In addition, a considerable number of viable cells was detected in $CaCO_3$-containing microgels after incubation in simulated intestinal fluids in comparison to $Mg(OH)_2$-loaded counterparts, which showed no viable cells.

Generally, food-grade hydrogels could be successfully used to encapsulate, protect, and deliver sensitive probiotic bacteria in functional foods, pharmaceuticals, and supplements.

2.3.6 MAILLARD REACTION PRODUCTS (MRPs) AS ENCAPSULATING MATERIALS

The Maillard reaction between reducing sugars and free amino-bearing compounds (e.g., proteins, peptides, and amino acids) generates various compounds from early Maillard reaction products (MRPs) to melanoidins. MRPs have some functionalities include antioxidant, antimicrobial, antihypertensive, antibrowning, and prebiotic activities. Recently, the Maillard reaction has gained a lot of research and industrial attention as a simple way to ameliorate techno-functional properties of food proteins. Protein-polysaccharide/oligosaccharide Maillard-based conjugates possess outstanding stability towards high temperature, acidic pHs, and high ionic strength in conjugation with remarkable oxidative stability (Nooshkam and Madadlou, 2016a, b; Nooshkam et al., 2018, 2019). Thus, MRPs are one of the latest promising food-born carriers to protect and deliver health-promoting molecules, such as probiotic microorganisms.

Whey protein concentrate (WPC) and isomaltooligosaccharide (IMO) were heated to obtain WPC-IMO conjugates and the WPC-IMO MRPs were then used to encapsulate *L. rhamnosus* through an emulsification-cold gelation technique. The obtained microsphere by MRPs had higher encapsulation efficiency (88.88%) in comparison to that produced from non-heated WPC-IMO mixture. It was reported that the network forming ability of WPC-IMO MRPs could lower acid diffusion rate to the corresponding system, resulting in an improved pH tolerance of *L. rhamnosus*. In addition, a good protection effect towards bile salt solution was provided by WPC-IMO physical mixtures and MRPs. However, there were no significant differences between the release behavior of the encapsulated probiotic and it was released quickly, likely due to the low cross-linking degree between protein and saccharide complexes (Liu et al., 2016).

The emulsification-cold gelation procedure was also employed to microencapsulate *L. rhamnosus* in WPC-IMO MRPs to modify its viability during white-brined cheese storage and GI transition (Liu et al., 2017). The encapsulation provided significantly higher viable cells in the cheese and after *in vitro* GI digestion. In addition, *B. longum* encapsulation by soy protein isolate (SPI)-carrageenan MRPs enhanced its stability towards prolong storage, pasteurization temperature, and GI conditions (Mao et al., 2018). These findings proved the ability of MRPs to form a strong, cohesive, and viscoelastic film around probiotics, which lowers the diffusing rate of acids and heat into microspheres. MRPs may also confer prebiotic effects to probiotic bacteria and increase their survivability under harsh conditions.

2.3.7 COMPLEX COACERVATES

Complex coacervation is based on mixing of two polymers with oppositely charges, which leads to the phase separation and generation of packed solid particles or liquid droplets. It is not only frequently used for flavor encapsulation, but also is a suitable technique for probiotic microencapsulation (Chávarri et al., 2012).

In a study, casein-pectin complex was applied to encapsulate *L. acidophilus* and *B. lactis* through complex coacervation-atomization technique to ensure high stability of the coacervated bacteria. The technique and the casein-pectin wall were efficient in protecting probiotic bacteria towards spray drying conditions and *in vitro* gastric digestion. *B. lactis* lost its survivability quickly during storage time, while *L. acidophilus* microencapsulated in the wall material maintained its viability for more prolonged storage period of 120 days at 7 and 37°C (Oliveira et al., 2007). Eratte et al. (2015) co-microencapsulated omega-3 fatty acid (tuna oil) and *L. casei* in whey protein isolate (WPI)-gum Arabic complex coacervation following spray and freeze-drying. *L. casei* had significantly higher viability in the co-microcapsules (i.e., WPI-gum Arabic) compared to WPI-probiotic-gum Arabic microcapsules, mainly due to oil droplet aggregation and subsequent cover by WPI-gum Arabic complex coacervates. This results in a better probiotic embedding in the complex coacervates and in turn improves its survivability (Eratte et al., 2017).

In general, the complex coacervation technology has some advantages compared to other techniques for probiotic encapsulation, such as a partially simple and low-cost process (i.e., it does not necessarily require organic solvents or high temperature) and the embedding of a large level of probiotics. However, it is known as a batch-scaled process and yields coacervates in aqueous solutions. Thus, an extra drying step should be applied to extend its shelf-life, which can be harmful to the probiotic cells (Chávarri et al., 2012).

2.3.8 ELECTROSPINNING TECHNIQUE

The electrospinning or electrospraying process is able to continuously generate submicron polymeric fibers by the action of an external electric field imposing on a polymer solution or melt (Dong et al., 2013). This technique has been recently used for probiotic encapsulation in nanofibers. In a study conducted by López-Rubio et al. (2009), the electrospinning method was used to encapsulate *B. animalis Bb12* in encapsulating polymer of polyvinyl alcohol (PVOH). The probiotic-incorporated fibers had a lower melting point and crystallinity and higher glass transition temperature compared to the probiotic-free PVOH counterparts. As well, *B. animalis* Bb12 encapsulated in electrospun PVOH fibers had significantly greater viable number during storage period of 40 days at ambient temperature and 130 days at cold conditions.

Studying the potential of soluble dietary fiber (SDF) in *L. acidophilus* nanoencapsulation through amending SDF solution with 8.0% POVH to form nanofibers using electrospinning technique, Fung et al. (2011) reported that the thermal behavior of the obtained nanofibers could provide thermal protection of probiotic microorganisms in heat-treated food products. Furthermore, about 78.69% of the bacteria remained viable under the process in conjugation with retaining its survivability during 21 days' refrigeration period. The technique has also been employed to encapsulate *B. animalis* Bb12 in WPI-pullulan fibers, and the viability of the probiotic strain was increased substantially (López-Rubio et al., 2012).

2.3.9 COLLOIDOSOMES

Colloidosome is a microcapsule with colloidal particles (1.0 nm to 1.0 μm dimension range) on its external shell. These colloidal particles form an ordered, stable, and regular structure on the colloidosome shell by self-assembly

(Rossier-Miranda et al., 2009). In general, the shell is fixed by heating the system above the glass transition temperature of the colloidal particles to fuse the particles together and so, shell formation (Dinsmore et al., 2002).

Colloidosomes have been used for encapsulation of probiotic bacteria. For instance, *L. crispatus* was successfully encapsulated polymeric colloidosomes (Keen et al., 2012). An aqueous suspension of *L. crispatus* and poly(methyl methacrylate-co-butyl acrylate) latex particles was firstly emulsified in an oil continuous phase, which was followed by adding low concentrations of salt and ethanol to the oil phase to fascinate latex particles aggregation at the emulsion droplet surfaces, leading to the formation of colloidosome shells. The encapsulated *L. crispatus* had the ability to metabolize glucose as a sugar source and generate acid slowly in comparison to the free bacteria, demonstrating transport limitation though the shell of colloidosome and lower cellular metabolism caused by encapsulation process. The colloidosome provided a significantly higher number of viable cells of the probiotic after gastric digestion compared to non-encapsulated bacteria.

2.3.10 IMPINGING AEROSOL TECHNOLOGY (IAT)

Impinging aerosol technology (IAT) utilizes two separate aerosols; one with calcium chloride solution and the other with microbial suspension in alginate solution (Figure 2.4). The calcium chloride solution and alginate solution are injected from the base and top of a cylinder, respectively, and alginate microbeads with <40 μm average diameter are then formed (Sohail et al., 2011). IAT is a suitable technology for the encapsulation of heat and solvent labile compounds as no solvent or heat is applied in this technique. As well, the technology provides a high production capacity and the obtained microbeads can be freeze or spray dried (Zhang et al., 2018).

In a study conducted by Sohail et al. (2011), the IAT-based microbeads and the extruded macrobeads (~2 mm diameter) presented similar protection to the probiotic *L. rhamnosus* GG towards acidic and bile salts conditions. The authors, in another research, investigated the microencapsulation effects on the survivability of bacteria *L. rhamnosus* GG and *L. acidophilus* NCFM, and their acidification in a food model system, orange juice, during storage periods of 9.0 days at 25°C and 35 days at 4°C. The non-encapsulated *L. rhamnosus* had superb viability in the model system at both temperatures; whilst, the free *L. acidophilus* experienced remarkable viability reduction. Although, the survivability of the bacteria did not manifestly enhance by the encapsulation, the process reduced the acidification rate at 4°C and 25°C.

Generally, *L. rhamnosus* had outstanding survivability in the orange juice and the encapsulation process has the ability to reduce negative acidification effects and possible adverse sensory effects of probiotics upon incorporating into orange juice or other fruit-based acidic food products (Sohail et al., 2012).

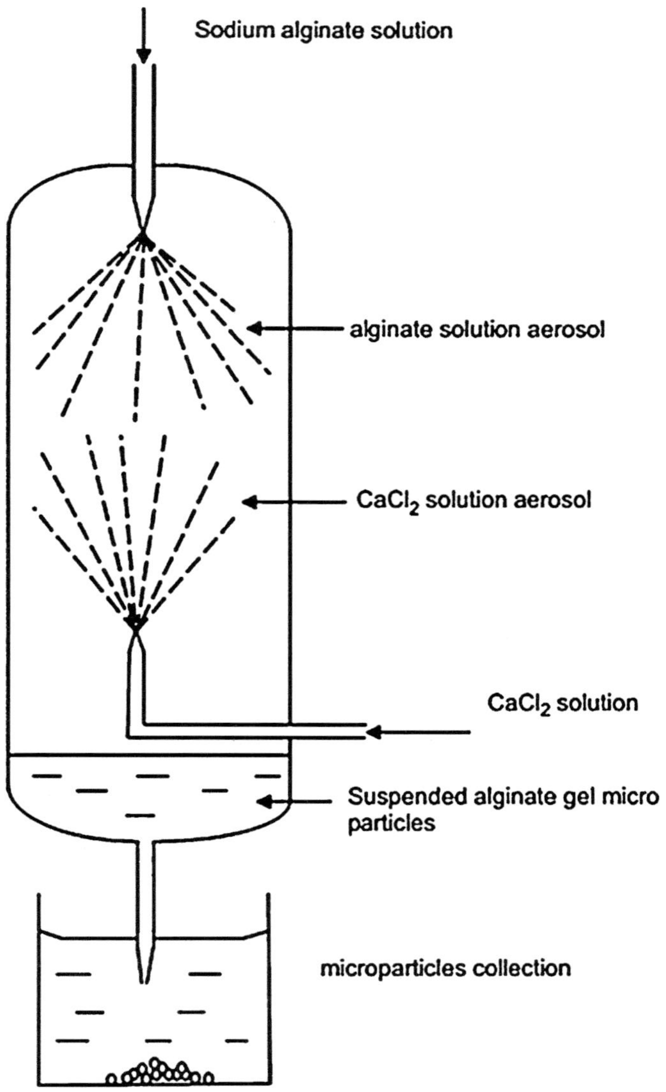

FIGURE 2.4 Schematic illustration of impinging aerosol technology (IAT).

Source: Reprinted with permission from Hariyadi et al., 2010). © Taylor & Francis.

2.4 STABILITY OF ENCAPSULATED PROBIOTICS IN FOODS AND GASTROINTESTINAL (GI) TRACT

2.4.1 *IN FOOD PRODUCTS*

The encapsulated probiotic might undergo some deteriorative stresses during processing, transportation, and storage of food products, such as (Manojlović et al., 2010):

- Mild heat treatments;
- Air incorporation during pumping, stirring, and homogenization;
- Freezing stress in frozen products;
- The addition of antimicrobial agents such as salt in cheese making;
- Drying process such as osmotic dehydration;
- Packaging (the ingress of oxygen through packaging material during storage);
- Inappropriate storage conditions such as the presence of oxygen and post-acidification in yogurt;
- The formation of large ice crystals during freezing and thawing of stored ice cream;
- The secretion of antimicrobial compounds from starter cultures during fermentation.

Culture companies have previously chosen the probiotic strains to resist the above-mentioned conditions. However, these companies have recently focused on the selection of probiotic strains based on their health-promoting and therapeutic effects. It is noteworthy that the viability of probiotic microorganisms is a very important property in the marketability of probiotic-based products, and the microencapsulation technology has currently been used to incorporate and protect viable cells into food products.

The survivability of probiotic bacteria in yogurt is low mainly due to its acidic nature (Kailaspathy and Rybka, 1997). The incorporation of encapsulated probiotics may enhance the viability of these health-promoting microorganisms in yogurts. The encapsulated probiotic bacteria in whey protein (Picot and Lacroix, 2004), carrageenan (Adhikari et al., 2000, 2003), and alginate-starch (Sultana et al., 2000; Godward and Kailasapathy, 2003) polymers had increased survivability in stirred and set yogurts during storage period. It was reported by Kailasapathy (2006) that the addition of encapsulated *L. acidophilus* and *B. lactis* in alginate-starch binary microcapsules into yogurts did not significantly change the overall sensory features of the product. Krasekoopt et al. (2006) reported that among encapsulation procedures tested, only chitosan-coated alginate beads had no improved protection for the probiotic strains in yogurt during refrigeration period of 4 weeks. Currently, *L. paracasei* entrapped in milk-based $W_1/O/W_2$ emulsions was added to yogurt and its effect on fermentation, structural characteristics, physicochemical properties, and bacterial growth kinetics of the final product was evaluated. The starter cultures were not affected during fermentation upon adding *L. paracasei* and or double emulsion. The encapsulated probiotic in the emulsion had increased viability during storage and after exposure to *in vitro* digestion conditions. Although the physicochemical and textural properties of the yogurt were altered by the incorporation of $W_1/O/W_2$ emulsion, these changes were in the acceptable range (El Kadri et al., 2018).

Other dairy products have also been incorporated with encapsulated probiotics to produce functional foods. Several studies have demonstrated a greater viability of probiotic bacteria encapsulated in carrageenan or alginate beads upon freezing in dairy desserts (Godward and Kailasapathy, 2003; Kebary et al., 1998; Shah and Ravula, 2000; Sheu et al., 1993). The incorporation of microencapsulated probiotics into milk and subsequent freezing to produce frozen ice milk did not lead to measurable effect on the sensory properties and overrun of the products with remarkably high probiotic survival of 90% (Sheu et al., 1993). It is noteworthy that the neutral conditions of dairy desserts and high fat level of ice cream are likely the main factors in providing an additional protection layer to the probiotic microorganisms. However, it seems that the addition of some cryoprotectants like glycerol is necessary to enhance the probiotic viability during the freezing step of the dairy desserts.

The acid-sensitive probiotics may have long-term survivability in cheddar cheese mainly due to its manifestly higher pH value (4.8–5.6) than yogurt (pH 3.7–4.3) and fermented milks, which provides a more stable medium for the probiotic strains (Stanton et al., 1998). The survival of probiotics can be favored in cheddar cheese as a result of metabolism of lactic acid bacteria (LAB) and subsequent anaerobic conditions during the ripening period (Van den Tempel et al., 2002). Enhanced survival of *L. paracasei* and *B. bifidum* in cheddar cheese, respectively, over 3.0- and 6.0-months ripening period has been reported in the literature (Gardiner et al., 2002; Dinakar and Mistry, 1994). Similar results were reported by Özer et al. (2009), who worked on the viability of encapsulated *B. bifidum* and *L. acidophilus* in white-brined cheese. In another research, it was shown that functional cheese containing encapsulated *Bifidobacterium* had similar texture, flavor, and appearance in comparison to the control one (Desmond et al., 2002; Dinakar and Mistry, 1994). Viability of entrapped *L. plantarum* in double emulsion during Oaxaca cheese (a Mexican pasta filata-type cheese) making, melting, and *in vitro* digestion was evaluated by Rodríguez-Huezo et al. (2014). The cheese containing encapsulated bacteria had higher viable cells after melting than the control sample. Furthermore, the exposure of probiotic-incorporated cheese to GI conditions of pH 2.3 and bile salts did not significantly affect the bacteria viability; whilst, the control cheese experienced a significant decrease in the cell viability.

In some dairy products such as powdered milk ones, the use of high temperatures and the possibility of osmotic degradation during the spray drying step might lower the viable number of probiotic cells. Therefore, the second drying step operating at lower temperatures followed by cooling to room temperature in conjugation with applying themoprotectants such as trehalose could provide improved viability (Meister et al., 1999; Conrad et al., 2000). Desmond et al. (2002) reported that the addition of soluble fiber gum acacia to milk-based medium before spray drying led to an increase in *L. paracasei* survivability during storage in comparison to milk powder alone.

There are some technological challenges in the incorporation of probiotics in meat-based products like dry fermented sausages, due to their low water activity, pH, curing salts, and the presence of other starter culture organisms. In a study, it was demonstrated that the encapsulated probiotic *L. reuteri* could be applied in dry fermented sausages in order to maintain the desirable content of the bacteria in the product at consumption time without changing its sensory characteristics (Muthukumarasamy and Holley, 2006).

Mayonnaise has also been incorporated with probiotic microorganisms. Non-encapsulated *B. infantis* and *B. bifidum* remained viable only for 14 days of refrigeration in mayonnaise. However, the encapsulated probiotics in calcium alginate beads were survived for 8 and 12 weeks, respectively. In addition, the product containing encapsulated probiotic strains had lower total microbial count and thiobarbituric acid and higher titratable acidity. The surviving Bifidobacteria were able to produce acidic compounds and reduce the oxidation process of the product. Surprisingly, the encapsulated probiotics provided an improved sensory feature of the functional mayonnaise (Khalil and Mansour, 1998).

The probiotic cells have been added to soy products (Wang et al., 2002) and fermented cereals (Laine et al., 2003). The microencapsulation in gellan-xanthan beads was applied to stabilize *Bifidobacteria* in traditional African-fermented corn product (McMaster et al., 2005). Similarly, the encapsulated probiotics possessed greater survivability during freezing storage of cranberry juice concentrate (Reid et al., 2007) and cold storage of apple and orange juices (Ding and Shah, 2008). Additionally, immobilized *L. acidophilus* cells in κ-carrageenan gel beads could resist the unfavorable low pH value in tomato juice and the better growth was observed in comparison to free cells. The encapsulated probiotic strain possessed higher viable cell counts (10^6 cfu/mL-gel) in the fermented tomato juice after 10 weeks' refrigeration storage, compared to free cells (10^4 cfu/mL-gel). Finally, the probiotic-contained fermented tomato juice presented better overall acceptability than that of free cells during cold storage (Tsen et al., 2008).

The high temperatures applied during processing of bakery products may have adverse effects on probiotic survival. Lactobacilli encapsulation in hard fat via spray coating did not provide an enhanced stability and survival of the probiotic during bread baking process (Belvis et al., 2006). Whey protein particles-based microencapsulation were however effective in improving the lactobacilli survival during heat treatment of biscuit making (Reid et al., 2007).

2.4.2 IN GASTROINTESTINAL (GI) TRACT

Encapsulation should provide an additional protection for probiotic microorganisms during passage through the upper part of the GIT to ensure desired health-promoting effects in the host intestines. After stomach transition, probiotic cells should resist the intestinal conditions. It was shown that incubation in simulated bile solution had generally lower detrimental effects on probiotic cells, in comparison to simulated gastric conditions with low pHs (Heidebach et al., 2012). The survival of encapsulated probiotics is mainly dependent on the concentration and type of encapsulants, the bead size, bacterial species, and initial cell numbers. In most cases, the encapsulation possessed good protection for the encapsulated probiotic under GI conditions (Cui et al., 2000; Lee and Heo, 2000; Sun and Griffiths, 2000; McMaster et al., 2005; Lian et al., 2003; Hou et al., 2003; Picot and Lacroix, 2004; Mandal et al., 2006; Ding and Shah, 2007). However, a few authors did not find protective effect of encapsulation for probiotic cells (Sultana et al., 2000; Reid et al., 2005; Leverrier et al., 2005).

2.5 CONCLUSION

Encapsulation technique has been recognized as an excellent way to increase the ability of probiotic bacteria to resist the harsh conditions of the digestive tract and to prolong the shelf-life of probiotic cells in food products. The encapsulated probiotics are protected from severe environments such as GI conditions, high content of organic acids, and freezing temperatures. The extrusion, spray drying, and emulsion are considered as the most common approaches among encapsulation methods to encapsulate probiotics. However, it seems that the development of some advanced encapsulation methods and the use of new coating material substitutes should be considered in future researches. It is noteworthy that clinical evidences regarding the health-enhancing effects of probiotics are accumulating, and therefore, the food and pharmaceutical industries are undergoing strong demands for formulating novel probiotic-based functional foods. In addition, new innovative ways of probiotic delivery and controlled release will be developed in upcoming researches.

KEYWORDS

- **coacervation**
- **colloidosome**
- **electrospinning**
- **emulsion**
- **emulsion gel**
- **encapsulation**
- **extrusion**
- **functional food**
- **hydrogel**
- **immobilization**
- **impinging aerosol technology**
- **Maillard reaction products**
- **microencapsulation**
- **probiotic encapsulation**
- **spray drying**

REFERENCES

Adhikari, K., Mustapha, A., & Grün, I. U., (2003). Survival and metabolic activity of microencapsulated *Bifidobacterium* longum in stirred yogurt. *Journal of Food Science, 68*(1), 275–280.

Adhikari, K., Mustapha, A., Grün, I. U., & Fernando, L., (2000). Viability of microencapsulated bifidobacteria in set yogurt during refrigerated storage. *Journal of Dairy Science, 83*(9), 1946–1951.

Anal, A. K., & Singh, H., (2007). Recent advances in microencapsulation of probiotics for industrial applications and targeted delivery. *Trends in Food Science & Technology, 18*(5), 240–251.

Anjani, K., Iyer, C., & Kailasapathy, K., (2004). Survival of co-encapsulated complementary probiotics and prebiotics in yoghurt. *Milchwissenschaft: Milk Science International, 59*(7–8), 396–399.

Arslan, S., Erbas, M., Tontul, I., & Topuz, A., (2015). Microencapsulation of probiotic *Saccharomyces cerevisiae* var. boulardii with different wall materials by spray drying. *LWT-Food Science and Technology, 63*(1), 685–690.

Belvis, J., Tompkins, T. A., Wallace, T. A., Casavant, L., Fortin, C., & Caron, C., (2006). Stability of probiotic bacteria in foodstuffs. In: *CIFST Meeting, Montréal* (Vol. 30).

Borges, S., Barbosa, J., Camilo, R., Carvalheira, A., Silva, J., Sousa, S., Gomes, A. M., et al., (2012). Effects of encapsulation on the viability of probiotic strains exposed to lethal conditions. *International Journal of Food Science & Technology, 47*(2), 416–421.

Burgain, J., Gaiani, C., Linder, M., & Scher, J., (2011). Encapsulation of probiotic living cells: From laboratory scale to industrial applications. *Journal of Food Engineering, 104*(4), 467–483.

Chávarri, M., Marañón, I., & Villarán, M. C., (2012). Encapsulation technology to protect probiotic bacteria. In: *Probiotics*. InTech.

Chávarri, M., Marañón, I., Ares, R., Ibáñez, F. C., Marzo, F., & Del, C. V. M., (2010). Microencapsulation of a probiotic and prebiotic in alginate-chitosan capsules improves survival in simulated gastro-intestinal conditions. *International Journal of Food Microbiology, 142*(1, 2), 185–189.

Cohen, R., Schwartz, B., Peri, I., & Shimoni, E., (2011). Improving bioavailability and stability of genistein by complexation with high-amylose corn starch. *Journal of Agricultural and Food Chemistry, 59*(14), 7932–7938.

Conrad, P. B., Miller, D. P., Cielenski, P. R., & De Pablo, J. J., (2000). Stabilization and preservation of *Lactobacillus acidophilus* in saccharide matrices. *Cryobiology, 41*(1), 17–24.

Corcoran, B. M., Ross, R. P., Fitzgerald, G. F., & Stanton, C., (2004). Comparative survival of probiotic lactobacilli spray-dried in the presence of prebiotic substances. *Journal of Applied Microbiology, 96*(5), 1024–1039.

Cui, J. H., Goh, J. S., Kim, P. H., Choi, S. H., & Lee, B. J., (2000). Survival and stability of bifidobacteria loaded in alginate poly-l-lysine microparticles. *International Journal of Pharmaceutics, 210*(1–2), 51–59.

Dafe, A., Etemadi, H., Dilmaghani, A., & Mahdavinia, G. R., (2017). Investigation of pectin/starch hydrogel as a carrier for oral delivery of probiotic bacteria. *International Journal of Biological Macromolecules, 97*, 536–543.

Desai, K. G. H., & Jin, P. H., (2005). Recent developments in microencapsulation of food ingredients. *Drying Technology, 23*(7), 1361–1394.

Desmond, C., Ross, R. P., O'callaghan, E., Fitzgerald, G., & Stanton, C., (2002). Improved survival of *Lactobacillus paracasei* NFBC 338 in spray-dried powders containing gum *Acacia*. *Journal of Applied Microbiology, 93*(6), 1003–1011.

Dickinson, E., (2012). Emulsion gels: The structuring of soft solids with protein-stabilized oil droplets. *Food Hydrocolloids, 28*(1), 224–241.

Dinakar, P., & Mistry, V. V., (1994). Growth and viability of *Bifidobacterium* bifidum in Cheddar cheese. *Journal of Dairy Science, 77*(10), 2854–2864.

Ding, W. K., & Shah, N. P., (2007). Acid, bile, and heat tolerance of free and microencapsulated probiotic bacteria. *Journal of Food Science, 72*(9), M446–M450.

Ding, W. K., & Shah, N. P., (2008). Survival of free and microencapsulated probiotic bacteria in orange and apple juices. *International Food Research Journal, 15*(2), 219–232.

Dinsmore, A. D., Hsu, M. F., Nikolaides, M. G., Marquez, M., Bausch, A. R., & Weitz, D. A., (2002). Colloidosomes: Selectively permeable capsules composed of colloidal particles. *Science, 298*(5595), 1006–1009.

Doherty, S. B., Gee, V. L., Ross, R. P., Stanton, C., Fitzgerald, G. F., & Brodkorb, A., (2011). Development and characterization of whey protein micro-beads as potential matrices for probiotic protection. *Food Hydrocolloids, 25*(6), 1604–1617.

Dong, Q. Y., Chen, M. Y., Xin, Y., Qin, X. Y., Cheng, Z., Shi, L. E., & Tang, Z. X., (2013). Alginate-based and protein-based materials for probiotics encapsulation: A review. *International Journal of Food Science & Technology, 48*(7), 1339–1351.

Dos, S. D. X., Casazza, A. A., Aliakbarian, B., Bedani, R., Saad, S. M. I., & Perego, P., (2019). Improved probiotic survival to *in vitro* gastrointestinal stress in a mousse containing *Lactobacillus acidophilus* La-5 microencapsulated with inulin by spray drying. *LWT, 99*, 404–410.

El Kadri, H., Lalou, S., Mantzouridou, F., & Gkatzionis, K., (2018). Utilization of water-in-oil-water (W1/O/W2) double emulsion in a set-type yogurt model for the delivery of probiotic *Lactobacillus paracasei*. *Food Research International, 107*, 325–336.

Eratte, D., Dowling, K., Barrow, C. J., & Adhikari, B. P., (2017). *In-vitro* digestion of probiotic bacteria and omega-3 oil co-microencapsulated in whey protein isolate-gum Arabic complex coacervates. *Food Chemistry, 227*, 129–136.

Eratte, D., McKnight, S., Gengenbach, T. R., Dowling, K., Barrow, C. J., & Adhikari, B. P., (2015). Co-encapsulation and characterization of omega-3 fatty acids and probiotic bacteria in whey protein isolate-gum Arabic complex coacervates. *Journal of Functional Foods, 19*, 882–892.

Esfanjani, A. F., Assadpour, E., & Jafari, S. M., (2018). Improving the bioavailability of phenolic compounds by loading them within lipid-based nanocarriers. *Trends in Food Science & Technology, 76*, 56–66.

Falco, C. Y., Falkman, P., Risbo, J., Cárdenas, M., & Medronho, B., (2017). Chitosan-dextran sulfate hydrogels as a potential carrier for probiotics. *Carbohydrate Polymers, 172*, 175–183.

Fang, Z., & Bhandari, B., (2010). Encapsulation of polyphenols: A review. *Trends in Food Science & Technology, 21*(10), 510–523.

Favaro-Trindade, C. S., Heinemann, R. J. B., & Pedroso, D. L., (2011). Developments in probiotic encapsulation. *CAB Rev., 6*, 1–8.

Fung, W. Y., Yuen, K. H., & Liong, M. T., (2011). Agro waste-based nanofibers as a probiotic encapsulant: Fabrication and characterization. *Journal of Agricultural and Food Chemistry, 59*(15), 8140–8147.

Gardiner, G. E., Bouchier, P., O'Sullivan, E., Kelly, J., Collins, J. K., Fitzgerald, G., Ross, R. P., & Stanton, C., (2002). A spray-dried culture for probiotic Cheddar cheese manufacture. *International Dairy Journal, 12*(9), 749–756.

Gbassi, G. K., & Vandamme, T., (2012). Probiotic encapsulation technology: From microencapsulation to release into the gut. *Pharmaceutics, 4*(1), 149–163.

Gebara, C., Chaves, K. S., Ribeiro, M. C. E., Souza, F. N., Grosso, C. R., & Gigante, M. L., (2013). Viability of *Lactobacillus acidophilus* La5 in pectin-whey protein microparticles during exposure to simulated gastrointestinal conditions. *Food Research International, 51*(2), 872–878.

Gerez, C. L., Font De, V. G., Gigante, M. L., & Grosso, C. R. F., (2012). Whey protein coating bead improves the survival of the probiotic *Lactobacillus rhamnosus* CRL 1505 to low pH. *Letters in Applied Microbiology, 54*(6), 552–556.

Godward, G., & Kailasapathy, K., (2003). Viability and survival of free, encapsulated and co-encapsulated probiotic bacteria in yoghurt. *Milchwissenschaft: Milk Science International, 58,* 396–399.

Gu, M., Zhang, Z., Pan, C., Goulette, T., Zhang, R., Hendricks, G., McClements, D. J., & Xiao, H., (2019). Encapsulation of *Bifidobacterium pseudocatenulatum* G7 in gastroprotective microgels: Improvement of the bacterial viability under simulated gastrointestinal conditions. *Food Hydrocolloids, 91*, 283–289.

Hariyadi, D. M., Lin, S. C. Y., Wang, Y., Bostrom, T., Turner, M. S., Bhandari, B., & Coombes, A. G., (2010). Diffusion loading and drug delivery characteristics of alginate gel microparticles produced by a novel impinging aerosols method. *Journal of Drug Targeting, 18*(10), 831–841.

Heidebach, T., Först, P., & Kulozik, U., (2009). Transglutaminase-induced caseinate gelatin for the microencapsulation of probiotic cells. *International Dairy Journal, 19*(2), 77–84.

Heidebach, T., Först, P., & Kulozik, U., (2012). Microencapsulation of probiotic cells for food applications. *Critical Reviews in Food Science and Nutrition, 52*(4), 291–311.

Hou, R. C. W., Lin, M. Y., Wang, M. M. C., & Tzen, J. T. C., (2003). Increase of viability of entrapped cells of *Lactobacillus delbrueckii* ssp. bulgaricus in artificial sesame oil emulsions. *Journal of Dairy Science, 86*(2), 424–428.

Jantzen, M., Göpel, A., & Beermann, C., (2013). Direct spray drying and microencapsulation of probiotic *Lactobacillus reuteri* from slurry fermentation with whey. *Journal of Applied Microbiology, 115*(4), 1029–1036.

Jiménez-Pranteda, M. L., Poncelet, D., Náder-Macías, M. E., Arcos, A., Aguilera, M., Monteoliva-Sánchez, M., & Ramos-Cormenzana, A., (2012). Stability of lactobacilli encapsulated in various microbial polymers. *Journal of Bioscience and Bioengineering, 113*(2), 179–184.

Kailasapathy, K., & Rybka, S. L., (1997). *Acidophilus* and *Bifidobacterium* spp.-their therapeutic potential and survival in yogurt. *Australian Journal of Dairy Technology, 52*(1), 28–35.

Kailasapathy, K., (2006). Survival of free and encapsulated probiotic bacteria and their effect on the sensory properties of yoghurt. *LWT-Food Science and Technology, 39*(10), 1221–1227.

Kailasapathy, K., Harmstorf, I., & Phillips, M., (2008). Survival of *Lactobacillus acidophilus* and *Bifidobacterium animalis* ssp. lactis in stirred fruit yogurts. *LWT-Food Science and Technology, 41*(7), 1317–1322.

Kebary, K. M. K., Hussein, S. A., & Badawi, R. M., (1998). Improving viability of *Bifidobacterium* and their effect on frozen ice milk. *Egyptian Journal of Dairy Science, 26*, 319–337.

Keen, P. H., Slater, N. K., & Routh, A. F., (2012). Encapsulation of lactic acid bacteria in colloidosomes. *Langmuir, 28*(46), 16007–16014.

Khalil, A. H., & Mansour, E. H., (1998). Alginate encapsulated bifidobacteria survival in mayonnaise. *Journal of Food Science, 63*(4), 702–705.

Krasaekoopt, W., Bhandari, B., & Deeth, H. C., (2006). Survival of probiotics encapsulated in chitosan-coated alginate beads in yoghurt from UHT-and conventionally treated milk during storage. *LWT-Food Science and Technology, 39*(2), 177–183.

Krasaekoopt, W., Bhandari, B., & Deeth, H., (2003). Evaluation of encapsulation techniques of probiotics for yoghurt. *International Dairy Journal, 13*(1), 3–13.

Laine, R., Salminen, S., Benno, Y., & Ouwehand, A. C., (2003). Performance of bifidobacteria in oat-based media. *International Journal of Food Microbiology, 83*(1), 105–109.

Lee, K. Y., & Heo, T. R., (2000). Survival of *Bifidobacterium* longum immobilized in calcium alginate beads in simulated gastric juices and bile salt solution. *Applied and Environmental Microbiology, 66*(2), 869–873.

Lee, Y. K., & Salminen, S., (2009). *Handbook of Probiotics and Prebiotics* (p. 596). John Wiley & Sons.

Leverrier, P., Fremont, Y., Rouault, A., Boyaval, P., & Jan, G., (2005). *In vitro* tolerance to digestive stresses of propionibacteria: Influence of food matrices. *Food Microbiology, 22*(1), 11–18.

Lian, W. C., Hsiao, H. C., & Chou, C. C., (2003). Viability of microencapsulated bifidobacteria in simulated gastric juice and bile solution. *International Journal of Food Microbiology, 86*(3), 293–301.

Liu, L., Chen, P., Zhao, W., Li, X., Wang, H., & Qu, X., (2017). Effect of microencapsulation with the Maillard reaction products of whey proteins and isomaltooligosaccharide on the survival rate of *Lactobacillus rhamnosus* in white brined cheese. *Food Control, 79*, 44–49.

Liu, L., Li, X., Zhu, Y., Bora, A. F. M., Zhao, Y., Du, L., Li, D., & Bi, W., (2016). Effect of microencapsulation with Maillard reaction products of whey proteins and isomaltooligosaccharide on the survival of *Lactobacillus rhamnosus*. *LWT-Food Science and Technology, 73*, 37–43.

López-Rubio, A., Sanchez, E., Sanz, Y., & Lagaron, J. M., (2009). Encapsulation of living bifidobacteria in ultrathin PVOH electro spun fibers. *Biomacromolecules, 10*(10), 2823–2829.

López-Rubio, A., Sanchez, E., Wilkanowicz, S., Sanz, Y., & Lagaron, J. M., (2012). Electrospinning as a useful technique for the encapsulation of living bifidobacteria in food hydrocolloids. *Food Hydrocolloids, 28*(1), 159–167.

Mandal, S., Puniya, A. K., & Singh, K., (2006). Effect of alginate concentrations on survival of microencapsulated *Lactobacillus casei* NCDC-298. *International Dairy Journal, 16*(10), 1190–1195.

Manojlović, V., Nedović, V. A., Kailasapathy, K., & Zuidam, N. J., (2010). Encapsulation of probiotics for use in food products. In: *Encapsulation Technologies for Active Food Ingredients and Food Processing* (p. 402). Springer.

Mao, L., Pan, Q., Hou, Z., Yuan, F., & Gao, Y., (2018). Development of soy protein isolate-carrageenan conjugates through Maillard reaction for the microencapsulation of *Bifidobacterium* longum. *Food Hydrocolloids, 84*, 489–497.

Mattila-Sandholm, T., Myllärinen, P., Crittenden, R., Mogensen, G., Fondén, R., & Saarela, M., (2002). Technological challenges for future probiotic foods. *International Dairy Journal, 12*(2, 3), 173–182.

McMaster, L. D., Kokott, S. A., Reid, S. J., & Abratt, V. R., (2005). Use of traditional African fermented beverages as delivery vehicles for *Bifidobacterium* lactis DSM 10140. *International Journal of Food Microbiology, 102*(2), 231–237.

Meister, N., Sutter, A., & Vikas, M., (1999). *Dehydrated Food Containing Lactic Acid Bacteria*. EP0924993B1.

Mortazavian, A. M., Azizi, A., Ehsani, M. R., Razavi, S. H., Mousavi, S. M., Sohrabvandi, S., & Reinheimer, J. A., (2008). Survival of encapsulated probiotic bacteria in Iranian yogurt drink (Doogh) after the product exposure to simulated gastrointestinal conditions. *Milchwissenschaft, 63*(4), 1–3.

Muthukumarasamy, P., & Holley, R. A., (2006). Microbiological and sensory quality of dry fermented sausages containing alginate-microencapsulated *Lactobacillus reuteri*. *International Journal of Food Microbiology, 111*(2), 164–169.

Naidu, A. S., Bidlack, W. R., & Clemens, R. A., (1999). Probiotic spectra of lactic acid bacteria (LAB). *Critical Reviews in Food Science and Nutrition, 39*(1), 13–126.

Nooshkam, M., & Madadlou, A., (2016a). Maillard conjugation of lactulose with potentially bioactive peptides. *Food Chemistry, 192*, 831–836.

Nooshkam, M., & Madadlou, A., (2016b). Microwave-assisted isomerization of lactose to lactulose and Maillard conjugation of lactulose and lactose with whey proteins and peptides. *Food Chemistry, 200*, 1–9.

Nooshkam, M., Babazadeh, A., & Jooyandeh, H., (2018). Lactulose: Properties, techno-functional food applications, and food-grade delivery system. *Trends in Food Science & Technology, 80*, 23–34.

Nooshkam, M., Varidi, M., & Bashash, M., (2019). The Maillard reaction products as food-born antioxidant and anti-browning agents in model and real food systems. *Food Chemistry, 275*, 644–660.

Nualkaekul, S., Cook, M. T., Khutoryanskiy, V. V., & Charalampopoulos, D., (2013). Influence of encapsulation and coating materials on the survival of *Lactobacillus plantarum* and *Bifidobacterium* longum in fruit juices. *Food Research International, 53*(1), 304–311.

O'riordan, K., Andrews, D., Buckle, K., & Conway, P., (2001). Evaluation of microencapsulation of a *Bifidobacterium* strain with starch as an approach to prolonging viability during storage. *Journal of Applied Microbiology, 91*(6), 1059–1066.

Oliveira, A. C., Moretti, T. S., Boschini, C., Baliero, J. C. C., Freitas, O. D., & Fávaro-Trindade, C. S., (2007). Stability of microencapsulated *B. lactis* (BI 01) and *L. acidophilus* (LAC 4) by complex coacervation followed by spray drying. *Journal of Microencapsulation, 24*(7), 685–693.

Özer, B., Kirmaci, H. A., Şenel, E., Atamer, M., & Hayaloğlu, A., (2009). Improving the viability of *Bifidobacterium bifidum* BB-12 and *Lactobacillus acidophilus* LA-5 in white-brined cheese by microencapsulation. *International Dairy Journal, 19*(1), 22–29.

Özer, B., Uzun, Y. S., & Kirmaci, H. A., (2008). Effect of microencapsulation on viability of *Lactobacillus acidophilus* LA-5 and *Bifidobacterium bifidum* BB-12 during Kasar cheese ripening. *International Journal of Dairy Technology, 61*(3), 237–244.

Paim, D. R., Costa, S. D., Walter, E. H., & Tonon, R. V., (2016). Microencapsulation of probiotic jussara (Euterpe edulis M.) juice by spray drying. *LWT, 74*, 21–25.

Pandey, S., Senthilguru, K., Uvanesh, K., Sagiri, S. S., Behera, B., Babu, N., Bhattacharyya, M. K., et al., (2016). Natural gum modified emulsion gel as single carrier for the oral delivery of probiotic-drug combination. *International Journal of Biological Macromolecules, 92*, 504–514.

Papagianni, M., & Anastasiadou, S., (2009). Encapsulation of *Pediococcus acidilactici* cells in corn and olive oil microcapsules emulsified by peptides and stabilized with xanthan in oil-in-water emulsions: Studies on cell viability under gastro-intestinal simulating conditions. *Enzyme and Microbial Technology, 45*(6, 7), 514–522.

Picone, C. S. F., Bueno, A. C., Michelon, M., & Cunha, R. L., (2017). Development of a probiotic delivery system based on gelation of water-in-oil emulsions. *LWT-Food Science and Technology, 86*, 62–68.

Picot, A., & Lacroix, C., (2004). Encapsulation of bifidobacteria in whey protein-based microcapsules and survival in simulated gastrointestinal conditions and in yoghurt. *International Dairy Journal, 14*(6), 505–515.

Pimentel-González, D. J., Campos-Montiel, R. G., Lobato-Calleros, C., Pedroza-Islas, R., & Vernon-Carter, E. J., (2009). Encapsulation of *Lactobacillus rhamnosus* in double emulsions formulated with sweet whey as emulsifier and survival in simulated gastrointestinal conditions. *Food Research International, 42*(2), 292–297.

Prasanna, P. H. P., & Charalampopoulos, D., (2018). Encapsulation of *Bifidobacterium longum* in alginate-dairy matrices and survival in simulated gastrointestinal conditions, refrigeration, cow milk and goat milk. *Food Bioscience, 21*, 72–79.

Prevost, H., & Divies, C., (1992). Cream fermentation by a mixed culture of lactococci entrapped in two-layer calcium alginate gel beads. *Biotechnology Letters, 14*(7), 583–588.

Reid, A. A., Champagne, C. P., Gardner, N., Fustier, P., & Vuillemard, J. C., (2007). Survival in food systems of *Lactobacillus rhamnosus* R011 microentrapped in whey protein gel particles. *Journal of Food Science, 72*(1), M031–M037.

Reid, A. A., Vuillemard, J. C., Britten, M., Arcand, Y., Farnworth, E., & Champagne, C. P., (2005). Microentrapment of probiotic bacteria in a Ca^{2+}-induced whey protein gel and effects on their viability in a dynamic gastrointestinal model. *Journal of Microencapsulation, 22*(6), 603–619.

Rodrigues, D., Sousa, S., Rocha-Santos, T., Silva, J. P., Lobo, J. S., Costa, P., Amaral, M. H., et al., (2011). Influence of L-cysteine, oxygen and relative humidity upon survival throughout storage of probiotic bacteria in whey protein-based microcapsules. *International Dairy Journal, 21*(11), 869–876.

Rodríguez-Huezo, M. E., Durán-Lugo, R., Prado-Barragán, L. A., Cruz-Sosa, F., Lobato-Calleros, C., Alvarez-Ramírez, J., & Vernon-Carter, E. J., (2007). Pre-selection of protective colloids for enhanced viability of *Bifidobacterium* bifidum following spray-drying and storage, and evaluation of aguamiel as thermoprotective prebiotic. *Food Research International, 40*(10), 1299–1306.

Rodríguez-Huezo, M. E., Estrada-Fernández, A. G., García-Almendárez, B. E., Ludena-Urquizo, F., Campos-Montiel, R. G., & Pimentel-González, D. J., (2014). Viability of *Lactobacillus plantarum* entrapped in double emulsion during Oaxaca cheese manufacture, melting and simulated intestinal conditions. *LWT-Food Science and Technology, 59*(2), 768–773.

Rossier-Miranda, F. J., Schroën, C. G. P. H., & Boom, R. M., (2009). Colloidosomes: Versatile microcapsules in perspective. *Colloids and Surfaces A: Physicochemical and Engineering Aspects, 343*(1–3), 43–49.

Sabikhi, L., Babu, R., Thompkinson, D. K., & Kapila, S., (2010). Resistance of microencapsulated *Lactobacillus acidophilus* LA1 to processing treatments and simulated gut conditions. *Food and Bioprocess Technology, 3*(4), 586–593.

Sandoval-Castilla, O., Lobato-Calleros, C., García-Galindo, H. S., Alvarez-Ramírez, J., & Vernon-Carter, E. J., (2010). Textural properties of alginate-pectin beads and survivability of entrapped *Lb. casei* in simulated gastrointestinal conditions and in yoghurt. *Food Research International, 43*(1), 111–117.

Santillo, A., Albenzio, M., Bevilacqua, A., Corbo, M. R., & Sevi, A., (2012). Encapsulation of probiotic bacteria in lamb rennet paste: Effects on the quality of Pecorino cheese. *Journal of Dairy Science, 95*(7), 3489–3500.

Semyonov, D., Ramon, O., Kaplun, Z., Levin-Brener, L., Gurevich, N., & Shimoni, E., (2010). Microencapsulation of *Lactobacillus paracasei* by spray freeze-drying. *Food Research International, 43*(1), 193–202.

Shah, N. P., & Ravula, R. R., (2000). Microencapsulation of probiotic bacteria and their survival in frozen fermented dairy desserts. *Australian Journal of Dairy Technology, 55*(3), 139–144.

Shah, N. P., (2000). Probiotic bacteria: selective enumeration and survival in dairy foods. *Journal of Dairy Science, 83*(4), 894–907.

Sheu, T. Y., Marshall, R. T., & Heymann, H., (1993). Improving survival of culture bacteria in frozen desserts by microentrapment. *Journal of Dairy Science, 76*(7), 1902–1907.

Shi, L. E., Li, Z. H., Li, D. T., Xu, M., Chen, H. Y., Zhang, Z. L., & Tang, Z. X., (2013). Encapsulation of probiotic *Lactobacillus bulgaricus* in alginate-milk microspheres and evaluation of the survival in simulated gastrointestinal conditions. *Journal of Food Engineering, 117*(1), 99–104.

Shima, M., Matsuo, T., Yamashita, M., & Adachi, S., (2009). Protection of *Lactobacillus acidophilus* from bile salts in a model intestinal juice by incorporation into the inner-water phase of a W/O/W emulsion. *Food Hydrocolloids, 23*(2), 281–285.

Shima, M., Morita, Y., Yamashita, M., & Adachi, S., (2006). Protection of *Lactobacillus acidophilus* from the low pH of a model gastric juice by incorporation in a W/O/W emulsion. *Food Hydrocolloids, 20*(8), 1164–1169.

Siuta-Cruce, P., & Goulet, J., (2001). Improving probiotic survival rates. *Food Technology, 55*(10), 36–42.

Sohail, A., Turner, M. S., Coombes, A., Bostrom, T., & Bhandari, B., (2011). Survivability of probiotics encapsulated in alginate gel microbeads using a novel impinging aerosols method. *International Journal of Food Microbiology, 145*(1), 162–168.

Sohail, A., Turner, M. S., Prabawati, E. K., Coombes, A. G., & Bhandari, B., (2012). Evaluation of *Lactobacillus rhamnosus* GG and *Lactobacillus acidophilus* NCFM encapsulated using a novel impinging aerosol method in fruit food products. *International Journal of Food Microbiology, 157*(2), 162–166.

Stanton, C., Gardiner, G., Lynch, P. B., Collins, J. K., Fitzgerald, G., & Ross, R. P., (1998). Probiotic cheese. *International Dairy Journal, 8*(5, 6), 491–496.

Su, L. C., Lin, C. W., & Chen, M. J., (2007). Development of an Oriental-style dairy product coagulated by microcapsules containing probiotics and filtrates from fermented rice. *International Journal of Dairy Technology, 60*(1), 49–54.

Su, R., Zhu, X. L., Fan, D. D., Mi, Y., Yang, C. Y., & Jia, X., (2011). Encapsulation of probiotic *Bifidobacterium* longum BIOMA 5920 with alginate-human-like collagen and evaluation of survival in simulated gastrointestinal conditions. *International Journal of Biological Macromolecules, 49*(5), 979–984.

Sultana, K., Godward, G., Reynolds, N., Arumugaswamy, R., Peiris, P., & Kailasapathy, K., (2000). Encapsulation of probiotic bacteria with alginate-starch and evaluation of survival in simulated gastrointestinal conditions and in yoghurt. *International Journal of Food Microbiology, 62*(1, 2), 47–55.

Sun, W., & Griffiths, M. W., (2000). Survival of bifidobacteria in yogurt and simulated gastric juice following immobilization in gellan-xanthan beads. *International Journal of Food Microbiology, 61*(1), 17–25.

Todorov, S. D., LeBlanc, J. G., & Franco, B. D., (2012). Evaluation of the probiotic potential and effect of encapsulation on survival for *Lactobacillus plantarum* ST16Pa isolated from papaya. *World Journal of Microbiology and Biotechnology, 28*(3), 973–984.

Tsen, J. H., Lin, Y. P., Huang, H. Y., & King, V. A. E., (2008). Studies on the fermentation of tomato juice by using κ-carrageenan immobilized *Lactobacillus acidophilus*. *Journal of Food Processing and Preservation, 32*(2), 178–189.

Urbanska, A. M., Bhathena, J., & Prakash, S., (2007). Live encapsulated *Lactobacillus acidophilus* cells in yogurt for therapeutic oral delivery: Preparation and *in vitro* analysis of alginate-chitosan microcapsules. *Canadian Journal of Physiology and Pharmacology, 85*(9), 884–893.

Van, D. T. T., Gundersen, J. K., & Nielsen, M. S., (2002). The microdistribution of oxygen in Danablu cheese measured by a microsensor during ripening. *International Journal of Food Microbiology, 75*(1, 2), 157–161.

Wang, S., Chen, Y., Liang, H., Chen, Y., Shi, M., Wu, J., Liu, X., et al., (2015). Intestine-specific delivery of hydrophobic bioactives from oxidized starch microspheres with an enhanced stability. *Journal of Agricultural and Food Chemistry, 63*(39), 8669–8675.

Wang, Y. C., Yu, R. C., & Chou, C. C., (2002). Growth and survival of bifidobacteria and lactic acid bacteria during the fermentation and storage of cultured soymilk drinks. *Food Microbiology, 19*(5), 501–508.

Yeung, T. W., Üçok, E. F., Tiani, K. A., McClements, D. J., & Sela, D. A., (2016). Microencapsulation in alginate and chitosan microgels to enhance viability of *Bifidobacterium* longum for oral delivery. *Frontiers in Microbiology, 7*, 494.

Ying, D., Schwander, S., Weerakkody, R., Sanguansri, L., Gantenbein-Demarchi, C., & Augustin, M. A., (2013). Microencapsulated *Lactobacillus rhamnosus* GG in whey protein and resistant starch matrices: Probiotic survival in fruit juice. *Journal of Functional Foods, 5*(1), 98–105.

Zhang, D., Zhang, M., & Gu, X., (2018). Seaweed-derived hydrocolloids as food coating and encapsulation agents. In: *Bioactive Seaweeds for Food Applications* (p. 322). Academic Press.

Zhang, Z., Zhang, R., Sun, Q., Park, Y., & McClements, D. J., (2017). Confocal fluorescence mapping of pH profile inside hydrogel beads (microgels) with controllable internal pH values. *Food Hydrocolloids, 65*, 198–205.

Zhao, R., Sun, J., Torley, P., Wang, D., & Niu, S., (2008). Measurement of particle diameter of *Lactobacillus acidophilus* microcapsule by spray drying and analysis on its microstructure. *World Journal of Microbiology and Biotechnology, 24*(8), 1349–1354.

Zhu, F., (2017). Encapsulation and delivery of food ingredients using starch based systems. *Food Chemistry, 229*, 542–552.

CHAPTER 3

Bioactive Peptides (BPs) as Functional Foods: Production Process, Techno-Functional Applications, Health-Promoting Effects, and Safety Issues

AFSHIN BABAZADEH,* MAJID NOOSHKAM, and MAHNAZ TABIBIAZAR

*Corresponding author. E-mail: babazadeh.afshin@gmail.com

ABSTRACT

Nowadays, bioactive peptides (BPs) are introduced as a new source of compounds with multiple functions in various fields of health sciences. These compounds are usually generated from food proteins, which mainly were used in pharmaceutical sciences. Thus, introducing their applications in food industries is a new era in the science community, which can bring new methods for disease treatment. These compounds are good candidates for producing functional foods with the food therapy purposes. Thus, this chapter will provide information about BPs and will discuss major challenges and concerns to cover the principles needed for preparing functional foods with special health-promoting features.

3.1 INTRODUCTION

Bioactive peptides (BPs) are one of the main components of 'Functional Foods.' BPs are defined as short-chain sequences (usually 2–20 amino acid residues) of the parent proteins, showing their effects on the human body during digestive hydrolysis (Tagliazucchi et al., 2017). It is reported that food proteins could be one of the main resource of BPs containing numerous physiological effects (Hartmann and Meisel, 2007; Hernández-Ledesma et al., 2004). In these products, the bioactive compounds such as

BPs are either added or enriched by food-grade carriers or modification of the usual manufacturing process (Andishmand et al., 2017; Babazadeh et al., 2016, 2017). Some of these products, however, can be traditional foods or newly formulated ones. Table 3.1 lists existing functional foods and food ingredients carrying BPs and Table 3.2 provides information about newly founded peptides with potential applications in functional foods.

TABLE 3.1 Available Bioactive Containing Functional Foods

Bioactive Peptide Sequence/Amino Acids	Functional Food	References
VPP, IPP from β- and k-CN	Sour milk, fermented milk, calcium-enriched	Nakamura et al. (1996)
Bonito-derived peptides	Soup	Gill et al. (1996)
Casein-derived dodecapeptide FFVAPFPEVFGK	Soft drink	Phelan et al. (2009)
Soy peptides bound to phospholipids	Drink powder	
aS1-CN (fragment 91–100): YLGYLEQLLR	Milk drink, confectionery	
RPKHPIKHQ and YPFPGPIPN	Gouda cheese	Fitzgerald and Murray (2006)
KKYNVPQL, VRYL, and VRGPFP	Manchego cheese	
LVYPFPGPINSLPQ	Crescenza cheese	
Casein phosphor-peptides (CPP)	Soft drink	Phelan et al. (2009)

According to previous fundamental studies within the last decade, new field of researches focusing on BPs derived from food products were carried out (Mohanty et al., 2016; Pessione and Cirrincione, 2016; Singh et al., 2014). Regarding various definitions, food-grade BPs are genuine or generated compounds of foods with regulatory activities, along with their nutritive functions. Therefore, food originated bioactive compounds could be used in functional/medicinal foods, dietary supplements, and conventional foods, whereby each group requires its own claims (Gobbetti et al., 2004).

TABLE 3.2 Determined Bioactive Peptides with Potential Applications Functional Foods

Bioactive Peptide Sequence/Amino Acids	Origin	Determination Method	Production Method	References
Gly-Pro-Leu, -Trp-Pro-Glu-Ala-Ala-Glu-Leu-Met-Met-Glu-Val-Asp-Pro, Gly-Asp-Leu-Gly-Lys-Thr-Thr-Thr-Val-Ser-Asn-Trp-Ser-Pro-Pro-Lys-Try-Lys-Asp-Thr-Pro, -Ile-Phe-Val-Pro-Ala-Phe, -Met-Glu-Gly-Ala-Gln-Glu-Ala-Gln-Gly-Asp.	Marine foods	–	Neutrase, Pepsin, Papain, Alcalase, α-Chymotrypsin, Collagenase	Kim and Wijesekara (2010)
KRQKYD, EKERERQ, VKKVLGNP, RMLGQTPTK, LGFPTTKTYFPHF, VVYPWT	Meat	LC-MS, RP-HPLC, MALDI,	Different enzymatic degradations	Lafarga and Hayes (2014)
DHPFLF, ELIN, ESIINF, ESI, RADHPF	Egg	RP-HPLC-MS/MS, RP-HPLC, MALDI-TOF MS, and μ-HPLC-ESI MS/MS	Alcalase, Pepsin, pancreatin, Chymotrypsin	Grootaert et al. (2017)
MRWRD, MRW, LRIPVA, IAYKPAG, YPLDL, YPLDLF, ELAAAC, LRDDF, MDN, FC, VW	Plant protein (Ribulose-1,5-bisphosphate carboxylase/oxygenase-(RuBisCO))	Chromatographic and electrodialysis with ultrafiltration membrane techniques	Maceration and buffer extraction, enzymatic hydrolysis	Udenigwe et al. (2017)
Ser-Asp-Ile-Thr-Arg-Pro-Gly-Gly-Asn-Met	*Palmaria palmata*	RP-HPLC, ESI-MS/MS	Proteolytic enzyme Corolase PP, solid-phase extraction	Harnedy et al. (2017)

TABLE 3.2 *(Continued)*

Bioactive Peptide Sequence/Amino Acids	Origin	Determination Method	Production Method	References
Pro-Pro-His-Met-Leu-Pro, Pro-Leu-Pro-Trp-Gly-Ala-Gly-Phe, Pro-Pro-His-Met-Gly-Gly-Pro, Pro-Leu-Pro-Leu-His-Met-Leu-Pro, Leu-Ser-Ser-Leu-Glu-Met-Gly-Ser-Leu-Gly-Ala-Leu-Phe-Val-Cys-Met	Pinto bean	LCMS LTQ Orbitrap, PEAKS studio	Protex	Ngoh and Gan (2017)
IPLH, IPIVLL	Mushroom foot	Gel chromatography and LC-MS/MS	Ultra-high-pressure	Zhao et al. (2017)
Different protein hydrolysates	Sheep cheese whey	SDS-PAGE electrophoresis	Bacterial proteases	Welsh et al. (2017)
Gly-Pro-Ala-Val, Val-Cys, Phe-Phe	Atlantic salmon (Salmo salar)	SDS-PAGE electrophoresis, reverse phase-HPLC, gel permeation HPLC	Alcalase 2.4 L, Flavourzyme 500 L, Corolase PP, Promod 144MG.	Neves et al. (2017)
Different mushroom protein hydrolysates	Mushroom (*Agaricus bisporus*)	SDS-PAGE electrophoresis and HPLC	Alcalase, pancreatin, flavourzyme	Komatsu et al. (2017)
FPGPIPK, IPPK, QPPQ	Buffalo milk	Exclusion chromatography and LC-MS	Papain, pepsin, or trypsin	Abdel-Hamid et al. (2017)
Glycine, alanine, leucine, lysine	Freshwater fish (*Cirrhinus mrigala*) muscle	Waters pico-tag method, Ft-IR, and gel permeation chromatography	Papain	Elavarasan et al. (2016)
Valine-proline-proline	Goats' milk	HPLC	Harmonized static *in vitro* digestion	Tagliazucchi et al. (2017)
Isoleucine-Proline-Proline	Goats' milk	HPLC		

Various BPs resources were reported; however, milk proteins are still the principal resources for a wide range of BPs. It is reported that milk proteins' primary structure requires proteolysis in order to be able to release their biofunctional effects. Proteolysis may occur during food processing or through gastrointestinal (GI) transit (Gobbetti et al., 2002). In general, there are three main procedures to characterize BPs: (i) isolating BPs from primary proteins via *in vitro* enzymatic digestions; (ii) or *in vivo* GI digestions; and (iii) synthetically generating peptides that have identical structures to those known BPs (Gobbetti et al., 2004; Meisel, 2001).

This chapter focuses on the current published data based on the background, experimental, clinical, and techno-functional food applications of BPs and their potential values in producing healthier and functional foods.

3.2 TYPES OF BIOACTIVE PEPTIDES (BPS)

Various types of BPs introduced during the last decade were mainly based on milk and milk products. However, the most important ones are expressed as follows.

3.2.1 ACE-INHIBITORY PEPTIDES

Many researchers reported the ACE-inhibitory peptides derived from some strains of lactic acid bacteria (LAB). These bacteria biosynthesize these BPs in fermented milks. However, there are few data regarding their presence in commercial fermented products (Meisel et al., 1997). Moreover, although the use of enzymes to generate BPs derived from milks has been already presented as a promising way of producing these BPs, their tolerability to GI enzymes is poorly understood (Gobbetti et al., 2002). ACE-inhibitory peptides can be generated from their inactive precursors, a protein or oligopeptide, during food processing or in GI digestion. In the case of food processing, a general example is milk fermentation using *S. cerevisiae* and *L. helveticus* (Hernández-Ledesma et al., 2004). The antihypertensive effects of these BPs have been well documented. There are various strains such as *L. delbrueckii* subsp. *bulgaricus* and *L. lactis* subsp. *cremoris*, which introduced as potential strains to produce ACE-inhibitory peptides containing milks. These BPs were also have been found in some ripened cheeses. These findings suggest that fermented products of milk such as cheese can be regarded as functional food in the ca BPs, which might have a great effect on hypertension prevention (Ferreira-Santos et al., 2017; Padghan et al., 2017).

3.2.2 ANTIHYPERTENSIVE PEPTIDES

Hypertension is one of the major risk factors in cardiovascular diseases occurrence. Thus, any reduction in blood pressure can be important for health. Nutritional factors besides pharmacological therapy can significantly prevent and cure hypertension. These BPs may be synthesized via different routes such as enzymatic hydrolysis of parent proteins during GI digestion or with microbial enzymes and fermentation of milk with cultures containing proteolytic starters (Phelan et al., 2009). Determining the structure of the peptide will make it possible to synthesize these peptides using different procedures such as recombinant DNA technology and chemical or enzymatic synthesis (Gill et al., 1996).

3.2.3 WHEY PROTEIN-DERIVED PEPTIDES

Whey protein derived antihypertensive peptides such as β-lactorphin (Tyr-Leu-Leu-Phe) and α-Lactorphin (Tyr-Gly-Leu-Phe) were derived from β-lactoglobulin (β-Lg) and α-lactalbumin (α-La), respectively, during enzymatic proteolysis (Antila et al., 1991). An *in vitro* study by Sipola et al. (2002) showed that both tetrapeptides were able to improve arterial functions. Furthermore, strong antihypertensive effects of β-lactosin B (Ala-Leu-Pro-Met, another tetrapeptide from β-Lg) on spontaneously hypertensive rats (SHR) was observed. Cheese originated whey proteins (β-Lg) treated with proteinase K produced tripeptide Ile-Pro-Ala, which showed good ability in decreasing blood pressure in SHR after a single-dose administration. Whey proteins' therapeutic applications can be found in Marshall and Saito studies (Jäkälä and Vapaatalo, 2010). Whey proteins and their derived peptides have been attracted less attention, particularly in the field of clinical researches. It was reported that supplementing with hydrolyzed whey proteins can decrease blood pressure o by 5.5 and 8.0 mmHg of diastolic and systolic blood pressure, respectively (Pins and Keenan, 2006). In contrast, fortification of milk drink with whey powder did not show a reduction in blood pressure in mild-hypertensive rats (Lee et al., 2007). Thus, more intervention investigations are required to assess the potential clinical benefits of the whey protein-derived BPs.

3.2.4 CASEIN-DERIVED PEPTIDES

Milk casein-derived peptides were first introduced for their antihypertensive effects using *L. helveticus* CP790 for milk fermentation (Yamamoto et al., 1994). It was reported that extracellular proteinase peptides generated substances with ACE-inhibitory effects from casein, which were in charge of the antihypertensive activity. They were identified as Ile-Pro-Pro and Val-Pro-Pro, which showed the IC_{50}-values of 5 and 9 μM, respectively. Bovine β-casein and κ-casein, especially their primary structure, are casein-based sources for these amino acid sequences (Nakamura et al., 1995). According to *in vitro* experiments, it has been found that IC_{50}-values of Val-Pro-Pro and Ile-Pro-Pro can be even less based on the substrate concentration (Lehtinen et al., 2010). Researchers were reported that feeding SHR with lyophilized sour milk (2.5%) containing diet in the long run can prevent the hypertension progress in SHR (Nakamura et al., 1996). Further findings from the ACE-inhibitory effects of the casein-derived peptides were reported by Masuda et al. (1996). Fermented milk products containing Val-Pro-Pro and Ile-Pro-Pro have significantly attenuated hypertension in rats. Pure tripeptides did not show strong antihypertensive effect in comparison to the milk products containing them. Although naturally occurring minerals (potassium and calcium) in the milk product might contributed in providing more antihypertensive effects, they did not individually attenuate the blood pressure development as the fermented milk products do (Jäkälä and Vapaatalo, 2010). Tripeptides-treated SHR showed high plasma renin activity (Sipola et al., 2002). Moreover, serum ACE activity was decreased by treatment with a fermented milk product containing both lacto-tripeptides and plant sterols (Jäkälä et al., 2009). Furthermore, lacto-tripeptide-containing fermented milk products decreased aldosterone and serum ACE levels in salt-loaded GK rats (Jauhiainen et al., 2009).

Studies on two fermented milk products with different lacto-tripeptides values indicated dose-dependent antihypertensive effect of lacto-tripeptides (Sipola et al., 2002). It was shown that termination of the treatment either with fermented milk product or pure tripeptides, will cause the gradually increase in blood pressure. Protection effects of Val-Pro-Pro and Ile-Pro-Pro on isolated SHR mesenteric arteries was also reported (Jäkälä and Vapaatalo, 2010). Researchers also showed significant antihypertensive effects of ß-casein-based BPs (LHLPLP, derived from *E. faecalis*-fermented milk) on SHR (Jäkälä and Vapaatalo, 2010). In most long-term clinical studies, the effect of lacto-tripeptides on blood pressure has been studied using sour milk prepared by fermenting skim milk with *L. helveticus* and/or *S. cerevisiae*. It is important to mention that the milk product's components such as minerals can also affect the peptide absorption and can play a role directly on blood pressure. Hence, the results from different treatments with different products cannot be compared directly, particularly when there are small effects on a biological variable.

Dose is another parameter that should be considered. It was reported that the lowest effective dose of lacto-tripeptides in humans is 3.07 mg per a day (Sano et al., 2005). The highest reported dose is 52.5 mg (Jauhiainen et al., 2005). Furthermore, it is important to consider that the blood pressure is taken at resting times or not. It is important to notice that neither in fermentation nor in enzymatic hydrolysis test products might contain different BPs with different amounts (Jäkälä and Vapaatalo, 2010).

3.2.5 WHEAT GERM PROTEIN HYDROLYSATES (WGPH)

Potent ACE inhibitory BPs of wheat germ proteins were successfully generated by Alcalase. The molecular mass of the main hydrolysates was less than 1,500 Da (ranging between 1,050 and 180 Da). Moreover, the antioxidant properties of several hydrolysates prepared at different periods of hydrolytic reactions were studied and the antioxidant activities of the hydrolysates with 6 h reaction time were high. The Alcalase based enzymatic hydrolysis of wheat germ protein provided a high proportion of BPs ranging from 1,500 Da to free amino acids. Freeze-dried WGPH exhibited nearly high free radical-scavenging and antioxidant activities. Nowadays, consumers highly demanding functional foods based on natural compounds, and therefore WGPH can be used as potent functional food ingredients in food industries. However, further detailed investigations on peptide fractions of WGPH along with assessing their *in vivo* antioxidant properties and related mechanisms are required (Zhu et al., 2006).

3.2.6 MEAT DERIVED PEPTIDES

The integrity and activity of proteins experience changes during rigor mortis, resolution of rigor mortis and long-term frozen storage and these are important factors in food processing. A large number of free amino acids and peptides are released during meat processing due to proteolytic mechanisms and chemical or physical treatment (Lafarga and Hayes, 2014).

It is reported that challenges in the characterization of natural BPs in processed meats depend on many factors, but one of the most important ones is the severe proteolysis in muscle proteins because of endogenous enzymes and microbial peptidases activities. In dry-fermented meats these activities not only influence flavor and taste but also the synthesizes of BPs. The identification and characterization progress in natural BPs activities is mostly depend on modern bioinformatics tools as well as updated protein databases such as proteomics and silico approaches. Finally, multiple reaction monitoring mass spectrometry methodology was announced as a strong tool for BPs' quantitation purposes (Mora et al., 2017).

3.2.7 *ANTIOXIDATIVE PEPTIDES*

3.2.7.1 *SOY GENERATED BPS*

Soybean proteins with antioxidative peptides, are comprised of 3 to 16 amino acids (Pihlanto, 2006). These proteins include leucine, valine or hydrophobic amino acids at the N-terminal sites, and tyrosine, histidine, or proline in the sequence. It was reported that histidine-containing peptides can act as an active-oxygen quencher, metal-ion chelator, and hydroxy radical scavenger. Chemical structure and nature of peptide fractions, which is depend on the specificity action of protease, can determine the capability of protein hydrolysates in preventing unwanted changes due to lipid oxidation. A good representative for soy protein hydrolysates' antioxidant activity is the peptide with a sequence of Leu-Leu-Pro-His-His (Chen et al., 1996, 1998).

3.2.7.2 *MILK GENERATED BPS*

Digestion of milk proteins can generate peptides with antioxidative activities. These Milk-derived antioxidative BPs are usually include 5 to 11 amino acids such as hydrophobic amino acids, tyrosine histidine, proline, or tryptophan in their sequence. Caseins have been exhibited antioxidant activity against TBARS (Wong and Kitts, 2003). It was reported that lipid autoxidation induced by lipoxygenase can be prevented by milk caseins. The explanation for this phenomenon, is free radicals quenching activity due to amino acid residues oxidation in casein. However, free amino acids cannot be replaced for casein as the antioxidant, signifying the main role of the primary structure of casein (Laakso, 1984). Free radical scavenging activity were isolated and identified from casein's peptic digestion. The peptide, Tyr-Phe-Tyr-Pro-Glu-Leu, showed promising superoxide anion radical scavenging activity, which was mainly attributed to the C-terminal dipeptide Glu-Leu (Suetsuna et al., 2000). Caseins' polar domains include phosphorylated serine residues and related specific sequences (SERP-SERP-SERP-Glu-Glu), which are effective chelators creating complexes with cations like calcium, iron, and zinc. Therefore, phosphorylated caseins or derived peptides can act as natural chelators in food emulsions and control lipid oxidation by binding to metals. Casein hydrolysates showed more effective inhibitory impacts on lipid oxidation than the enriched Casein phosphor-peptides (CPP) at same phosphorous content. The mechanism for antioxidant activities of casein hydrolysates and CPP has further been studied using phosphatidylcholine-based liposomes (Díaz and Decker, 2004). Results showed efficient inhibitory effects of them on TBARS development, however, high quantities of CPP were observed to act as prooxidant agents, while casein hydrolysates only showed antioxidant activities. Casein hydrolysates and their low molecular weight casein hydrolysates' role in peroxyl radical scavenging activities were better than enriched CPP, which might be attributed to their amino acid composition. Casein hydrolysates contain higher concentrations of free radical scavenging amino acids such as lysine, histidine, tyrosine, and proline than CPP. CPP seems to act as an antioxidant with chain-breaking properties in the presence of AAPH, which produces peroxyl radicals. Same results were found with the hydrophilic free radical quenching activities (Kitts and Weiler, 2003). Casein calcium peptides (CCPs ~3 kDa) exhibited dose-dependently strong antioxidant activity in the b-carotene bleaching method. CCPs also revealed scavenging activity against radicals of hydroxyl, DPPH, and superoxide (Sakanaka et al., 2005). *In vitro* effects of different bovine caseins on the lipoxygenase activity showed that a subtilisin and tryptic b-casein digests of whole casein kept their inhibitory features. In this research, the highest inhibitory effects on oxidation of linoleic acid were observed in the fraction involving b-casein f(169–176) with a trace amount of b-casein f(33–48). The structure-function relationship between the amino acid sequence, antioxidant capacity, and effectiveness were examined with different procedures v synthetic peptides. Results showed that, although β-Casein f(177–183) was a highly potent antioxidant. It was almost inactive against enzymatic oxidation of linoleic acid (Rival et al., 2001a, b). Other substances, such as vitamin E, were reported to play an antioxidant role without having significant inhibitory effects on linoleic acid oxidation. It was reported that non-enzymatic and enzymatic lipid peroxidation was inhibited by casein-derived peptides, due to their act as a desired target over free radicals of fatty acids (Pihlanto, 2006).

3.2.7.3 MARINE GENERATED BPs

Marine organisms are highly potential sources of various bioactive compounds with numerous biological effects making them to be considered as a potent source of novel bioactive substances (Aneiros and Garateix, 2004; Barrow and Shahidi, 2009). These BPs can be produced by their inherent enzymatic hydrolysis (Sheih et al., 2009). However, it was reported that in food sauces produced from fermented marine, enzymatic hydrolysis by microorganisms can generate the BPs. These BPs can be purified without any further hydrolysis. It was stated that marine gelatin derived BPs are able to pose potent antioxidant effects. Thus, marine-derived BPs may have great potential for use as natural compounds with antioxidative properties and also can act as a substitute for synthetic antioxidants (Je et al., 2005).

3.2.8 ANTIMICROBIAL PEPTIDES

The total antibacterial properties of milk are much higher than the sum of the individual contributions of immunoglobulin and non-immunoglobulin defense proteins. Along with their synergistic effects, the presence of natural bactericidal peptides beside to generated ones from inactive protein precursors may be another parameter. Antibacterial and antifungal milk proteins such as basic glycopeptides and lactoferrin, and lactoferrin peptides (ex. lactoferricin B) showed bacteria/fungicidal activity against various strains of *Streptococcus C. albicans, and S. aureus* (Lahav et al., 1971; Wakabayashi et al., 1996). Other studies also showed that combination of lactoferrin-related peptides with triazoles, prevented the growth of azole-resistant *C. albicans* hyphae (Wakabayashi et al., 1998; Clare and Swaisgood, 2000).

3.2.9 CASEIN PHOSPHOR-PEPTIDES (CPP)

CPP have been recognized after releasing of trypsin from αs and β caseins. Almost all CPP include three serine phosphate clusters and two glutamic acid residues, forming soluble organophosphate salts, and possibly acting as a mineral carrier particularly calcium. Different degrees of phosphorylation can be seen in these fractions. A direct relationship between mineral chelating ability and the degree of phosphorylation has been described as follows; αs2-CN > αs1-CN > β-CN > κ-CN (non-uniform phosphoserine clusters distribution). It was further illustrated that calcium-binding CPP are mostly hardy hydrolyzed with gut enzymes and most frequently founded in calcium phosphate complexes (Reynolds et al., 1994). This complex formation provides the reason for improved absorption of calcium through the distal small intestines of animals fed with casein diets compared to control animals fed with soy-based diets (Kitts et al., 1992).

3.2.10 IMMUNOMODULATORY PEPTIDES

Immunomodulatory peptides obtained from milk affect both cell proliferation responses and the immune system. Bradykinin, an immune-enhancing hormone, can be inactivated by β-casokinins, which inhibits ACE enzymes. Thus, this chain of actins indirectly generates a total immunostimulatory response. Casein hydrolysates derivative peptides revealed an increase in phagocytotic activity of human macrophages against aging red blood cells (Migliore-Samour et al., 1989). *In vivo* studies on rats revealed immunostimulatory activity of hexapeptide (obtained by hydrolysis of human β-CN) against *Klebsiella pneumoniae* (Parker et al., 1984). In yogurt production, BPs reduced cell proliferation in IEC-6 and Caco-2 cells. This report can make it clear, how yogurt use has been related with a reduced level of colon cancer incidence (Clare and Swaisgood, 2000; Ganjam et al., 1997).

3.2.11 OPIOID MILK PEPTIDES

Fragments of β-CN involve the major opioid peptides group, called β-casomorphins. β-casomorphins have exogenous origin and morphine-like features; however, they can be produced by hydrolysis of bovine αs1-CN fractions by pepsin (Meisel and Schlimme, 1990). Human β-CN fractions have been shown to have similar peptides. The Y-P-F sequence, which is a popular sequence in bovine β-casomorphin, was also reported to exist in the primary structure of human β-CN (Greenberg et al., 1984). In general, peptides derived from α- and β-CN generate agonist reactions, whereas κ-CN derived one's cause antagonist impacts. It is worth to mention that BPs are produced from most of the major proteins in either human or bovine milk (Clare and Swaisgood, 2000).

3.3 PRODUCTION PROCESS OF BIOACTIVE PEPTIDES (BPS)

BPs can be synthesized in various conditions and procedures, which the most important ones are reported here (Table 3.2).

3.3.1 ANGIOTENSIN-CONVERTING ENZYME (ACE) INHIBITION

ACE is known as a multifunctional enzyme linked with the rennin-angiotensin system. Peripheral blood pressure is regulated by ACE, where it catalyzes both inactivation of the vasodilator bradykinin and creation of the vasoconstrictor angiotensin II. Therefore, ACE inhibition in the organism can cause a reduced levels of blood pressure, and it can also influence several inmmunodefense regulatory systems. This procedure is one of the most important and known methods for producing ACE-inhibitory peptides (Hernández-Ledesma et al., 2004).

3.3.2 GASTROINTESTINAL (GI) DIGESTION

Milk proteins hydrolysis during various stages of GI digestion can produce physiologically active peptides (Parrot et al., 2003). In the gastrointestinal tract (GIT), proteinases (i.e., pepsin, trypsin, and chymotrypsin) are able to hydrolyze ingested proteins into the peptides with various lengths. However, there are some proteins that are resistant against proteinases and remain intact in GI conditions. Subsequently, further digestion on some of the generated peptides to generate amino acids or short chain peptides can be carried out with brush-border peptidases placed at the surface of epithelial cells (Jäkälä and Vapaatalo, 2010).

3.3.3 FERMENTATION

Fermentation process of milk using dairy starter cultures can generate BPs with potential proteolytic activities (Jäkälä and Vapaatalo, 2010). In this regard, LAB are able to utilize milk proteins especially caseins, as a growth source. For example, *Lactococcus lactis* and *Lactobacillus helveticus* strains have been used for producing BPs (Fitzgerald and Murray, 2006).

3.3.4 ENZYMATIC HYDROLYSIS

GI enzymes and also enzyme complexes (e.g., Alcalase™) can be used to produce BPs from whole proteins (Wu and Ding, 2002). LAB's proteolytic enzymes can be used to produce BPs from caseins that were obtained from various species. Increasing functional food production yields can be achieved by application of commercially accessible microbial-derived proteinases and cost-effective ultrafiltration membrane reactors (Korhonen and Pihlanto, 2003). It is reported that hydrolysis by papain, pepsin or trypsin enhanced the ACE-inhibitory activity and radical scavenging capacity (Abdel-Hamid et al., 2017). Alcalase is an alkaline protease that can be used for the generation of BPs and can also be used for synthesizing of protein hydrolysates with improved functional and nutritional characteristics in comparison to original proteins, but also for. It was stated that Alcalase is able to solubilize 85% of wheat germ proteins with favorable techno-functional properties (Claver and Zhou, 2005).

3.4 TECHNO-FUNCTIONAL APPLICATIONS OF BIOACTIVE PEPTIDES (BPS)

3.4.1 ANTIOXIDANT ACTIVITY

Antioxidant activity of the protein hydrolysates might be attributed to their inherent amino acid sequences. The results imply that the derived peptides from milk proteins can enhance antioxidant features preventing oxidation reactions during functional food processing. Casein hydrolysates, enriched CPP and low molecular weight casein hydrolysates showed to be good TBARS inhibitors in cooked ground beef. Although TBARS inhibitory ability of enriched CPP was highest, hydrolysates showed better antioxidant activities. Since the catalytic activity of iron can be increased by cooking, enriched CPP can be more effective antioxidants in cooked muscle foods due to their strong chelating activity (Díaz and Decker, 2004). Application of CCP in ground beef homogenates also showed high antioxidant activity. Incorporation 2% of CCP inhibited lipid oxidation (~70%) in homogenates. These results indicate that BPs can also be potent antioxidants in meat processing technologies and also can help to prevent off-flavor formation besides increasing shelf life (Sakanaka et al., 2005).

3.4.2 APPLICATION IN MEAT PRODUCTS

Whey proteins are already used as functional components in meat-based products and can act as potent antioxidants in these products. For this purpose, the antioxidant activity of some whey hydrolysates was evaluated in cooked meat pork patties (Peña-Ramos and Xiong, 2003). The results revealed that

2% of whey protein isolates (WPIs) (and their hydrolytic derivatives) not only decreased the cooking loss but also reduced lipid oxidation during refrigerated storage. Hydrolysis of whey proteins with protamex enhanced their capability to inhibit the formation of conjugated dienes or hydroperoxides (early-stage lipid oxidation) as well as slowing down the hydroperoxides degradations causing TBARS formation. Therefore, partially hydrolyzation of whey proteins seems to produce more promising compounds for quality enhancement in processed muscle foods.

3.4.3 FOAMING AND EMULSIFYING PROPERTIES

Limited enzymatic hydrolysis (protease) was applied for producing gluten hydrolysates. The emulsifying and foaming properties of retentates, hydrolysates, and permeates analysis displayed hydrolysates foaming capacity; however, the foams were not stable. Permeates produced goof foaming features at pH of 6.5 but a very short lifetime. Permeates exhibited no emulsifying features. Retentates generated good stability foams, which were more efficient than entire hydrolysates to stabilize emulsions. They provided strong resistance to coalescence. Techno-functional properties of retentates were only influenced by ionic strength and pH slightly (Popineau et al., 2002).

3.5 EFFECTS OF PROCESSING PARAMETERS ON BIOACTIVE PEPTIDES (BPS)

3.5.1 EFFECTS OF OXIDIZING CONDITIONS

Endogenous production of oxidative compounds (peroxides and free radicals) during food processing and applying oxidizing agents into the food industry (e.g., aseptic packaging with hydrogen peroxide) might cause oxidation of amino acids. In this case, peroxides can easily oxidize the methionine to methionine sulfoxide, which later can be oxidized to homocysteine acid and methionine sulfone. These two compounds are biologically unavailable. Under acidic conditions such as stomach, methionine can be regenerated from Methionine sulfoxide. Similarly, mono/disulfoxides forms of L-cystine showed bioavailability once they reduced to L-cystine (in the body conditions). Acidic and mild oxidizing environments are able to oxidize Trp. The presence of riboflavin, a photo-sensitizer, in dairy products makes them more susceptible to photo-oxidation. Cys, Trp, His, Met, and Tyr are just some amino acids susceptible to this type of oxidation (Mestdagh et al., 2005). It is enough clear that these types of oxidative changes can cause an increment in the molecular structure changes of antihypertensive and ACE-inhibitory peptides, which consequently may alter the bioactivity of BPs (López-Fandiño et al., 2006).

3.5.2 EFFECTS OF HEATING PROCESSING

Thermal processing at high pH values (alkaline conditions) is able to make adjustments in the structure of amino acids including Arg, Lys, Ser, and Thr. In these conditions Arg is prone to decompose to ornithine. One reason for these changes might be attributed to the racemization phenomenon, in which L-amino acids partially racemized into D-amino acids. This phenomenon is highly occurred in amino acids containing strong electron donor activity in their side chain, i.e., Asp, Phe, Glu, Cys, Asn, Ser, and Thr. It was stated that heating processes induced this phenomenon in roasted bovine serum albumin and casein. Furthermore, L-amino acids can be converted to D-amino acids by microorganisms-based enzymes such as transaminases, epimerases, and amino acid oxidases (Friedman, 1999b). Therefore, fermented milk and ripened cheeses which get the benefits of microorganisms are a good source of D-amino acids. Dehydroalanine residues, highly reactive substances, can be generated from phosphoserine, Cys, Ala, and cystine during high temperature (200°C)-neutral pH or normal heat-alkaline pH conditions. Dehydroalanine residues can react with different amino acids such as Cys, Lys, His, and ornithine and is able to change these amino acids to lanthionine, lysinoalanine, histidinoalanine, and ornithoalanine, respectively. It was reported that thermally treated milk products will contain high levels of lysinoalanine amino acid due to the presence of precursors such as cysteine (in whey proteins) and serine phosphate groups (in casein) (Friedman, 1999a). These changes may influence the sensitivity of peptides to the proteolysis, which consequently can change the bioactive sites of the peptides.

3.5.2.1 EFFECTS OF MAILLARD REACTION

The Maillard reaction, as one of the main heating reactions, is a common occurring phenomenon in heated products. It is known as a non-enzymatic browning, and could have significant effects on nutritional aspects of amino acids. The reaction engages amino acids and reduced sugars. The main point in Maillard reactions is the speed of reaction, which goes up faster with elevated temperatures. The results are decomposed amino acids and sugars condensing them into brown compounds such as melanoidins (Nooshkam and Madadlou, 2016a, b). High levels of Lys-rich proteins and lactose in some food products such as milk make them sensitive to this reaction. Therefore, they are more prone to the amino acid loss during food processing. Severe heat treatments such as in-can sterilization, can cause more amounts of available amino acids loss in comparison with new technologies such as high temperature-short time (HTST) and or ultra-high temperature (UHT) processes (Ma et al., 2017).

3.5.3 EFFECTS OF SPRAY DRYING ON CASEIN HYDROLYSATES

Bitterness and hygroscopicity are two important features hindering the use of protein hydrolysates in food systems. Spray drying technology was mentioned as an alternative for decreasing these issues. In a study, masking bitter taste of casein hydrolysate was investigated using spray drying and different combinations of soy protein isolate (SPI) and gelatin as carriers. Scanning electron microscopy analysis exhibited almost all spherical particles, some particles have pores and many with concavities. In this study, all formulations were water and oil compatible with less hygroscopicity levels than free casein hydrolysate. The free hydrolysate exhibited that the Tg at aw of 0.83 was lower than 25, indicating more stability of developed formulations at aw of 0.65. The sensory panel, who tasted in the powder form, reported less bitter ($P < 0.05$) tastes for developed formulations compared to free casein hydrolysate. Therefore, the results suggested that spray drying of mixtures of SPI, gelatin, and casein hydrolysate was successful to reduce the bitterness of casein hydrolysate in free from. Thus, spray drying expands the feasibilities of casein hydrolysates application in food-producing processes (Favaro-Trindade et al., 2010).

3.5.4 EFFECTS OF OVEN AND FREEZE DRYING

Freeze and oven-dried fish hydrolysate proteins (OD-FPH and FD-FPH, respectively) were studied to check the effects of these processes on the BPs derived from fresh fish muscle. The results of amino acid profiles revealed a higher percentage of hydrophilic and hydrophobic residues in FD-FPH and OD-FPH samples, respectively. Fourier transform infrared (FT-IR) spectra showed random coil and b-sheet structures in OD-FPH and FD-FPH samples, respectively. The ACE-inhibitory activity of FD-FPH was less stable (during digestion with pancreatin and pepsin) than that of OD-FPH samples. OD-FPH had a brownish-yellow color caused Maillard reaction. Peptides' secondary structures in the hydrolysates were changed, especially by the oven drying process. Thermal aggregations caused slightly higher molecular weight of OD-FPH than FD-FPH samples. OD-FPH effects on the ACE inhibitory activity was comparable to that of the FD-FPH and suggested better stability against progressive digestion by pancreatin and pepsin. These results indicates that oven drying (80 ± 2°C) can be used in the generation of FPHs without any remarkable loss in ACE inhibitory activity (Elavarasan et al., 2016).

3.6 SENSORY EVALUATIONS OF BIOACTIVE PEPTIDES (BPs)

Sensory properties especially the taste plays an important role in consumer. Peptides can show various tastes from sweet to umami (Temussi, 2012). Low molecular weight peptides with hydrophobic amino acid residues (e.g., Pro and Phe) can cause bitter taste and, consequently will influence consumer satisfactoriness (Siro et al., 2008). Therefore, for oral consumption of peptide-based foods, taste should be carefully considered. Various methods were reported for removing/ reducing the bitterness of small BPs and protein hydrolysates such as activated carbon treatments; transpeptidation, adding amino acids (Asp and Glu), intact proteins, and cyclodextrin (FitzGerald and O'cuinn, 2006). One of the concerns about these methods is that they mostly include removal or addition of molecules to or from the peptides. Hence, optimization is greatly required to these processes for specific peptides and protein hydrolysates to confirm that debittering process does not cause any significant alterations in profiles of molecular that might cause loss of stability or activity (Lacroix and Li-Chan, 2012). Moreover, microencapsulation via spray drying has been showed to be successfully masking the undesirable taste of bitter peptides. The findings of processes debittering peptides without significant changes on their physiological functions and molecular structural integrity will be an innovation in designing peptide-based medicinal-functional foods that can meet the consumers satisfactory criteria. Additionally, concerns should be raised to characterize the chemosensory features of newly developed/produced BPs and evaluation of consumer sensitivity for different bitterness levels of functional quantities within the intended BPs delivery carriers (Udenigwe, 2014).

3.7 HEALTH PROMOTING AND BENEFICIAL EFFECTS OF BPs

BPs are also involved in non-food matrices to supply certain health-improving features. In this regard, CPPs, and amorphous calcium phosphate that were added in toothpaste (Prospec MI™ Paste, GC Tooth Mousse™), mouth rinsing solution, or chewing gum (Recaldent™, Trident™) (Hartmann and Meisel, 2007). Some analytical methods for determining the antioxidative activity of BPs under *in vitro* situations can also be used to evaluate their protective roles in human intervention investigations. Strong attempts are used to explain the different risk factors of oxidative DNA damage, cancer, and other age-related diseases (Hoelzl et al., 2005). Some human and animal investigations recommend that fermented milk products could have potent antioxidative roles. Kullisaar et al. (2003) reported that using fermented goat milk prolonged the lipoprotein fraction resistance to oxidation, and subsequently enhanced anti-atherogenicity in healthy subjects by decreasing the levels of 8-isoprostanes, oxidized LDL, peroxidized lipoproteins, and the glutathione redox ratio, and improved overall antioxidative activity (Kullisaar et al., 2003). The release of numerous BPs from milk proteins via microbial proteolysis has also been reported (Gobbetti et al., 2004).

3.8 ABSORPTION MECHANISMS OF BPs

Paracellular diffusion and carrier-mediated transport were reported to be absorbance pathway of BPs (Shimizu, 2004). Oligopeptides and short tripeptides are actively transported via the paracellular route and a PepT1 transporter, respectively. Three different absorption models for antihypertensive peptides, Val-Pro-Pro, and Ile-Pro-Pro, were investigated. The results showed that small amounts of them are transported in intact form across the intestinal epithelium barrier (van der Pijl et al., 2008). Main transport mechanisms for these tripeptides were passive diffusion and paracellular transport. A human investigation showed that Leu-Pro-Pro and Ile-Pro-Pro were shown to be measurable from plasma after ingestion of yogurt enriched with tripeptides (Foltz et al., 2007). Remarkably, detectable levels of Ile-Pro-Pro in plasma were also found after ingestion of a placebo yogurt beverage without addition of tripeptides. This might be because of the production of Ile-Pro-Pro from milk proteins by intestinal brush border peptidases or luminal (van der Pijl et al., 2008). Half-lives for elimination and absorption were only a few minutes. Subsequently, there is a high feasibility that these peptides can get to target sites of action intact and also exhibit systemic effects. In the case of Ile-Pro-Pro derived from casein, after oral administration of a single dose to rats, radiolabeled tripeptide absorbed in partially intact form through the GIT. An *in vitro* study showed that Ile-Pro-Pro did not bind to plasma proteins (i.e., albumin) (Jauhiainen et al., 2007). As peptides are quickly metabolized to their amino acids by peptidases located at brush border membrane, bioavailability, and absorption remains very low after oral administration. However, there is data indicating that at least intact form of di/tripeptides might be absorbed following by entering the circulation and exhibit systemic effects (Miguel et al., 2008).

3.9 SAFETY ISSUE AND HUMAN CONCERNS

The European Food Safety Authority (EFSA), which make scientific instructions to the European Commission (EC), indicated that published data about lacto-tripeptides have stated clear antihypertensive effects in different animal hypertension models, however, human data is more inconsistent. As almost clinical investigations are done by industries, not all data regarding the product manufacturing or composition is accessible. As the interest on health-promoting and disease-preventing features of foods has been increasing, there must be more emphasis on the legal regulations associated with the health claims of these products. Authorities around the world have had to establish systematic methods for reviewing and assessing scientific data (Jäkälä and Vapaatalo, 2010).

3.10 PROSPECTIVE FUTURE AND RESEARCH OPPORTUNITIES

The development of any functional food products with health benefits beyond their basic nutrition needs an in-depth understanding of the mechanisms causing these benefits. In the case of fermented milks with hypotensive features, it is necessary that detailed information should be provided about the biological mechanisms and how cardiovascular control can be modulated by substances in fermented dairy products. In regard to peptides with ACE inhibitory activities, a detailed information of the mechanisms contributed to the interaction between the existing peptides/amino acids and active site of the enzyme. The use of enrichment procedures such as chromatographic isolation and membrane techniques may also be an interest area in the future of high potent BPs extractions from fermented milks for further use as functional food substances. Many BPs derived from milk proteins can be found in the same regions in the intact protein called strategic zones. These regions are considered might be partially protected from breakdowns caused by proteolysis. Therefore, it can be possible to design new fermented dairy foods providing synergistic effects due to their multifunctional bioactivities. In summary, milk includes numerous peptide sequences with various modulatory and regulatory processes such as hormone secretion (casomorphins), immune nutrient uptake (casomorphins and phosphopeptides), neurological information transmission (casokinins), and defense (casomorphins, casokinins, and immunopeptides). The recovery for non-denatured, biologically active whey proteins from cheese manufacturing by-products are already industrially developed. In this case, caseins are produced on a large scale for being used in food products, i.e., cottage cheese). Therefore, it should be reasonably affordable to generate BPs from them for use as natural dietary supplements, safe, healthy, and efficient preservatives. This field of study can promisingly contribute to generate novel milk-based biologically active peptides that will significantly have effects on the food industries.

3.11 SUMMARY AND CONCLUSION

One of the main objectives for newly formulated foods is to ensure that people are not misinformed, and health claims truly endorse healthier choices. Indeed, nutritional quality and status of a new food formulation should not be masked by incorrect health claims. The problem in the newly formulated functional foods is that in almost all cases small group of people are involved in the research. Therefore, the population of investigations should be large enough to strongly prove the efficacy of these types of medicinal-functional foods. We suggest that more preclinical studies should be taken with different research groups until coming to the results with high consistency and repeatability. With these investigations good hypotheses with well-defined methods can provide valuable outcomes about the mechanisms of action. In the case of BPs, EFSA considered the evidence on the antihypertensive effect of lacto-tripeptides to be insufficient. Although several clinical investigations revealed positive results, a few researches claiming no effect on blood pressure. However, statements such as helps to control and or/to keep blood pressure at healthy levels might be contributed to the lactotripeptide-containing foods in the future.

KEYWORDS

- **antihypertensive peptides**
- **bioactive peptides**
- **casein phosphor-peptides**
- **food processing**
- **functional foods**
- **immunomodulatory peptides**
- **milk proteins**
- **protein hydrolysates**
- **soy proteins**
- **wheat proteins**

REFERENCES

Abdel-Hamid, M., Otte, J., De Gobba, C., Osman, A., & Hamad, E., (2017). Angiotensin I-converting enzyme inhibitory activity and antioxidant capacity of bioactive peptides derived from enzymatic hydrolysis of buffalo milk proteins. *International Dairy Journal, 66*, 91–98.

Andishmand, H., Hamishehkar, H., Mohammadifar, M. A., Babazadeh, A., Taghvimi, A., & Tabibiazar, M., (2017). A colon targeted delivery system for resveratrol enriching in pH responsive-model. *Pharmaceutical Sciences, 23*(1), 42–49.

Aneiros, A., & Garateix, A., (2004). Bioactive peptides from marine sources: Pharmacological properties and isolation procedures. *Journal of Chromatography B, 803*(1), 41–53.

Antila, P., Paakkari, I., Järvinen, A., Mattila, M. J., Laukkanen, M., Pihlanto-Leppälä, A., Mäntsälä, P., & Hellman, J., (1991). Opioid peptides derived from *in-vitro* proteolysis of bovine whey proteins. *International Dairy Journal, 1*(4), 215–229.

Babazadeh, A., Ghanbarzadeh, B., & Hamishehkar, H., (2016). Novel nanostructured lipid carriers as a promising food-grade delivery system for rutin. *Journal of Functional Foods, 26*, 167–175.

Babazadeh, A., Ghanbarzadeh, B., & Hamishehkar, H., (2017). Formulation of food-grade nanostructured lipid carrier (NLC) for potential applications in medicinal-functional foods. *Journal of Drug Delivery Science and Technology, 39*, 50–58.

Barrow, C., & Shahidi, F., (2007). *Marine Nutraceuticals and Functional Foods* (p. 512). CRC Press.

Chen, H. M., Muramoto, K., Yamauchi, F., & Nokihara, K., (1996). Antioxidant activity of designed peptides based on the antioxidative peptide isolated from digests of a soybean protein. *Journal of Agricultural and Food Chemistry, 44*(9), 2619–2623.
Chen, H. M., Muramoto, K., Yamauchi, F., Fujimoto, K., & Nokihara, K., (1998). Antioxidative properties of histidine-containing peptides designed from peptide fragments found in the digests of a soybean protein. *Journal of Agricultural and Food Chemistry, 46*(1), 49–53.
Clare, D. A., & Swaisgood, H. E., (2000). Bioactive milk peptides: A prospectus. *Journal of Dairy Science, 83*(6), 1187–1195.
Claver, I. P., & Zhou, H., (2005). Enzymatic hydrolysis of defatted wheat germ by proteases and the effect on the functional properties of resulting protein hydrolysates. *Journal of Food Biochemistry, 29*(1), 13–26.
Díaz, M., & Decker, E. A., (2004). Antioxidant mechanisms of caseinophosphopeptides and casein hydrolysates and their application in ground beef. *Journal of Agricultural and Food Chemistry, 52*(26), 8208–8213.
Elavarasan, K., Shamasundar, B. A., Badii, F., & Howell, N., (2016). Angiotensin I-converting enzyme (ACE) inhibitory activity and structural properties of oven-and freeze-dried protein hydrolysate from freshwater fish (Cirrhinus mrigala). *Food Chemistry, 206*, 210–216.
Favaro-Trindade, C. S., Santana, A. S., Monterrey-Quintero, E. S., Trindade, M. A., & Netto, F. M., (2010). The use of spray drying technology to reduce the bitter taste of casein hydrolysate. *Food Hydrocolloids, 24*(4), 336–340.
Ferreira-Santos, P., Carrón, R., Recio, I., Sevilla, M. Á., & Montero, M. J., (2017). Effects of milk casein hydrolyzate supplemented with phytosterols on hypertension and lipid profile in hypercholesterolemic hypertensive rats. *Journal of Functional Foods, 28*, 168–176.
Fitzgerald, R. J., & Murray, B. A., (2006). Bioactive peptides and lactic fermentations. *International Journal of Dairy Technology, 59*(2), 118–125.
Fitzgerald, R. J., & O'cuinn, G., (2006). Enzymatic debittering of food protein hydrolysates. *Biotechnology Advances, 24*(2), 234–237.
Foltz, M., Meynen, E. E., Bianco, V., Van, P. C., Koning, T. M., & Kloek, J., (2007). Angiotensin-converting enzyme inhibitory peptides from a lactotripeptide-enriched milk beverage are absorbed intact into the circulation. *The Journal of Nutrition, 137*(4), 953–958.
Friedman, M., (1999a). Chemistry, biochemistry, nutrition, and microbiology of lysinoalanine, lanthionine, and histidinoalanine in food and other proteins. *Journal of Agricultural and Food Chemistry, 47*(4), 1295–1319.
Friedman, M., (1999b). Chemistry, nutrition, and microbiology of D-amino acids. *Journal of Agricultural and Food Chemistry, 47*(9), 3457–3479.
Ganjam, L. S., Thornton, W. H., Marshall, R. T., & MacDonald, R. S., (1997). Antiproliferative effects of yogurt fractions obtained by membrane dialysis on cultured mammalian intestinal cells. *Journal of Dairy Science, 80*(10), 2325–2329.
Gill, I., López-Fandiño, R., Jorba, X., & Vulfson, E. N., (1996). Biologically active peptides and enzymatic approaches to their production. *Enzyme and Microbial Technology, 18*(3), 162–183.
Gobbetti, M., Minervini, F., & Rizzello, C. G., (2004). Angiotensin I-converting-enzyme-inhibitory and antimicrobial bioactive peptides. *International Journal of Dairy Technology, 57*(2, 3), 173–188.
Gobbetti, M., Stepaniak, L., De Angelis, M., Corsetti, A., & Di Cagno, R., (2002). Latent bioactive peptides in milk proteins: Proteolytic activation and significance in dairy processing. *Critical Reviews in Food Science and Nutrition, 42*(3), 223–239.

Greenberg, R., Groves, M. L., & Dower, H. J., (1984). Human beta-casein. Amino acid sequence and identification of phosphorylation sites. *Journal of Biological Chemistry, 259*(8), 5132–5138.

Grootaert, C., Matthijs, B., Voorspoels, S., Possemiers, S., Smagghe, G., & Van, C. J., (2017). Egg-derived bioactive peptides with ACE-inhibitory properties: A literature update. *Food & Function, 8*(11), 3847–3855.

Harnedy, P. A., O'Keeffe, M. B., & FitzGerald, R. J., (2017). Fractionation and identification of antioxidant peptides from an enzymatically hydrolyzed *Palmaria palmata* protein isolate. *Food Research International, 100*, 416–422.

Hartmann, R., & Meisel, H., (2007). Food-derived peptides with biological activity: From research to food applications. *Current Opinion in Biotechnology, 18*(2), 163–169.

Hernández-Ledesma, B., Amigo, L., Ramos, M., & Recio, I., (2004). Angiotensin-converting enzyme inhibitory activity in commercial fermented products. Formation of peptides under simulated gastrointestinal digestion. *Journal of Agricultural and Food Chemistry, 52*(6), 1504–1510.

Hoelzl, C., Bichler, J., Ferk, F., Simic, T., Nersesyan, A., Elbling, L., Ehrlich, V., et al., (2005). Methods for the detection of antioxidants which prevent age-related diseases: A critical review with particular emphasis on human intervention studies. *Journal of Physiology and Pharmacology, 56*, 49.

Jäkälä, P., & Vapaatalo, H., (2010). Antihypertensive peptides from milk proteins. *Pharmaceuticals, 3*(1), 251–272.

Jäkälä, P., Hakala, A., Turpeinen, A. M., Korpela, R., & Vapaatalo, H., (2009). Casein-derived bioactive tripeptides Ile-Pro-Pro and Val-Pro-Pro attenuate the development of hypertension and improve endothelial function in salt-loaded Goto-Kakizaki rats. *Journal of Functional Foods, 1*(4), 366–374.

Jauhiainen, T., Pilvi, T., Cheng, Z. J., Kautiainen, H., Müller, D. N., Vapaatalo, H., Korpela, R., & Mervaala, E., (2009). Milk products containing bioactive tripeptides have an antihypertensive effect in double transgenic rats (dTGR) harboring human renin and human angiotensinogen genes. *Journal of Nutrition and Metabolism, 2010*, 1–6.

Jauhiainen, T., Vapaatalo, H., Poussa, T., Kyrönpalo, S., Rasmussen, M., & Korpela, R., (2005). *Lactobacillus helveticus* fermented milk lowers blood pressure in hypertensive subjects in 24-h ambulatory blood pressure measurement. *American Journal of Hypertension, 18*(12), 1600–1605.

Jauhiainen, T., Wuolle, K., Vapaatalo, H., Kerojoki, O., Nurmela, K., Lowrie, C., & Korpela, R., (2007). Oral absorption, tissue distribution and excretion of a radiolabeled analog of a milk-derived antihypertensive peptide, Ile-Pro-Pro, in rats. *International Dairy Journal, 17*(10), 1216–1223.

Je, J. Y., Park, J. Y., Jung, W. K., Park, P. J., & Kim, S. K., (2005). Isolation of angiotensin I converting enzyme (ACE) inhibitor from fermented oyster sauce, *Crassostrea gigas*. *Food Chemistry, 90*(4), 809–814.

Kim, S. K., & Wijesekara, I., (2010). Development and biological activities of marine-derived bioactive peptides: A review. *Journal of Functional Foods, 2*(1), 1–9.

Kimatu, B. M., Zhao, L., Biao, Y., Ma, G., Yang, W., Pei, F., & Hu, Q., (2017). Antioxidant potential of edible mushroom (Agaricus bisporus) protein hydrolysates and their ultrafiltration fractions. *Food Chemistry, 230*, 58–67.

Kitts, D. D., & Weiler, K., (2003). Bioactive proteins and peptides from food sources. Applications of bioprocesses used in isolation and recovery. *Current Pharmaceutical Design, 9*(16), 1309–1323.

Kitts, D. D., Yuan, Y. V., Nagasawa, T., & Moriyama, Y., (1992). Effect of casein, casein phosphopeptides and calcium intake on ileal 45 Ca disappearance and temporal systolic blood pressure in spontaneously hypertensive rats. *British Journal of Nutrition, 68*(3), 765–781.

Korhonen, H., & Pihlanto, A., (2003). Food-derived bioactive peptides-opportunities for designing future foods. *Current Pharmaceutical Design, 9*(16), 1297–1308.

Kullisaar, T., Songisepp, E., Mikelsaar, M., Zilmer, K., Vihalemm, T., & Zilmer, M., (2003). Antioxidative probiotic fermented goats' milk decreases oxidative stress-mediated atherogenicity in human subjects. *British Journal of Nutrition, 90*(2), 449–456.

Laakso, S., (1984). Inhibition of lipid peroxidation by casein Evidence of molecular encapsulation of 1, 4-pentadiene fatty acids. *Biochimica et Biophysica Acta (BBA)-Lipids and Lipid Metabolism, 792*(1), 11–15.

Lacroix, I. M., & Li-Chan, E. C., (2012). Evaluation of the potential of dietary proteins as precursors of dipeptidyl peptidase (DPP)-IV inhibitors by an in silico approach. *Journal of Functional Foods, 4*(2), 403–422.

Lafarga, T., & Hayes, M., (2014). Bioactive peptides from meat muscle and by-products: Generation, functionality and application as functional ingredients. *Meat Science, 98*(2), 227–239.

Lahav, E., Edelsten, D., Sode-Mogensen, M. T., & Sofer, E., (1971). Properties of basic glycopeptides released from cow milk protein by heat. *Milchwissenschaft, 26*, 489–495.

Lee, Y. M., Skurk, T., Hennig, M., & Hauner, H., (2007). Effect of a milk drink supplemented with whey peptides on blood pressure in patients with mild hypertension. *European Journal of Nutrition, 46*(1), 21–27.

Lehtinen, R., Jauhiainen, T., Kankuri, E., Lindstedt, K., Kovanen, P. T., Kerojoki, O., Korpela, R., & Vapaatalo, H., (2010). Effects of milk casein-derived tripeptides Ile-Pro-Pro, Val-Pro-Pro, and Leu-Pro-Pro on enzymes processing vasoactive precursors *in vitro*. *Arzneimittelforschung, 60*(04), 182–185.

López-Fandiño, R., Otte, J., & Van, C. J., (2006). Physiological, chemical and technological aspects of milk-protein-derived peptides with antihypertensive and ACE-inhibitory activity. *International Dairy Journal, 16*(11), 1277–1293.

Ma, X. J., Gao, J. Y., Tong, P., Li, X., & Chen, H. B., (2017). Tracking the behavior of Maillard browning in lysine/arginine-sugar model systems under high hydrostatic pressure. *Journal of the Science of Food and Agriculture, 97*(15), 5168–5175.

Masuda, O., Nakamura, Y., & Takano, T., (1996). Antihypertensive peptides are present in the aorta after oral administration of sour milk containing these peptides to spontaneously hypertensive rats. *The Journal of Nutrition, 126*(12), 3063–3068.

Meisel, H., & Schlimme, E., (1990). Milk proteins: Precursors of bioactive peptides. *Trends in Food Science & Technology, 1*, 41–43.

Meisel, H., (2001). Bioactive peptides from milk proteins: A perspective for consumers and producers. *Australian Journal of Dairy Technology, 56*(2), 83.

Meisel, H., Goepfert, A., & Günther, S., (1997). ACE-inhibitory activities in milk products. *Milchwissenschaft, 52*, 307–311.

Mestdagh, F., De Meulenaer, B., De Clippeleer, J., Devlieghere, F., & Huyghebaert, A., (2005). Protective influence of several packaging materials on light oxidation of milk. *Journal of Dairy Science, 88*(2), 499–510.

Migliore-Samour, D., Floc'h, F., & Jolles, P., (1989). Biologically active casein peptides implicated in immunomodulation. *Journal of Dairy Research, 56*(3), 357–362.

Miguel, M., Dávalos, A., Manso, M. A., De La Peña, G., Lasunción, M. A., & López-Fandiño, R., (2008). Transepithelial transport across Caco-2 cell monolayers of antihypertensive egg-derived peptides. PepT1-mediated flux of Tyr-Pro-Ile. *Molecular Nutrition & Food Research, 52*(12), 1507–1513.

Mohanty, D. P., Mohapatra, S., Misra, S., & Sahu, P. S., (2016). Milk derived bioactive peptides and their impact on human health: A review. *Saudi Journal of Biological Sciences, 23*(5), 577–583.

Mora, L., Gallego, M., Reig, M., & Toldrá, F., (2017). Challenges in the quantitation of naturally generated bioactive peptides in processed meats. *Trends in Food Science & Technology, 69*, 306–314.

Nakamura, Y., Masuda, O., & Takano, T., (1996). Decrease of tissue angiotensin I-converting enzyme activity upon feeding sour milk in spontaneously hypertensive rats. *Bioscience, Biotechnology, and Biochemistry, 60*(3), 488, 489.

Nakamura, Y., Yamamoto, N., Sakai, K., & Takano, T., (1995). Antihypertensive effect of sour milk and peptides isolated from it that are inhibitors to angiotensin I-converting enzyme. *Journal of Dairy Science, 78*(6), 1253–1257.

Neves, A. C., Harnedy, P. A., O'Keeffe, M. B., & FitzGerald, R. J., (2017). Bioactive peptides from *Atlantic salmon* (Salmo salar) with angiotensin-converting enzyme and dipeptidyl peptidase IV inhibitory, and antioxidant activities. *Food Chemistry, 218*, 396–405.

Ngoh, Y. Y., & Gan, C. Y., (2018). Identification of Pinto bean peptides with inhibitory effects on α-amylase and angiotensin-converting enzyme (ACE) activities using an integrated bioinformatics-assisted approach. *Food Chemistry, 267*, 124–131.

Nooshkam, M., & Madadlou, A., (2016a). Microwave-assisted isomerization of lactose to lactulose and Maillard conjugation of lactulose and lactose with whey proteins and peptides. *Food Chemistry, 200*, 1–9.

Nooshkam, M., & Madadlou, A., (2016b). Maillard conjugation of lactulose with potentially bioactive peptides. *Food Chemistry, 192*, 831–836.

Padghan, P. V., Mann, B., Sharma, R., Bajaj, R., & Saini, P., (2017). Production of angiotensin-I-converting-enzyme-inhibitory peptides in fermented milks (Lassi) fermented by *Lactobacillus acidophilus* with consideration of incubation period and simmering treatment. *International Journal of Peptide Research and Therapeutics, 23*(1), 69–79.

Parker, F., Migliore-Samour, D., Floch, F., Zerial, A., Werner, G. H., Jollès, J., Casaretto, M., Zahn, H., & Jollès, P., (1984). Immuno stimulating hexapeptide from human casein: Amino acid sequence, synthesis and biological properties. *European Journal of Biochemistry, 145*(3), 677–682.

Parrot, S., Degraeve, P., Curia, C., & Martial-Gros, A., (2003). *In vitro* study on digestion of peptides in Emmental cheese: Analytical evaluation and influence on angiotensin I converting enzyme inhibitory peptides. *Food/Nahrung, 47*(2), 87–94.

Peña-Ramos, E. A., & Xiong, Y. L., (2003). Whey and soy protein hydrolysates inhibit lipid oxidation in cooked pork patties. *Meat Science, 64*(3), 259–263.

Pessione, E., & Cirrincione, S., (2016). Bioactive molecules released in food by lactic acid bacteria: Encrypted peptides and biogenic amines. *Frontiers in Microbiology, 7*, 876.

Phelan, M., Aherne, A., FitzGerald, R. J., & O'Brien, N. M., (2009). Casein-derived bioactive peptides: Biological effects, industrial uses, safety aspects and regulatory status. *International Dairy Journal, 19*(11), 643–654.

Pihlanto, A., (2006). Antioxidative peptides derived from milk proteins. *International Dairy Journal, 16*(11), 1306–1314.

Pins, J. J., & Keenan, J. M., (2006). Effects of whey peptides on cardiovascular disease risk factors. *The Journal of Clinical Hypertension, 8*(11), 775–782.

Popineau, Y., Huchet, B., Larré, C., & Bérot, S., (2002). Foaming and emulsifying properties of fractions of gluten peptides obtained by limited enzymatic hydrolysis and ultrafiltration. *Journal of Cereal Science, 35*(3), 327–335.

Reynolds, E. C., Riley, P. F., & Adamson, N. J., (1994). A selective precipitation purification procedure for multiple phosphoryl-containing peptides and methods for their identification. *Analytical Biochemistry, 217*(2), 277–284.

Rival, S. G., Boeriu, C. G., & Wichers, H. J., (2001a). Caseins and casein hydrolysates. 2. Antioxidative properties and relevance to lipoxygenase inhibition. *Journal of Agricultural and Food Chemistry, 49*(1), 295–302.

Rival, S. G., Fornaroli, S., Boeriu, C. G., & Wichers, H. J., (2001b). Caseins and casein hydrolysates. 1. Lipoxygenase inhibitory properties. *Journal of Agricultural and Food Chemistry, 49*(1), 287–294.

Sakanaka, S., Tachibana, Y., Ishihara, N., & Juneja, L. R., (2005). Antioxidant properties of casein calcium peptides and their effects on lipid oxidation in beef homogenates. *Journal of Agricultural and Food Chemistry, 53*(2), 464–468.

Sano, J., Ohki, K., Higuchi, T., Aihara, K., Mizuno, S., Kajimoto, O., Nakagawa, S., et al., (2005). Effect of casein hydrolysate, prepared with protease derived from *Aspergillus oryzae*, on subjects with high-normal blood pressure or mild hypertension. *Journal of Medicinal Food, 8*(4), 423–430.

Sheih, I. C., Wu, T. K., & Fang, T. J., (2009). Antioxidant properties of a new antioxidative peptide from algae protein waste hydrolysate in different oxidation systems. *Bioresource Technology, 100*(13), 3419–3425.

Shimizu, M., (2004). Food-derived peptides and intestinal functions. *Biofactors, 21*(1–4), 43–47.

Singh, B. P., Vij, S., & Hati, S., (2014). Functional significance of bioactive peptides derived from soybean. *Peptides, 54*, 171–179.

Sipola, M., Finckenberg, P., Korpela, R., Vapaatalo, H., & Nurminen, M. L., (2002). Effect of long-term intake of milk products on blood pressure in hypertensive rats. *Journal of Dairy Research, 69*(1), 103–111.

Siro, I., Kápolna, E., Kápolna, B., & Lugasi, A., (2008). Functional food. Product development, marketing and consumer acceptance: A review. *Appetite, 51*(3), 456–467.

Suetsuna, K., Ukeda, H., & Ochi, H., (2000). Isolation and characterization of free radical scavenging activities peptides derived from casein. *The Journal of Nutritional Biochemistry, 11*(3), 128–131.

Tagliazucchi, D., Shamsia, S., Helal, A., & Conte, A., (2017). Angiotensin-converting enzyme inhibitory peptides from goats' milk released by *in vitro* gastrointestinal digestion. *International Dairy Journal, 71*, 6–16.

Temussi, P. A., (2012). The good taste of peptides. *Journal of Peptide Science, 18*(2), 73–82.

Udenigwe, C. C., (2014). Bioinformatics approaches, prospects and challenges of food bioactive peptide research. *Trends in Food Science & Technology, 36*(2), 137–143.

Udenigwe, C. C., Okolie, C. L., Qian, H., Ohanenye, I. C., Agyei, D., & Aluko, R. E., (2017). Ribulose-1, 5-bisphosphate carboxylase as a sustainable and promising plant source of bioactive peptides for food applications. *Trends in Food Science & Technology, 69*, 74–82.

Van, D. P. P. C., Kies, A. K., Ten, H. G. A., Duchateau, G. S., & Deutz, N. E., (2008). Pharmacokinetics of proline-rich tripeptides in the pig. *Peptides, 29*(12), 2196–2202.

Wakabayashi, H., Abe, S., Okutomi, T., Tansho, S., Kawase, K., & Yamaguchi, H., (1996). Cooperative anti-Candida effects of lactoferrin or its peptides in combination with azole antifungal agents. *Microbiology and Immunology, 40*(11), 821–825.

Wakabayashi, H., Abe, S., Teraguchi, S., Hayasawa, H., & Yamaguchi, H., (1998). Inhibition of hyphal growth of azole-resistant strains of *Candida albicans* by triazole antifungal agents in the presence of lactoferrin-related compounds. *Antimicrobial Agents and Chemotherapy, 42*(7), 1587–1591.

Welsh, G., Ryder, K., Brewster, J., Walker, C., Mros, S., Bekhit, A. E. D. A., McConnell, M., & Carne, A., (2017). Comparison of bioactive peptides prepared from sheep cheese whey using a food-grade bacterial and a fungal protease preparation. *International Journal of Food Science & Technology, 52*(5), 1252–1259.

Wong, P. Y. Y., & Kitts, D. D., (2003). Chemistry of buttermilk solid antioxidant activity. *Journal of Dairy Science, 86*(5), 1541–1547.

Wu, J., & Ding, X., (2002). Characterization of inhibition and stability of soy-protein-derived angiotensin I-converting enzyme inhibitory peptides. *Food Research International, 35*(4), 367–375.

Wu, J., Aluko, R. E., & Nakai, S., (2006). Structural requirements of angiotensin I-converting enzyme inhibitory peptides: Quantitative structure-activity relationship study of di-and tripeptides. *Journal of Agricultural and Food Chemistry, 54*(3), 732–738.

Yamamoto, N., Akino, A., & Takano, T., (1994). Antihypertensive effect of the peptides derived from casein by an extracellular proteinase from *Lactobacillus helveticus* CP790. *Journal of Dairy Science, 77*(4), 917–922.

Zhao, R. J., Huo, C. Y., Qian, Y., Ren, D. F., & Lu, J., (2017). Ultra-high-pressure processing improves proteolysis and release of bioactive peptides with activation activities on alcohol metabolic enzymes *in vitro* from mushroom foot protein. *Food Chemistry, 231*, 25–32.

Zhu, K., Zhou, H., & Qian, H., (2006). Antioxidant and free radical-scavenging activities of wheat germ protein hydrolysates (WGPH) prepared with Alcalase. *Process Biochemistry, 41*(6), 1296–1302.

Part II
Probiotics and Potential Health Benefits

CHAPTER 4

Probiotics in Fruits and Vegetables: Challenges, Legislation Issues, and Potential Health Benefits

MAMTA THAKUR,* DEEPAK KUMAR VERMA,* SUDHANSHU BILLORIA, H. W. DESHPANDE, AMI R. PATEL, and GEETANJALI KAUSHIK

**Corresponding authors. E-mails: thakurmamtafoodtech@gmail.com; mamta.ft@gmail.com (M. Thakur), deepak.verma@agfe.iitkgp.ernet.in; rajadkv@rediffmail.com (D. K. Verma)*

ABSTRACT

These days consumers are highly selective about their food and their interest has been shifting towards the nutrients-balance diets that offer the physiological and health benefits. Consequently, the scientists are focusing on the functional foods among which the probiotics are highly beneficial to maintain the balance of gut microbiota thus preventing the chronic diseases. The genera of *Lactobacillus* and *Bifidobacterium* are frequently employed by the food processing industry and are mostly incorporated in dairy substrates but the increasing fashion of vegan, the demand for low-fat and low-cholesterol foods, and cases of lactose intolerance caters the ever-expanding demand for non-dairy probiotic products. Therefore, the fruit and vegetable are the best matrices due to the presence of essential nutrients, bioactive components, pleasing taste-profiles. However, the viability of probiotics is a serious challenge to reflect their functionality in the gastrointestinal tract (GIT) which is strongly dependent on several processing and storage factors. The present contribution overviews the details of the mechanism of probiotics action, their selection criteria and potential in non-dairy substrates mainly fruits and vegetables. This review also highlights the fruits and vegetables as a potential substrate for the development of probiotic products.

4.1 INTRODUCTION

Today, consumers are more conscious of the correlation between food and well-being that has resulted in the huge demand for healthy foods and beverages. However, the concept of health-improving foods is not recent; 2,400 years ago, the Hippocrates said, "*Let food be thy medicine and medicine be thy food.*" These days, healthy foods typically refer to the functional foods that were introduced by the Japanese government in the mid-1980s and defined as, "any food or food ingredient which on consumption imparts the positive effects to the highly advanced body functions simultaneously offering the basic nutritional effects" (Cenci and Chingwaru, 2010). The foods which are inherently possessing or prepared using the bioactive components such as dietary fiber, oligosaccharides, phytochemicals, and "friendly" bacteria or probiotics are the prominent examples of functional or health foods (Jankovic et al., 2010) among which probiotics are expeditiously growing and constitute nearly 60–70% of the functional foods (Tripathi and Giri, 2014).

The role of probiotics in health maintenance and improvement is not new. Eli Metchnikoff, the Nobel Prize winner in Medicine, in 1908 firstly noted the effect of what is now called probiotic. He associated the health and long-life of the Bulgarian villagers with the intake of bacteria like *Lactobacillus bulgaricus* and *Streptococcus thermophilus* found in yogurt (Anukam and Reid, 2007). Basically, the term 'probiotic' is composed of the Latin word 'pro' meaning 'for' and Greek word 'biotic' that refers to 'life.' Probiotics are defined by FAO/WHO as, "live microorganisms, which when administered in adequate amounts, confer a health benefit on the host" (FAO/WHO, 2001). This definition strictly emphasized the consumption of live microorganisms' insufficient high number, typically greater than 10^8–10^9 CFU g^{-1} of the product during consumption (Prado et al., 2008). The majority of probiotic microorganisms employed today in the food sector are lactic acid bacteria (LAB) and *Bifidobacteria* species that have "generally recognized as safe" (GRAS) status. The probiotic cultures can be obtained in the capsules form but are chiefly sold by their addition in the fermented foods particularly dairy products (Heller, 2001). While choosing probiotic bacteria for any kind of industrial manufacturing processes; their functional, safety, and technological features must be kept in mind. Any bacterial strain is typically not thought to be a probiotic until its health benefits are validated clinically (Reid, 2008).

The use of dairy-based substrates is an ideal medium for probiotics growth and survival but their market is decreasing recently due to the milk fat content, possible allergens (casein), cholesterol content and increasing trend of vegan. Among the non-dairy food matrices, the fruits and vegetables are preferred because they are nutrient-dense, healthy, and refreshing but the survival of probiotics is a major challenge. The main objective of this chapter is to demonstrate the several kinds of fruits and vegetable-based probiotic products, their health potential, and major challenges encountered during processing and storage of probiotic products.

4.2 MECHANISMS OF PROBIOTICS' ACTION

Generally, the probiotics are linked with several health benefits including mitigation of the lactose-intolerance symptoms, alleviation of symptoms of the antibiotic treatment of *Helicobacter pylori*, alleviation of atopic dermatitis symptoms in children and inhibiting the risk of allergy in infancy, prevention of dental caries, treatment of viral as well as antibiotic-associated diarrhea, alleviation of inflammable bowel disease (IBD) and irritable bowel syndrome (IBS) symptoms, and improving the immune response (Saarela et al., 2000; Nase et al., 2001; Kumar et al., 2013). The potential probiotic mechanisms of action as described by Kim et al. (2016) are depicted in Figure 4.1. In addition to these, the distinct advantage of probiotics is the growth and development of microbiota in the gastrointestinal tract (GIT) to maintain a suitable balance between the beneficial bacteria and pathogens (Oelschlaeger, 2010). The good or friendly bacteria, i.e., probiotics are thus mainly consumed through food to restore and flourish the natural microflora of the gut after the antibiotic therapy (Johnston et al., 2006). Another important function is balancing the action of pathogenic bacteria encountered due to the contaminated or spoiled food and environmental conditions. The probiotic cultures, therefore, must restrain the growth of pathogenic microbes like *Campylobacter jejuni*, *Clostridium perfringens*, *Salmonella Enteritidis*, *E. coli*, and several species of *Staphylococcus*, *Yersinia*, and *Shigella*, successfully (Carter et al., 2017; Chingwaru and Vidmar, 2017; Hussain et al., 2017). Moreover, the B-complex vitamins (vit.-B_1, vit.-B_2, vit.-B_3, vit.-B_6, vit.-B_8, vit.-B_9 and vit.-B_{12}) are synthesized by probiotic microorganism like *Lactobacillus reuteri*, *L. plantarum*, *Bifidobacterium pseudocatenulatum*, and *B. adolescentis* that also produce several enzymes (lipase, esterase, and co-enzymes A, Q, NADP, and NAD), improve the assimilation of minerals and vitamins, trigger the amino acids and organic acids production, and improve the effectiveness of the immunological system (Nova et al., 2007; Gu et al., 2015; Markowiak and Śliżewska, 2017). Colonization of probiotics over the intestinal tract does not always necessary to impart the beneficial effects. Few microorganisms such as *B. longum* may become a segment of GI flora, while few, for instance, *L. casei* influences the GIT indirectly during transient path (Ohland and Macnaughton, 2010). Therefore, the potency of probiotics to exert health effects is strongly determined by their mechanism of action that is as in the four types (Figure 4.2).

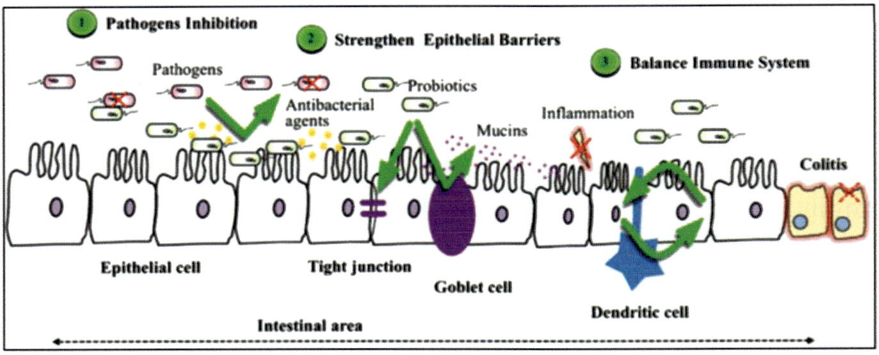

FIGURE 4.1 Potential mechanisms of action of probiotics.

Source: Reprint with permission from Kim et al, 2016. © Springer Nature.

The first two mechanisms are no doubt linked with the effect of probiotics on other microbes, thus, significant in the prevention and treatment of various infectious diseases to maintain the balance of GI microbiota. The probiotic cultures may aggregate with other microorganisms (one among their mechanisms) and discharge several guts-protective compounds such as glutamine, arginine, conjugated linoleic acids, and small fatty acids to generate a defensive barrier for the inhibition of pathogens to colonize the epithelial cells that are important in immune response (Eizaguirre et al., 2002; Mangell et al., 2002). Thus, the pathogens like *C. botulinum* can be repressed using the probiotics while competing for restrained nutrients or binding sites, essential for the receptors or fermentation.

This mechanism, in turn, imparts a positive effect on the health and/or immune system of the host (Lebeer et al., 2008). Also, the adherence of probiotics over the epithelial cells can stimulate a signaling jet that results in the immunological harmonization. The immunological modulation activated by probiotics can be explained on the basis of enhanced interferon and immunoglobulins (Ig) synthesis and stimulation of lymphocytes and macrophages activity. On the other hand, the discharge of few soluble compounds can generate either direct or indirect actuation of the immunological cells that is beneficial in inhibiting and treating the infectious diseases together with the chronic alimentary tract inflammation and sometimes eradication of the cancer cells also (Collado et al., 2007; Oelschlaeger, 2010). After adherence to the epithelial cells, numerous biological actions like the discharge of cytokines and chemokines occurs which in turn make their secondary action like triggering of the mucosal and systemic host immunity (Delcenserie et al., 2005). The probiotic cultures obstruct the invading microbes from epithelium infiltration either by generating or releasing the peptides. Numerous *in-vitro* studies demonstrated the potential of low-molecular-weight compounds like short-chain fatty acids and hydroperoxide synthesized by probiotics for restraining the pathogens' multiplication. Further, the LAB along with *Bifidobacterium* genus species can produce antibacterial peptides, antibiotics, class III bacteriocins, de-conjugated bile acids, etc., that exhibit the intense antimicrobial action.

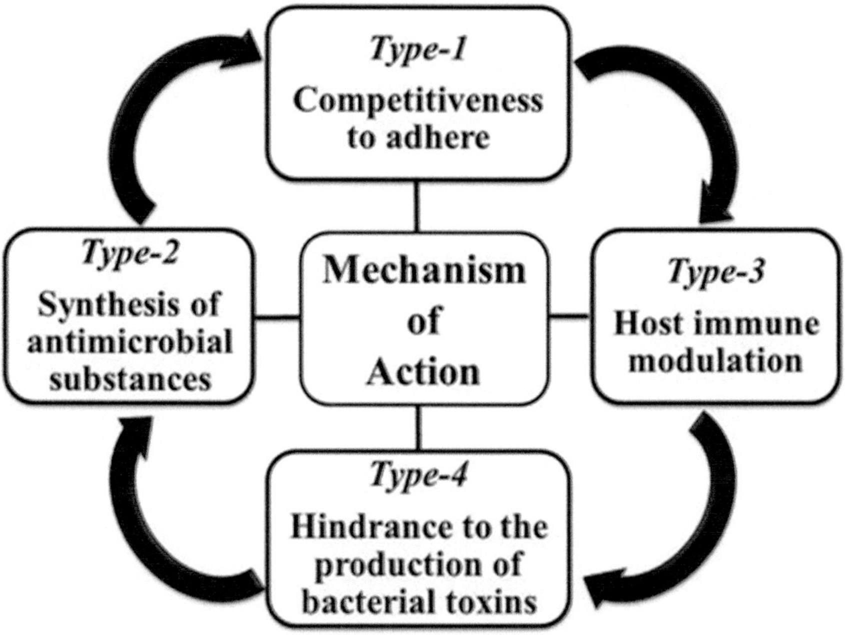

FIGURE 4.2 Mechanism of probiotic action in a stepwise manner.

However, more investigations are essential for illustrating the mechanism of probiotic cells about how will they gain resistance against the self-produced metabolic products (Begley et al., 2006; Oelschlaeger, 2010). Perhaps it may be the favorable fact for the LAB that they do not require iron in their indigenous habitat that is an essential mineral for almost all bacteria. The performance of other microorganisms is influenced by *L. delbrueckii* by restricting the iron hydroxide to its cell surface, accordingly turning it inaccessible for other microorganisms (Weinberg, 1997; Elli et al., 2000). Lastly, the synthesis of bacterial toxins can be prevented either deactivating or eliminating the toxins from the body through the process of adsorption as some bacterial strains may attach with the toxins thus minimizing to the absorption of intestinal toxins (Schatzmayr et al., 2006). But such detoxifying characteristics are particular to the kind of probiotic strains. Therefore, more efforts should be taken to isolate and identify such strains. Diarrhea can be resisted by some probiotic strains because of their potential to safeguard the body from toxins (Oelschlaeger, 2010).

4.3 PROBIOTIC CULTURES

Several probiotic microorganisms are used globally for producing the probiotic products among which most belong to the genera: *Lactobacillus* and *Bifidobacterium*. Numerous strains of *Lactobacillus* and *Bifidobacterium* are employed commercially which are enlisted in Table 4.1. Beyond these, the microorganisms from other genera like *Enterococcus*, *Streptococcus*, *Pediococcus*, and *Leuconostoc* also exhibit the probiotic effects (Table 4.1). *Saccharomyces baulardii*-yeast is being used commercially as a probiotic. The probiotics can be used in single or mixed cultures. *Bifidobacterium lactis* and *Lactobacillus acidophilus* commonly employed probiotics are often integrated with *Streptococcus thermophiles* as they are deficient in necessary proteolytic activity (Fijan, 2014; Casarotti et al., 2014; Yerlikaya, 2014).

TABLE 4.1 Examples of Widely Used Probiotic Cultures

Probiotic Micrograms	**Examples of Probiotic Cultures**
Lactobacillus species	*L. acidophilus, L. amylovorus, L. brevis, L. bulgaricus, L. casei, L. crispatus, L. fermentum, L. gasseri, L. helveticus, L. johnsonii, L. johnsonii, L. lactis, L. paracasei, L. pentosus, L. plantarum, L. reuteri, L. rhamnosus*
Bifidobacterium species	*B. adolescentis, B. animalis, B. bifidum, B. brevis, B. infantis, B. lactis, B. longum*
Other lactic acid bacteria (LAB)	*Enterococcus fraecium, Lactococcus cremoriss, L. lactis, Leuconostoc dextranicum, Pediococcus acidilactici, Propionibacterium freudenreichii, Streptococcus thermophillus*
Non-lactic acid bacteria (N-LAB)	*Bacillus cereus, B. clausii, Clostridium botyricum, Escherichia coli Nissle 1917, Saccharomyces boulardi*

4.4 INCREASING MARKET OF NON-DAIRY PROBIOTIC IN TRADE

Nowadays, probiotic bacteria are available in a variety of food products, dietary supplements, and drugs. However, the most common probiotic products found in the global markets are dairy-based from the ancient times and yogurt only contributes to the nearly 78% of global probiotics sale (Granato et al., 2010). Dairy no doubt is a very nutritious substrate for probiotic microorganisms fulfilling their nutritional requirements, however, increasing trend of the vegan lifestyles, the issues of lactose intolerance and milk protein allergies, and the demand for low-fat and low-cholesterol foods have created a growing demand for non-dairy based foods, such as fruits and vegetables, coconut milk, nutrition bars, fruit drinks, soy products, and cereal-based products to deliver probiotics in the diet (Ranadheera et al., 2010; Peres et al., 2012). Some non-dairy probiotic beverages especially based on cereals are not recent and numerous traditional beverages are found globally. Table 4.2 illustrates the current research activities to develop non-dairy based probiotic beverages and also include the traditional non-dairy probiotic beverages from different parts of the world (excluding the fruits and vegetables that will be explained in this section).

TABLE 4.2 Examples of Non-Dairy based Probiotic Beverages (Excluding the Fruits and Vegetables)

Product Name	Probiotic Cultures	Substrate(s) Used	References
Adai	Lactic acid bacteria	Cereal, legume	Farnworth (2005)
Atole	Lactic acid bacteria	Maize	Escamilla-Hurtado et al. (1993)
Ben-Saalga	Lactic acid bacteria	Pearl millet	Tou et al. (2006)
Boza	*L. plantarum, L. brevis, L. rhamnosus, L. fermentum,* and *Leuc. mesenteroides* subsp. *dextranium*	Wheat, rye, millet, maize, and other cereals	Gotcheva et al. (2000)
Bushera	*Lactobacillus, Lactococcus, Leuconostoc, Enterococcus, and Streptococcus*	Sorghum	Muyanja et al. (2003)
Chhang	*L. plantarum, L. casei, Enterococcus faecium* and *Pediococcus pentosaceus*	Rice	Thakur et al. (2015)
Chicha	Lactic acid bacteria	Rice	Puerari et al. (2015)
Dosa	*Leuc. mesenteroides, L. fermentum,* and *S. cerevisiae*	Rice and Bengal gram	Soni et al. (1986)

TABLE 4.2 *(Continued)*

Product Name	Probiotic Cultures	Substrate(s) Used	References
Idli	*Leuc. mesenteroides*, LAB, and yeast	Cereal, legume	Aidoo et al. (2006)
Kenkey	*L. casei, L. lactis, L. plantarum, L. brevis, L. acidophilus, L. fermentum, L. casei*, and yeast	Maize	Olasupo et al. (1997)
Kishk	Lactic acid bacteria	Cereal and milk	Tamime and McNulty (1999)
Kisra	*Lactobacillus* sp., and *L. brevis*	Sorghum	Mohammed et al. (1991)
Koko	*L. fermentum*, and *L. salivarius*	Millet	Lei and Jacobsen (2004)
Mahewu	*B. lactis*	Maize	McMaster et al. (2005)
Ngari	Lactic acid bacteria	Fish	Thapa et al. (2004)
Ogi	*L. plantarum, L. fermentum, Leuc. mesenteroides*, and *S. cerevisiae*	Maize	Omemu et al. (2007)
Omegisool	*L. pentosus SW02*, *L. plantarum SW03*, *L. sakei SW04*, *L. plantarum SW06*, *L. plantarum SW07*, *Pediococcus pentosaceus SW01*, and *Pediococcus acidilactici SW05*	Omegik cake (millet)	Oh and Jung (2015)
Pozol	*Streptococcus bovis*, *Streptococcus macedonicus*, *Lactococcus lactis*, and *Enterococcus sulfureus*	Maize	Díaz-Ruiz et al. (2003)
PROVIVA	*Lactobacillus plantarum* 299v	Oat	Prado et al. (2008)
Togwa	Lactic acid bacteria	Maize flour and finger millet malt	Oi and Kitabatake (2003)
Som-fug	Lactic acid bacteria	Fish	Riebroy et al. (2007)
Tarhana	*Streptococcus thermophilus, L. bulgaricus*, and *L. plantarum*	Parboiled wheat meal and yogurt	Ozdemir et al. (2007)
Tempeh	Lactic acid bacteria and *L. plantarum*	Maize, sorghum cassava	Onyango et al. (2003, 2004)
Yosa	LAB and *Bifidobacteria*	Oat bran	Martensson et al. (2002)

PROVIVA is the first-ever developed non-dairy probiotic product that is typically fermented oat gruel containing *L. plantarum 299v* (Prado et al., 2008). Recently, chocolate is also being exploited as a novel substrate for probiotics delivery of microencapsulated *Bifidobacteria* and LAB (Possemiers, 2010). Fermented coconut water is also available in supermarkets as a functional drink. Such kind of beverages may be an economic substitute to deliver the probiotic cultures benefitting the host. However, the incorporation of probiotic cultures into non-dairy food matrices is also a great challenge in itself because of the viability and varied response against the processing and storage conditions (Vinderola and Reinheimer, 2003).

4.5 FRUITS AND VEGETABLES: A HEALTHY APPROACH FOR BIOACTIVE COMPONENTS

The fruits and vegetables are no doubt the significantly nutritious, colorful, and flavorful constituents of the balanced diet that are rich in minerals, vitamins (such as A, E, B_1, B_6, B_9, C), dietary fiber, and phytochemicals (Dhandevi and Jeewon, 2015). They are designated as the low-calorie food because of the high water and low-fat content. On a general basis, they contain 70–80% moisture content (MC) (except melons and leafy vegetables that have nearly 92–95% moisture and tubers crops (cassava, corms, and yam) containing about 50% moisture), 2–40% carbohydrates (lowest in cucurbits and highest in cassava), 1–2% proteins, less than 1% lipids (except olives-15% and avocado-20% lipid content), and considerable amount of dietary fiber, essential vitamins, minerals, pigments, organic acids, and polyphenols. Among the nutrients, the ascorbic acid is most sensitive during processing while other nutrients like minerals, vitamins, and polyphenolic compounds are typically stable (Dias, 2012; Yahia et al., 2009). Carbohydrates are present in the form of simple sugars like glucose and fructose that are easily digested and assimilated in the body. Several organizations such as WHO, FAO, and USDA believe the intake of fruits and vegetables as a serious health concern and recommends their consumption globally to meet the nutritional requirements (Ragaert et al., 2004; Su and Arab, 2006).

Fruits and vegetables must comprise about 50% of the meal to maintain health better (Tiwari and Cummins, 2013). Several bioactive or functional components are found in fruits and vegetables (Table 4.3) like carotenoids, flavonoids, phytoestrogens, phenolic acids, etc., that contributes towards the several health benefits (Rickman et al., 2007). Due to the richness of phyto-nutraceuticals, the fruits and vegetables are potent antioxidants thus minimizing the probability of chronic diseases in the following ways: (i) protection from free radical injury; (ii) modification of metabolic activities; (iii) carcinogens detoxification; and (iv) alteration of tumor cells too (Craig and Beck, 1999; Southon, 2000; Dias and Ryder, 2011).

TABLE 4.3 Bioactive Components in Fruits and Vegetables Having Potential Health Benefits

Bioactive Component(s)	Source(s)	Potential Health Benefits
Flavonoids		
Anthocyanins: Cyanidin, Pelargonidin, Malvidin, Delphinidin	Red grapes, cherries, and berries	Boost the cellular antioxidant protection; reinforce the preservation of healthy brain functions
Flavanols: Epicatechins, Catechins, Epigallocatechin	Grapes and apples	Strengthen the maintenance of heart health
Flavanones: Hesperetin, Naringenin	Citrus fruits	Neutralization of the free radicals and lowering the risks of cancer
Proanthocyanidins and Procyanidins	Strawberries, grapes, cranberries, apples, and red wine	Strengthen the maintenance of urinary tract and heart health
Flavonols: Quercetin, Isorhamnetin, Kaempferol, Myricetin	Broccoli, tea, onions, and apples	Neutralization of the free radicals and boost the cellular antioxidant protection
Phytoestrogen		
Lignans	Many vegetables	Lowering cancer and heart disease
Carotenoids		
β-carotene	Sweet potatoes, pumpkin, carrots, tomatoes, and spinach	Conversion into vitamin A in the body, neutralization of the free radicals and boost the cellular antioxidant protection
Lutein	Carrots, asparagus, spinach, broccoli, and citrus fruits	Healthy vision
Lycopene	Watermelon, tomato, and red/pink grapefruit	Lowering the prostate cancer
Zea-xanthin	Citrus	Healthy vision
Dietary Fiber		
Soluble fiber	Peas, citrus fruits, apples, and beans	Lowering the risk of CHD and other types of cancer

TABLE 4.3 *(Continued)*

Bioactive Component(s)	Source(s)	Potential Health Benefits
Insoluble fiber	Fruit skins	Strengthen the maintenance of digestive health and lowering the risk of cancers
Isothiocyanates		
Sulforaphane	Broccoli sprouts, cabbage, cauliflower, and broccoli	Boost the detoxification of undesirable compounds as well as cellular antioxidant protection
Phenolic Acids		
Ferulic acid, caffeic acid	Pears, citrus fruits, apples, and some vegetables	Boost the cellular antioxidant protection and reinforce the maintenance of healthy eye and heart
Sulfides		
Dithiolthiones	Cruciferous vegetables	Boost the detoxification of undesirable components and maintain the healthy immune system
Allyl methyl trisulfide, diallyl sulfide	Onions, garlic, and leeks	Boost the detoxification of undesirable components and maintain the healthy immune system, heart, and digestive system
Minerals		
Calcium	Spinach	Lowering the risk of osteoporosis
Magnesium	Spinach, pumpkin seeds	Strengthen the maintenance of muscle and nerve function, immune, and bone health
Potassium	Leafy greens, banana, citrus juices, and potatoes	Lowering the risk of high blood pressure and stroke
Prebiotics		
Fructo-oligosaccharides	Banana, onions, garlic, leeks, and some fruits	Strengthen the maintenance of digestive health along with calcium absorption
Vitamins		
Vitamin A	Spinach, carrots, and sweet potato	Strengthen the maintenance of eye, immune, and bone health and contributes to cell integration

TABLE 4.3 *(Continued)*

Bioactive Component(s)	Source(s)	Potential Health Benefits
Thiamin	Peas	Strengthen the maintenance of mental function and regulate metabolism
Riboflavin	Green leafy vegetables	Encourages the cell growth and regulate the metabolism
Pantothenic acid	Sweet potato	Regulate the metabolism and hormone synthesis
Folate or folic acid	Green leafy vegetables, and citrus fruits	Lowering of the woman's risk of having a child with a brain or spinal cord defect; strengthen the immune health
Ascorbic acid	Kiwi, sweet pepper, guava, strawberries, and citrus fruits	Neutralization of the free radicals and supports the maintenance of bone and immune health

Fruits and vegetables pose a positive effect on cardiovascular health due to their anti-inflammatory, antioxidant, blood pressure regulatory, and anti-platelet characteristics. Therefore, they are always considered to be related with maintenance and improvement of vision, GI health, overall sound health, and lessened hazards of diabetes, stroke, cancer, gastric ulcer, anemia, rheumatoid arthritis, and several other chronic diseases (Prior and Cao, 2000; Dias, 2012).

4.6 WHY FRUITS AND VEGETABLES ARE A PROMISING SUBSTRATE FOR PROBIOTICS?

Generally, the plant tissues are composed of a multi-phase system that contains the complex microstructure encompassing the internal cells, intercellular spaces, capillaries, and small openings or pores. Among the four plant tissues namely parenchyma (storage), vascular (conducting), protecting, and supporting tissues; the edible part of fruits and vegetables are concentrated in parenchyma cells ($\approx$ 50–500 μm) thus carrying the essential nutrients to human growth and development. These cells appear from the middle lamella, primary cell wall, and plasma membrane to a fine-layer of parietal cytoplasm that carries several organelles like chloroplasts, mito-chondria, endoplasmic reticulum, plastids, etc., and restrained through the tonoplast membrane containing the one or more vacuoles. These vacuoles comprise the solution synthesized by organic acids, pigments, salts, and flavors, accountable to the cell osmotic potential (Aguilera and Stanley, 1991). The plasmodesmata-the channels linking the cytoplasm of adjoining cells can only infiltrate the plant cells walls that are otherwise an effective physical hindrance for the microorganisms due to the size of the very small pore (≈ 0.1–5.0 μm) restricting the bacterial entry (Alzamora et al., 2005). Nevertheless, the microbes may invade the plant tissue through tissue injury and later on, stoma may get entangled and adhere to the trichomes. On the other side, the openings present on the plant tissues may allow the discharge of nutrients that can be consumed by microorganisms for their survival and growth (Reina et al., 2002). The mature plant cells may enter the bacteria through parenchymal pores (≈ 50–500 μm), for instance, the microbes access in apples is common through parenchymal cells that constitute the 20–25% of the volume of apple (Alzamora et al., 2005). Moreover, the pores or intercellular spaces may be augmented along with the deterioration of the middle lamella and alteration in the composition of cell wall during the senescence that permits the entrance to bacteria (Castillejo et al., 2004).

The microorganisms are usually found in the holes, pores, or any other deformities like the cut or damaged tissues existing on the whole fruit surface and such sites support the adherence and protection of the microbes. Also, the plant tissues frequently generate such circumstances that may encourage microbial internalization (Sapers, 2001). The minimal processing of vegetables that includes several operations like cutting, peeling, slicing, etc., may enhance the nutrients (sugars, vitamins, minerals, etc.), discharge thus generating suitable conditions for the growth of microorganisms. And such properties of processed fruits and vegetables make them an ideal substrate for probiotics because they are rich in growth factors, fiber, and antioxidants (Rößle et al., 2010; Soccol et al., 2010).

In the past, few studies have been conducted regarding the vacuum impregnation of probiotics using fruits and vegetables (mainly apples) containing cellulose and showed the positive results (Betoret et al., 2003; Alzamora et al., 2005). The cellulose that is not digested by the GI system acts as a safeguard to the probiotic cultures across the GIT. Therefore, the integrity of probiotics like *L. casei* may be fully retained on their way to the colon thus aiding the host (Kourkoutas et al., 2006). In case of fruits like banana, apple, melon, and guava, the probiotic cultures were held strongly to the fruit surface leading to the favorable synergy as observed using the scanning electron microscopy (Martins et al., 2013). The fermented vegetable products from artichokes and olives allow to the survival of the probiotics comparable to those for dairy products and sometimes better than the dairy products which may be connected to makeup and microstructure of plants ensuring the protection and survival of the microbes. Moreover, the resistant cell wall of LAB permits the bacterial adaptation to the surrounding conditions like low moisture, antimicrobials presence, etc., of the plant cells (Masuda et al., 2010; Ranadheera et al., 2010).

Thus, after reviewing the literature, we can conclude that the fruits and vegetables have a great potential for survival and growth of the probiotics due to the inherent properties of plant surface ridges, natural oligosaccharides as prebiotic molecules, the richness of nutrients and survival from acidic conditions of the stomach (Ranadheera et al., 2010). However, more efforts are needed to explain the details of microbes' internalization in plant tissues mainly the manner how the microorganisms get fitted into the pores like lenticels or stomata and bruised tissues present on the fruit and vegetable surface.

4.7 POTENTIAL OF PROBIOTICS IN FRUITS AND VEGETABLES

Generally, the majority of probiotic microorganisms are delivered in the form of fermented milk or yogurt throughout the globe. However, there is recently an explosion of interest in the development of probiotic products based on non-dairy matrices especially fruit and vegetables (Ranadheera et al., 2010). Fresh fruits and vegetables seem to be ideal substrates for probiotics because they are rich in vitamins, minerals, antioxidants, and fibers and do not associate with any kind of dairy allergens that might inhibit the consumption by few-section of the population. They consist of large amount of sugars that are critical for the probiotic culture multiplication along with the quick transit across the severe acidic conditions in the stomach leading to the high survival rate of probiotic cells (Pimentel et al., 2015; Vijaya et al., 2015). Moreover, the fruit juices and beverages being refreshing and nutritious have delightful taste profiles enjoyed by all age groups. Thus, an ever-growing global market has been expected for probiotic products because of their health benefits that result in increased consumer demand. Some fruits employed in the probiotic-based preparations involve the acerola, acai, apple, blueberry, cranberry, strawberries, blackcurrant, mango, pomegranate, kiwifruits, guarana, bilberries, grapes, peach, cherries, feijoa, and plums (Sun-Waterhouse, 2011), given in Table 4.4. Numerous probiotic strains like *L. casei, L. acidophilus, L. rhamnosus, L. plantarum, L. fermentum, B. longum, B. bifidum*, etc., are extensively employed to develop several fruits and vegetables-based products particularly fruit and vegetable juices drinks, purees, etc., because they can tolerate moderate to severe acidic conditions (Thakur et al., 2015).

TABLE 4.4 Recently Developed Fruits and Vegetables based Probiotic Products

Name of Probiotic Products	Probiotic Culture	References
Apple and orange concentrate	*L. acidophilus* and *B. bifidum*	Marhamatizadeh et al. (2012)
Apple drink	*L. casei*	Ellendersen et al. (2012)
Apple juice	*L. paracasei* ssp. *paracasei*	Pimentel et al. (2015)
Beet juice	*L. acidophilus LA39, L. casei A4, L. delbrueckii D7,* and *L. plantarum C3*	Yoon et al. (2005)
Beetroot juice	*L. acidophilus NCDO1748*	Rakin et al. (2007)
A beverage containing drumstick leaves and beetroot juice	*L. plantarum* and *E. hirae*	Vanajakshi et al. (2015)
Cabbage juice	*L. plantarum C3, L. casei A4,* and *L. delbrueckii D7*	Yoon et al. (2006)
Carrot juice	*B. lactis Bb-12, B. bifidum B7.1,* and *B3.2*	Kun et al. (2007)
	L. acidophilus NCDO1748	Rakin et al. (2007)
Cashew apple juice	*L. casei NRRL B-442*	Pereira et al. (2011)
Cataloupe juice	*L. casei NRRL B-442*	Fonteles et al. (2011)
Cranberry juice	*L. casei DN 114 001, L. rhamnosus GG, L. paracasei NFBC 43338,* and *B. lactis Bb-12*	Sheehan et al. (2007)
Fruit cocktail (carrot, celery, and apple)	*L. acidophilus LA-5*	Nicolesco and Buruleanu (2010)
Fruit smoothies	*L. acidophilus LA-5* and *B. animalis ssp. lactis BB-12*	Rodgers (2007)
Honeydew melon juice	*L. casei NCIMB 4114*	Saw et al. (2011)
Litchi juice	*L. casei*	Zheng et al. (2014)

TABLE 4.4 *(Continued)*

Name of Probiotic Products	Probiotic Culture	References
Minimally processed symbiotic apples	*L. rhamnosus GG*	Rößle et al. (2010)
Orange juice	*L. casei DN 114 001, L. rhamnosus GG, L. paracasei NFBC 43338,* and *B. lactis Bb-12*	Sheehan et al. (2007)
Pear juice	*L. acidophilus*	Ankolekar et al. (2012)
Pineapple juice	*L. casei DN 114 001, L. rhamnosus GG, L. paracasei NFBC 43338,* and *B. lactis Bb-12*	Sheehan et al. (2007)
	L. casei NRRL B442	Costa et al. (2013)
Plum juice	*L. kefiranofaciens, Candida kefir,* and *S. boluradii*	Sheela and Suganya (2012)
Pomegranate beverage	*L. bulgaricus* and *L. plantarum*	Thakur and Sharma (2017)
Tomato juice	*L. acidophilus LA39, L. plantarum C3, L. casei A4 and L. delbrueckii D7*	Yoon et al. (2004)
–	*B. breve, B. longum and B. infantis*	Koh et al. (2010)
Turşu	*L. plantarum BC 7321*	Çetin (2011)

On the other hand, the dairy sector is also introducing hybrid probiotic products including the fruits and vegetables in order to make the final product more nutritious, flavorful, and convenient. Fruit rich fermented milk and yogurt containing the fruit cuts, pulp or even the peel flour have been explored as the nutrient as well as prebiotic source flourishing the GIT microbiota (Khurana and Kanawjia, 2007; Espírito-Santo et al., 2011). A study showed that the addition of *Lactobacillus plantarum* into the fruit juices has resulted in an unacceptable concentration of flavors like sour or savory and aroma like milky or perfumery. The conventional orange juice was preferred rather than one containing the probiotic bacteria on the basis of sensory properties, however, the consumers choose to drink the probiotic-containing orange juice compared to plain orange juice when they were informed about its health benefits (Luckow and Delahunty, 2004). The tropical fruit juices may be incorporated to cover the off-flavors generated due to probiotics (Luckow et al., 2006).

4.7.1 MARKET SIZE, TRENDS, AND GROWTH PERSPECTIVE IN FRUITS AND VEGETABLES BASED PROBIOTICS

Today, the probiotic microorganisms find applications in the market in the form of food and beverages, animal feed, and dietary supplements and the total market size of probiotics was worth of USD 45.64 billion in the FY 2017 that is supposed to rise with more than 7% compound annual growth rate (CAGR) to achieve the USD 65 billion till the FY 2024 (Global Market Insights Inc., 2017). The Asia Pacific dominates the global probiotic industry by the highest share of 40% or more and the fast-growing market in countries like Japan, China, and India contribute a lot to this scenario. *Lactobacilli* and *Bifidobacteria* are the frequently employed probiotic cultures in the food and beverage applications. The main players in the probiotics industry are Danone, Danisco A/S, Chr Hansen, Lifeway Foods Inc., Arla Foods Inc., AMUL, Yakult Honsha Co. Ltd., Nestle, BioGaia, Lallemand Inc, and Probi. Out of the total market size, the food and beverage-based probiotics accounted for the largest share of 85% followed by 10% animal feed-based probiotics and 5% dietary supplement probiotics. The main drivers for the probiotics market include: (1) Increasing consumers' choice for natural products with health benefits; (2) Growing health concerns; (3) Increasing disposable income as well as standards of living; (4) Growing investment in the research and developments; and (5) Rising awareness about the consumption of the probiotic.

The food and beverage probiotics segment include the dairy products as well as non-dairy products such as fruits and vegetables, baked products, fermented meat products, cereals, and dry food among which the dairy-based probiotics held the largest share in FY 2015. This segment is supposed to be dominated by dairy products; however, the probiotic industry is facing the trend of changing consumer preference from dairy probiotics to non-dairy based probiotic products due to the increasing concern about vegetarianism, milk cholesterol content and lactose intolerance. It is expected that the annual growth rate of fruit and vegetable-based probiotics will be almost twice than the dairy products during 2016–2024 (Grand View Research, Inc., 2016). According to Espírito-Santo et al. (2011), probiotic foods containing fruit ingredients are increasingly preferred by consumers. However, until today it is very difficult to gather information about the market of fruits and vegetable-based probiotic products as this segment is completely brand new. Even in some developing countries like India, there is dairy-based probiotics in the market dominated by yogurt, curd, and buttermilk.

Indian probiotic industry is still in its infancy stage and presently accounts for only a small fraction, i.e., less than 1% of the total world market turnover in the probiotic industry. But the probiotic industry is evolving at a steady pace with conditions set for tremendous growth in near future. Indian probiotic food market is dominated by North India, followed by South and Central India, and is expected to grow with 19.80% CAGR during the FY 2014–2019 in the terms of revenue with a handful of players like Amul, Mother Dairy, Yakult Danone, and Nestle India (TechSci Research, 2014). The segment comprising the probiotic-based drugs and dietary supplements are governed by the Tablets India, Dr. Reddy's Laboratories, and USV India whereas the Unique Biotech, Zeus Biotech, and Polchem Hygiene Laboratories are main giants in the animal probiotics segment. Probiotics products in India are generally sold in two forms: milk and fermented milk products with the former occupying 62% of the market share and the latter having 38% market share. AMUL is the main player in the sale of probiotic ice-cream and yogurt; however, Yakult is leading the market of probiotic drinks than Mother Dairy (Sharma et al., 2013). However, this is very unfortunate that fruits, being the delicious and rich source of nutrients, are not preferred for development of probiotic drinks in India instead of being the second-largest producer of fruits and vegetables in the world. The research and development activities should be undertaken by the food scientists, research institutes, and food companies to explore the fruit and vegetables-based probiotics so that diversity can be added in Indian probiotic market for meeting the growing demand of probiotics in upcoming years.

4.8 PROBIOTIC PRODUCTS FROM FRUITS AND VEGETABLES AND THEIR CHARACTERISTICS

Several kinds of fruits and vegetables-based probiotics are being exploited consistently to meet the increasing consumer demands for a healthy diet that should be free from animal products, cholesterol, and dairy allergens. However, it is very challenging to develop the non-dairy probiotics especially fruit and vegetables-based products due to their high perishable nature and they must be healthy and delicious simultaneously (Shah, 2007). Many factors are essential while developing new products such as: (i) Selection of suitable raw materials; (ii) Standardization of the basic product formulation; (iii) Optimum use of ingredients to ensure the perfect physicochemical as well as organoleptic features, chemical stability, prolonged shelf-life, and rational price also; and (iv) Ideal storage conditions for extended product life.

Because the probiotics market worldwide is loaded with dairy products, therefore it is very hard for the fruits and vegetable probiotics to penetrate the already established market. But innovation is the key to every kind of success today and the researchers throughout the globe are working hard for developing novel fruit-based probiotic products (Granato et al., 2010). The first probiotic fruit drink named Gefilus® developed by Valio Ltd. Finland contains *L. rhamnosus GG®* and has a shelf-life of 35 days under the refrigerated conditions. Biola® is another probiotic juice that is formulated by Tine, Norway who purchased the *L. rhamnosus GG®* from its industry partner Valio Ltd. Biogaia, Sweden had developed the world's first-ever 100% probiotic juice under the brand name Rela® containing *L. reuteri MM53* (Mishra and Mishra, 2012). Their commercial viability strongly relies on the health benefits, taste, flavor, appearance, and also price of final product.

Till today, the different kinds of probiotic products from fruits and vegetables are developed using the pomegranate, orange, pineapple, banana, guarana, cranberry, sweet lime, mango, acerola, grapes, cashew apple, carrot, cherries, beet, olive, strawberry, oranges, blueberry, blackcurrant, acai, bilberries, kiwifruits, feijoa, peach, and plums, etc., which can be classified under the following categories: 1) Probiotic beverages; 2) Probiotic purees; 3) Probiotic smoothies; 4) Probiotic pickles; 5) Minimally processed probiotic fruits; and 6) Other probiotic products.

4.8.1 PROBIOTIC BEVERAGES

Most of the probiotic products containing fruits and vegetables are available in the form of beverages due to the convenience and possibility to fulfill the consumer demands for container size, shape, contents, and appearance, ease of distribution and better storage for refrigerated and shelf-stable products and huge opportunity to incorporate the desirable nutrients and bioactive components (Kausar et al., 2012).

4.8.1.1 FRUIT-BASED BEVERAGES

Hardaliye-a traditional non-alcoholic beverage from Turkey is prepared from the grapes, benzoic acids, and crushed mustard seeds (so that oil can be released) together with the incorporation of LAB. The liberated mustard seed oil furnishes the distinct the flavor and aroma to end product along with inhibiting the yeast growth while benzoic acid is used to prevent the synthesis of an alcohol from yeast because higher alcohol production may turn the product alcoholic changing its unique characteristics. Hardaliye has moderately acidic pH (3.21–3.97) and rich in antioxidant compounds and phytochemicals based on the production method and grape cultivar. The LAB employed include *L. paracasei* subsp. *paracasei*, *L. casei* subsp. *pseudoplantarum*, *L. pontis*, *L. sanfranciscensis*, *L. vaccinostercus*, *L. brevis*, and *L. acetotolerans* that can produce lactic acid nearly 0.27% reducing its pH to 3.7 and making it non-alcoholic beverage (Güven and Aksoy, 2009; Malganji et al., 2015; Coskun, 2017).

Probiotic pomegranate beverage is prepared using LAB cultures (*L. plantarum*, *L. delbrueckii*, *L. acidophilus,* and *L. paracasei*) those consume citric acid mainly as a food source rather than glucose and fructose present in pomegranate juice containing nearly 85% water. The pomegranate juice contains the antioxidants higher than found in the red wine and green tea. This beverage is considered highly nutritious and healthy as it contains the fundamental essential nutrients from pomegranate juice and health benefits from probiotics. The shelf-life study of probiotic pomegranate beverage revealed the optimum growth and survival of *L. plantarum* and *L. delbrueckii* (10^7–10^9 CFU/ml) at 4°C during the initial 14 days but remaining probiotic bacteria lost their viability (Mousavi et al., 2011; Thakur and Sharma, 2017).

The litchi fruit-a wonderful source of ascorbic acid (400–900 mg kg^{-1}) is liked by consumers in the form of juice. Litchi juice can be fermented using *L. plantarum*, *L. Rhamnosus*, *L. casei,* and *L. acidophilus* under 37°C for 72 h in the literature. Due to more sugar consumption by *L. plantarum* and *L. rhamnosus*, the pH is reduced greatly during the first fermentation phase and thus exhibited the sound cell multiplication to 14 days but after 28 days, the cell count reduced drastically. The incorporation of probiotics typically increases the antioxidant levels in litchi juice thus enhancing the health benefits of litchi. The probiotic litchi juice treated with high hydrostatic pressure (HHP) retained a better overall appearance, taste, and flavor (Kalita et al., 2015; Zheng et al., 2014).

The probiotic apple beverage prepared using *L. paracasei* ssp. *paracasei* and prebiotic oligofructose (as a sugar substitute too) contains more acidity and turbidity compared to the pure apple juice but is equivalently liked by the consumers. This beverage has a shelf-life of 4 weeks under refrigeration when packed in glass containers that do not affect the product properties but also maintain the culture viability thoroughly (Pimentel et al., 2015).

Peach juice fermented with 24 h old culture of *L. delbrueckii* exhibits a nutritious beverage and consist of 10×10^9 CFU mL^{-1} just after 48 h of fermentation at 30°C. When stored under refrigeration conditions, the cell count remains 1.72×10^7 CFU mL^{-1} after 4 weeks (Pakbin et al., 2014).

The orange and pineapple juice are also a promising substrate for the growth of probiotics. Three bacterial strains of LAB viz. *L. rhamnosus GG*, *L. casei DN-114001* and *B. lactis Bb-12* are found to withstand the low juice pH thus exhibiting the high amount of viable cells while cold storage. In addition to these, Sheehan et al. (2007) discovered a novel bacterial strain *L. paracasei NFBC43338* that is capable to survive in the optimum concentrations ($>10^7$ CFU ml^{-1}) under the low pH and temperature conditions for minimum 12 weeks (84 days). However, these strains exhibit very poor viability in the cranberry juice due to high levels of benzoic acid (34 mg L^{-1})—a phenolic acid and extreme low juice pH of 2.5. Till now, the most acid susceptible probiotics are recognized as *L. salivarius* strains *UCC118* and *UCC500* that can survive in the low amount (10^6 CFU ml^{-1}) for only one week in the orange juice while in pineapple juice; their concentration is below 10^5 CFU ml^{-1}. The non-sweetened sonicated pineapple juice contains the viable cells of *L. casei NRRL B442* above 10^6 CFU ml^{-1} at pH 4.2 for 42 days, however, the addition of sucrose (10% w/v) to the juice reduces the survival of probiotic cells above 10^6 CFU ml^{-1} for 28 days only. This can be explained due to the increased amount of sugar that in turn raised the acidity and also the osmotic pressure which leads to the high death rate of microorganisms in sweetened juice (Sheehan et al., 2007; Costa et al., 2013). Similarly, the viability of *L. casei NRRL B442* in non-sweetened cashew apple juice is 8.72×10^7 CFU ml^{-1} for 42 days under the refrigeration that meets the recommended dose of probiotic cells to offer optimum benefits (Pereira et al., 2011).

The blackcurrant juice containing the probiotic culture of *L. plantarum* was preferred by the consumers on the basis of nutritional properties, appearance, flavor, aroma, and consistency in a survey as compared to the standard blackcurrant juice (Luckow and Delahunty, 2004).

Mango juice is fermented using strains of *L. acidophilus (MTCC10307)*, *L. delbrueckii (MTCC911)*, *L. plantarum (MTCC9511)* and *L. casei* for 72 h at 30°C to which lowers the pH from 4.5 to 3.2. Sugar is being utilized most by the *L. plantarum* than other microorganisms and the cells count remains at 1.0×10^7 CFU ml^{-1} during the entire storage that demonstrate the mango juice as a promising substrate to produce the probiotic beverage (Reddy et al., 2015).

The cantaloupe melon juice is rich in the sugars such as fructose and sucrose that enables the survival of *L. casei B-442* in high amounts without the presence of any other sugars. This juice comprises of high levels of amino acids like glutamic acid, aspartic acid, alanine, and arginine that are beneficial for multiplication of the probiotic cultures (Fonteles et al., 2012).

4.8.1.2 VEGETABLE-BASED BEVERAGES

Several kinds of native strains of LAB have been isolated and identified from fresh vegetables (Yoon et al., 2006; Bisakowski et al., 2007) and also from their lactic acid fermentation (Czyzowska et al., 2006). Being rich in phytochemicals, carbohydrates, minerals, and vitamins, the vegetables are necessary to build-up, repair the body, and sustain the alkaline reserve. The minimal processing of vegetables like peeling, cutting, etc., release the nutrients like sugars, vitamins, minerals, etc., that provide an optimum environment for the growth of the probiotics, as discussed previously.

The probiotic beet juice is prepared using the probiotic strains of *L. acidophilus*, *L. plantarum*, *L. casei*, and *L. delbrueckii* that synthesize the lactic acid lowering the pH from 6.3 to 4.5 just after 2 h fermentation at 30°C. All the four strains exhibit a good survival rate at 4°C for 4 weeks varying from 10^6 CFU ml^{-1} and 10^8 CFU ml^{-1}. The tomato juice also appears to be a promising raw material to prepare the probiotic juice using *L. acidophilus LA39*, *L. plantarum C3*, *L. casei A4,* and *L. delbrueckii D7* where the rate the viability of probiotics is comparable to probiotic beet juice. Similarly, the cabbage juice is also found suitable for the cultivation of probiotic bacteria that are used in case of beet juice and tomato juice also. The count of probiotic cells reaches to 10^9 CFU ml^{-1} just after the 48 h of fermentation. Further, the two strains of lactobacilli-*L. rhamnosus* and *L. bulgaricus* have good viability at low pH ($\approx$ 3.5) during the storage for 30 days. However, in another study, the Lactobacilli cell count lasts constantly under the refrigerated storage for more than 12 weeks besides the *L. plantarum Lp-115* whose viability is reduced to below 10^8 CFU ml^{-1} after storage. On the other side, the Bifidobacterium strains like *B. lactis Bb-12* and *B. bifidum B7.1* and *B3.2* have the potential to grow in the carrot juice without any kind of nutrient supplementation, but both the cultures of *Bifidobacteria* dropped drastically and not detected at the end of 8 weeks. The difference between these two studies is the kind of probiotic culture used because the *Bifidobacteria* consume 20% of glucose and 10% sucrose only beside the other carbohydrates present. Therefore, the nutrient reserves finish soon for further survival of *Bifidobacteria* cells (Yoon et al., 2004–2006; Kun et al., 2008).

Kanji-a carrot and beetroot based probiotic drink of India is prepared by the fermentation of LAB such as *L. casei, L. plantarum*, and *L. acidophilus*. This probiotic drink is known as a vitaminized drink due to high levels of tocopherol, β-carotene, and ascorbic acid. It also offers a wide range of therapeutic effects in addition to working as coolant (Panghal et al., 2018).

A novel beverage containing drumstick (*Moringa oleifera*) leaves and beetroot juice are prepared that is fermented for 48 h using the probiotic bacteria: *L. plantarum* and *Enterococcus hirae*. This beverage has the antibacterial activity against the *pathogenic bacteria due to the* drumstick leaves *and* supplies antioxidants compounds on account of beetroot juice and fermentation reduced the raffinose level to 60%. It consists of a sufficient amount of calcium and iron; therefore, it can be proposed for the consumption of malnourished people especially. It has a shelf-life of 30 days under the refrigeration conditions (Vanajakshi et al., 2015).

Kombucha-a fermented green or black tea drink is prepared by the fermentation of antioxidants rich tea leaves with probiotics such as *Saccharomyces cerevisiae, S. bisporus, S. ludwigii, Zygo saccharomyces*, and *Torulopsis* sp. (Jayabalan et al., 2014).

In Nepal and North-east India, the ripe cucumbers are sun-dried and fermented in the bamboo containers for 3–5 days to prepare a traditional probiotic drink called *Khalpi*. The natural fermentation of *Khalpi* includes the probiotic culture of *Pediococcus pentosaceus* and *L. plantarum* (Dahal et al., 2005).

4.8.2 PROBIOTIC PUREES

An innovative product from banana has been prepared using the immobilized probiotic cultures. A novel probiotic banana puree was prepared through the immobilization of *L. acidophilus* in the Ca-alginate and k-carrageenan matrix that protect the probiotic cultures effectively and minimize the drying losses during dehydration (Tsen et al., 2003, 2007).

4.8.3 PROBIOTIC PICKLES

The pickle preparation is impossible without LAB microorganisms-the primary agent for carrying natural vegetable fermentation; however, each culture acts according to the kind of raw materials, fermentation conditions, pH, and temperature (Roberts and Kidd, 2005; Yoon et al., 2006; Bisakowski et al., 2007). Frequently used strains in lactic acid fermentation belong to the *Leuconostoc* (*Leuc. mesenteroides*) and *Lactobacillus* (*L. plantarum, L. paracasei/casei, L. delbrueckii* and *L. brevis*) genus (Czyżowska et al., 2006; Yan et al., 2008). *Leuconostoc* species no doubt grows well on the vegetable juices but also provide conditions for the multiplication of probiotic cultures like *Bifidobacteria* and *Lactobacilli* (Sanz et al., 2006).

Peach pickle named *Yan-taozih* is prepared in Taiwan using the major cultures of *L. lactis, L. brevis, Weissella cibaria, W. paramesenteroides*, together with small amount of *Leuconostoc mesenteroides*, and *Enterococcus faecalis*. This pickle can be stored for more than 10 months at 6–10°C (Chen et al., 2013).

The few Lactobacilli strains which are acid and salt-tolerant can also be employed to ferment the table olives that may enhance the shelf-life of the final product. Thus, olives can be a potential carrier matrix for the probiotic microorganisms (Peres et al., 2012).

Kimchi-a popular probiotic product of East-Asia is prepared using cabbage, onion, garlic, red pepper, ginger, and radish that develop probiotics such as *Leuconostoc, Enterococcus, Pediococcus, Streptococcus,* and *Lactococcus* at the suitable pH, acidity, and temperature (Nam and Ahn, 2016). It contains the significant amount of β-carotene, vitamin C, chlorophyll, and dietary fiber together with antioxidant, antimutagenic, and angiotensin-converting enzyme inhibition activities thus protecting from several diseases (Oh et al., 2005).

Likewise, *Gundruk*-a pickled side dish of South-Asia mainly India and Nepal are prepared from cabbage, mustard, radish, and cauliflower using the probiotic cultures of *Lactobacillus* spp. and *Pediococcus* (Dahal et al., 2005). *Paocai*-widely eaten in China is prepared using the vegetables like celery, cabbage, radish, and cucumber with the incorporation of probiotics like *L. plantarum, L. pentosus, L. lactis, L. brevis, L. fermentum* and *Leuconostoc mesenteroides* (Feng et al., 2012).

4.8.4 MINIMALLY PROCESSED FRUITS

The fresh-cut fruits act as a matrix for the immobilization of probiotic cells as reported by Kourkoutas et al. (2005). The quince and apple pieces when immobilized with *L. casei* typically enhance the survival of probiotics as

products (Bruno and Shah, 2003). The room temperature (20°C) storage of dried probiotic products decreases the count of viable cells significantly.

4.9.2.2.6 Packaging

The viability of probiotic cells may be affected by following several factors like (Korbekandi et al., 2011): (i) thickness and kind of packaging material; (ii) light and gas (oxygen, carbon dioxide, and water vapors) barrier properties; and (iii) kind of packaging technique (modified, vacuum, active or intelligent packaging). The gas barrier properties of packaging materials are influenced by relative humidity and temperature of surroundings that may also affect the survival (Cruz et al., 2007). Most of the probiotic products are typically stored in the plastic packages of poor oxygen barrier that presents a serious concern about the viability of probiotic cells. Although the glass bottles are superior in maintaining the survival of the probiotics due to the very low oxygen transfer rate; but are not preferred due to their heavyweight, high cost, and fragile nature (Jayamanne and Adams, 2004). The plastic packaging materials may allow the entrance of oxygen molecules beyond the oxygen level in a product. Vacuum packaging of the product in the package of low oxygen permeability containing oxygen scavengers and antioxidants (catechins) is the best way to restrict the entry of dissolved into the product.

4.10 POTENTIAL HEALTH BENEFITS OF PROBIOTICS

The probiotic microorganisms have numerous advantages for human health such as improvement of lactose absorption, treatment, and prevention of diarrhea and regulation of the immune system. They may have hyper-cholesterol and anti-tumor effects and preventive action against constipation, urogenital infections, and allergy. However, the potential benefits from probiotics are strongly dependent on the particular strains and their dosage rather than the entire microbial family.

The lactose intolerance that is caused by the deficiency of lactase enzyme in the digestive tract exhibits the symptoms like abdominal pain and distension, flatus, and diarrhea that may occur between 30 min and 2 h after the ingestion of lactose (Li et al., 2012). The probiotic cultures mainly *L. bulgaricus* and *S. thermophilus* observed to enhance the lactose digestion by the microbial lactase activity consuming the lactose while reaching it in the intestine. If the food product has slower passage time, the probiotics will get more time for the digestion of lactose. The mice fed with *L. lactis MG1363/FGZW* (strain expressing food-grade β-galactosidase) than naïve mice showed the significant alleviation of diarrhea symptoms within 6 h post-challenge and suppressed intestinal motility after lactose challenge, although there was no significant increase of β-galactosidase activity in the small intestine (Li et al., 2012). Also, probiotics are helpful in the prevention of several kinds of diarrhea specifically the infantile and antibiotic-related diarrhea (Parvez et al., 2006) because they compete with the contagious microbes for the food and space in GIT.

The metabolic products of probiotic cultures inhibit the pathogens, for instance, bulgarican and acidophilin are the metabolic products of *L. casei, L. acidophilus,* and *L. bulgaricus* and have antimicrobial effects (Hemaiswarya et al., 2013). Thus, probiotic cultures are like a boon to the immune-compromised patients those take antibiotics on daily basis. The lactobacilli, for example, *L. acidophilus* exhibit the hypocholesterolemic effects by consuming the cholesterol in the bile presence. Some studies showed that free bile salts and cholesterol precipitate together in acidic conditions under the *L. acidophilus* existence thus, reducing the serum cholesterol (Pereira and Gibson, 2002). It has been shown experimentally that probiotics have the potential to reduce the growth of cancers by minimizing the concentration of colon enzymes such as nitroreductase, β-glucuronidase, and azoreductase that synthesizes the carcinogenic substances from procarcinogens. In another way, the probiotic culture can directly reduce the procarcinogens through the consumption of nitrites and lower the secondary bile salts levels (Oatley et al., 2000; Sanders et al., 2005). The probiotics exhibit the potential to modulate the immune response and prevent the onset of several allergic diseases. In a double-blind, randomized, placebo-controlled trial, *L. rhamnosus GG* was given to pregnant women for four weeks prior to delivery, then to newborns at high risk of allergy for six months with the result that there was a significant reduction in early atopic disease (Kalliomäki et al., 2001).

In a study, it was shown that stomach ulcers may be inhibited by lactobacilli because of the struggle of *L. acidophilus* cells with the *Heliobacter pylori*-a risk factor of stomach ulcers for the attachment sites thus reducing its cells those adhere to the cell wall (Michetti et al., 1999).

Constipation is a common problem in elderly subjects and probiotics have been suggested to improve intestinal motility and reduce the fecal enzyme activity. In a study, the consumers receiving the *L. rhamnosus/P. freudenreichii*-supplemented juice exhibited nearly 24% increase in defecation frequency as compared to the non-supplemented juice and juice supplemented with *Lactobacillus reuteri* without any reduction in laxative use (Ouwehand et al., 2002).

Crohn's disease (CD) and ulcerative colitis (UC) are the two major forms of inflammatory bowel disease (IBD) that lead to high morbidity and health care costs. Complex interactions between the immune system, enteric commensal bacteria, and host genotype are thought to underlie the development of IBD although the precise etiology of this group of diseases is still unknown. Studies far have shown that the intestinal microbiota drives the development of the gut immune system and can induce immune homeostasis as well as contribute to the development of IBD. Probiotics which deliver some of the beneficial immunomodulatory effects of the commensal gut microbiota and induce immune homeostasis have been proposed as a suitable treatment for mild to moderate IBD (Reiff and Kelly, 2010).

According to Reid (2001), several 100 million women are affected by urinary tract infection (UTI) annually. Uropathogenic *E. coli* originating in the bowel is the major responsible agent for about 85% of the cases. There is evidence, including the randomized controlled data to suggest that once-weekly vaginal capsules of freeze-dried *Lactobacillus* strains *GR-1* and *B-54* and once-daily oral capsule of *Lactobacillus* strains *GR-1* and *RC-14* may lead to re-establishment of lactobacilli-dominated vaginal flora and reduce the risk of UTI recurrences.

4.11 LEGISLATIVE ISSUES

There is a strong need for the establishment of probiotics as an ingredient of food on a global basis. The issues related to probiotics such as their safety, effectiveness, fraud, health claims, labeling, etc., must be addressed by an autonomous regulatory organization. The beneficial effects and adverse health consequences (if any) of the probiotics must be monitored through the surveillance systems. The food containing probiotic cells must be advertised widely particularly to the people having the risk of mortality and morbidity.

The probiotic foods from a legislative perspective may come under any of four food classes mentioned by Food and Drug Administration (FDA) as conventional foods, dietary supplements, medical foods, and foods for special dietary use (Berner and O'Donnell, 1998). But till now, there is no definite acknowledgment of positive health effects linked with the probiotics added in food products. The issues of probiotics safety are handled differently according to the country's regulations and there is a lack of any kind of safety guidelines on an international basis. Depending on the intended use of a probiotic, whether as a food/food ingredient a dietary supplement or a drug, regulatory requirements differ greatly among different countries. For most countries, if a probiotic is to be used as a drug, then it must undergo the regulatory process as a drug, which is similar to that of any new therapeutic agent. The probiotic drug safety and efficacy for its intended use must be evaluated and approved before marketing. But, if a probiotic is to be used as a food or dietary supplement, it is considered as foods, and then these products do not need any safety evaluation or approval before being marketed. However, there is an urgent need for harmonization of these regulatory standards on probiotic bacteria at the international level to ensure the safety and efficacy of probiotic products for their effective utilization in different countries around the world. While evaluating the probiotics for safety, several factors like infectivity, pathogenicity, and virulence must be considered that should include the inherent characteristics, toxicity, and metabolic activity of microbes (Donohue and Salminen, 1996).

In India, the market for probiotic products has been swiftly expanding, therefore, the Department of Biotechnology, Government of India and Indian Council of Medical Research formulated the probiotics' regulations in 2011 that describe the essential requirements for considering any product or strain as a probiotic. These requirements typically include the genus, species, and strain identification; *in-vivo* testing of probiotics safety and efficacy in animal and human models, *in-vitro* screening of probiotics strains, the dose of probiotic strains, viable cell count during storage and consumption, labeling requirements, etc. (Indian Council of Medical Research, 2011). Recently in 2016 Food Safety and Standards Authority of India (FSSAI) had formulated the standards of health supplements, nutraceuticals, food for special dietary use, food for special medical purpose, functional food and novel food that was implemented from 1 January 2018 throughout the country and specific provisions were included for the food with added probiotic ingredients (FSSAI, 2016).

4.12 FUTURE OUTLOOK

Due to the presence of lactose content, high cholesterol levels, and allergens (milk protein), the consumption of probiotic dairy products is limited that makes the fruits and vegetables based probiotic products a next category or good substitute to the individuals which are restricted to the diet containing cholesterol, milk proteins, and lactose. As discussed above, the fruits and vegetables as liked by all age groups contain a significant amount of nutrients and are a promising substrate for the probiotics showing the significant cell survival. Currently the probiotic products containing fruits and vegetables can be seen in global supermarkets, however, India is lacking in this. There is a strong need to develop such probiotic products in India as it ranks 2nd in the global production of fruits and vegetables but the processing level is only limited to 2%. The lower shelf-life of fruits requires their instant processing to develop probiotic products for the reduction of post-harvest losses and improving their market value and also availability.

For improving the acceptance of probiotic juices, the pleasing aroma and volatile components may be incorporated in a process called "masking" that would mask the presence of the probiotic in products. Another major challenge-the shelf-life of probiotic products is limited to few days to one month in maximum cases that are a big concern in probiotics consumption. Scientists must develop techniques to increase the shelf-life of probiotics. The direct liquid inoculation system in which the live bacterial cultures are added just before filling the cartons must be commercially employed. Several technologies like vacuum impregnation, immobilization, microencapsulation, and fortification with prebiotics or bioactive ingredients exerting the protective effect to acidic conditions have been used for improving the survival of probiotic cultures together with the incorporation of bioactive components. Efforts must be taken by the researchers to maintain the viability of probiotic cultures during the interruption to cold chain or ambient conditions. Further, the *in-vivo* clinical trials should be carried out for determining the adherence and viability of probiotics in the intestine to assess the degree of positive effects. Genetic engineering may also act as a blessing to dress or improve an organism with probiotics features.

4.13 SUMMARY AND CONCLUSIONS

The research focusing the non-dairy probiotic products are still at a nascent stage with huge possibilities globally to explore the novel microbial strains along with the budding substrates. The fruits and vegetable-based products are normally linked with well-being, antioxidants, minerals, and vitamins along with fiber and will be liked by people of all ages and the addition of probiotics is as a supplement to the fruits and vegetable product thus enhancing their health-promoting effects. However, the sensory acceptance and viability of probiotic cells from storage to the GIT in the fruits and vegetables-based substrates is a great challenge. Much more efforts are needed to analyze the probiotic strains and find novel substrates to engineer the products as well as processes so that the nutritious and delicious probiotic foods with high consumer demand can be developed.

KEYWORDS

- **bacteriocins**
- ***Bifidobacteria***
- **bioactive components**
- **functional foods**
- **health benefits**
- **lactic acid bacteria**
- **microencapsulation**
- **probiotic beverages**
- **shelf-life**

REFERENCES

Aggregation and Coaggregation Abilities, (2007). *European Food Research and Technology, 226*, 1065–1073.

Aguilera, J. M., & Stanley, D. W., (1991). *Microstructural Principles of Food Engineering and Processing*. Elsevier Applied Science, New York.

Aidoo, K. E., Nout, M. J. R., & Sarkar, P. K., (2006). Occurrence and function of yeast in Asian indigenous fermented foods. *FEMS Yeast Research, 6,* 30–39.

Akin, M. B., Akin, M. S., & Kirmaci, Z., (2007). Effects of inulin and sugar levels on the viability of yogurt and probiotic bacteria and the physical and sensory characteristics in probiotic ice-cream. *Food Chemistry, 104*, 93–99.

Alzamora, S. M., Salvatori, D., Tapia, M. S., López-Malo, A., Welti-Chanes, J., & Fito, P., (2005). Novel functional foods from vegetable matrices impregnated with biologically active compounds. *Journal of Food Engineering, 67*, 205–214.

Ankolekar, C., Pinto, M., Greene, D., & Shetty, K., (2012). *In-vitro* bioassay based screening of antihyperglycemia and antihypertensive activities of *Lactobacillus acidophilus* fermented pear juice. *Innovative Food Science and Emerging Technologies, 13*, 221–230.

Anukam, K. C., & Reid, G., (2007). Probiotics: 100 years (1907–2007) after Elie Metchnikoff's observation. In: Méndez-Vilas, A., (ed.), *Communicating Current Research and Educational Topics and Trends in Applied Microbiology* (pp. 466–474). FORMATEX, Spain.

Begley, M., Hill, C., & Gahan, C. G. M., (2006). Bile salt hydrolase activity in probiotics. *Applied and Environmental Microbiology, 72*, 1729–1738.

Berner, L. A., & O'Donnell, J. A., (1998). Functional foods and health claims legislation: Applications to dairy foods. *International Dairy Journal, 8*, 355–362.

Betoret, N., Puente, L., Díaz, M. J., Pagán, M. J., García, M. J., Gras, M. L., Martínez-Monzó, J., & Fito, P., (2003). Development of probiotic enriched dried fruits by vacuum impregnation. *Journal of Food Engineering, 56*, 273–277.

Bisakowski, B., Atwal, A. S., Gardner, N., & Champagne, C. P., (2007). Effect of lactic acid fermentation of onions (*Allium cepa*) on the composition of flavonol glucosides. *International Journal of Food Science and Technology, 42*, 783–789.

Boylston, T. D., Vinderola, C. G., Ghoddusi, H. B., & Reinheimer, J. A., (2004). Incorporation of bifidobacteria into cheeses: Challenges and rewards. *International Dairy Journal, 14*, 375–387.

Bruno, F. A., & Shah, N. P., (2003). Viability of two freeze-dried strains of *Bifidobacterium* and commercial preparations at various temperatures during prolonged storage. *Journal of Food Science, 68*, 2336–2339.

Carter, A., Adams, M., La Ragione, R. M., & WoodWard, M. J., (2017). Colonization of poultry by *Salmonella* enteritidis S1400 is reduced by combined administration of *Lactobacillus salivarius* 59 and *Enterococcus faecium* PXN-33. *Veterinary Microbiology, 199*, 100–107.

Carvalho, A. S., Silva, J., Ho, P., Teixeira, P., Malcata, F. X., & Gibbs, P., (2002). Survival of freeze-dried *Lactobacillus plantarum* and *Lactobacillus rhamnosus* during storage in the presence of protectants. *Biotechnology Letters, 24*, 1587–1591.

Casarotti, S. N., Monteiro, D. A., Moretti, M. M. S., & Penna, A. L. B., (2014). Influence of the combination of probiotic cultures during fermentation and storage of fermented milk. *Food Research International, 59*, 67–75.

Castillejo, C., Fuente, J. I., Iannetta, P., Botella, M. A., & Valpuesta, V., (2004). Pectin esterase gene family in strawberry fruit: Study of FaPE1, a ripening-specific isoform. *Journal of Experimental Botany, 55*, 909–918.

Cenci, A., & Chingwaru, W., (2010). The role of functional foods, nutraceuticals, and food supplements in intestinal health. *Nutrients, 2*, 611–625.

Çetin, B., (2011). Production of probiotic mixed pickles (Turşu) and microbiological properties. *African Journal of Biotechnology, 10*, 14926–14931.

Champagne, C. P., & Gardner, N. J., (2005). Challenges in the addition of probiotic cultures to food. *Critical Reviews in Food Science and Nutrition, 45*, 61–84.

Champagne, C. P., & Gardner, N. J., (2008). Effect of storage in a fruit drink on subsequent survival of probiotic lactobacilli to gastrointestinal stresses. *Food Research International, 41*, 539–543.

Champagne, C. P., Ross, R. P., Saarela, M., Hansen, K. F., & Charalampopoulos, D., (2011). Recommendations for the viability assessment of probiotics as concentrated cultures and in food matrices. *International Journal of Food Microbiology, 149*, 185–193.

Chen, Y., Wu, H., Pan, S., Lin, B., Lin, Y., Tung, W., Li, Y., Chiang, C., & Yanagida, F., (2013). Isolation and characterization of lactic acid bacteria from Yan-Taozih (pickled peaches) in Taiwan. *Annals of Microbiology, 63*, 607–614.

Chingwaru, W., & Vidmar, J., (2017). Potential of Zimbabwean commercial probiotic products and strains of *Lactobacillus plantarum* as prophylaxis and therapy against diarrhoea caused by *Escherichia coli* in children. *Asian Pacific Journal of Tropical Medicine, 10*, 57–63.

Collado, M. C., Meriluoto, J., & Salminen, S., (2007). Interactions between pathogens and lactic acid bacteria: Cell-surface properties and interactions with pathogens. *Journal of Food Science, 72*, M89–93.

Coskun, F., (2017). A traditional Turkish fermented non-alcoholic grape-based beverage,"hardaliye". *Beverages,* 3, 2. doi: 10.3390/beverages3010002.

Costa, M. G. M., Fonteles, T. V., Jesus, A. L. T., & Rodrigues, S., (2013). Sonicated pineapple juice as substrate for *L. casei* cultivation for probiotic beverage development: Process optimization and product stability. *Food Chemistry, 139*, 261–266.

Craig, W., & Beck, L., (1999). Phytochemicals: Health protective effects. *Canadian Journal of Dietetic Practice and Research, 60*, 78–84.

Cruz, A. G., Faria, J. A. F., & Van, D. A. G. F., (2007). Packaging system and probiotic dairy foods. *Food Research International, 40*, 951–956.

Czyżowska, A., Klewicka, E., & Libudzisz, Z., (2006). The influence of lactic acid fermentation process of red beet juice on the stability of biologically active colorants. *European Food Research Technology, 223*, 110–116.

Dahal, N. R., Karki, T. B., Swamylingappa, B., Li, Q., & Gu, G., (2005). Traditional foods and beverages of Nepal—A review. *Food Reviews International, 21*, 1–25.

Delcenserie, V., Martel, D., Lamoureux, M., Amiot, J., Boutin, Y., & Roy, D., (2005). Immunomodulatory effects of probiotics in the intestinal tract. *Current Issues in Molecular Biology, 10*, 37–54.

Dhandevi, P. E. M., & Jeewon, R., (2015). Fruit and vegetable intake: Benefits and progress of nutrition education interventions- narrative review article. *Iranian Journal of Public Health, 44,* 1309–1321.

Dias, J. S., & Ryder, E., (2011). World vegetable industry: Production, breeding, trends. *Horticultural Reviews, 38*, 299–356.

Dias, J. S., (2012). Nutritional quality and health benefits of vegetables: A review. *Food and Nutrition Sciences, 3*, 1354–1374.

Díaz-Ruiz, G., Guyot, J. P., Ruiz-Teran, F., Morlon-Guyot, J., & Wacher, C., (2003). Microbial and physiological characterization of weakly amylolytic but fast-growing lactic acid bacteria: A functional role in supporting microbial diversity in Pozol, a Mexican fermented maize beverage. *Applied and Environmental Microbiology, 69*, 4367–4374.

Donohue, D. C., & Salminen, S. J., (1996). Safety of probiotic bacteria. *Asia Pacific Journal of Clinical Nutrition, 5*, 25–28.

Eizaguirre, I., Urkia, N. G., Asensio, A. B., Zubillaga, I., Zubillaga, P., Vidales, C., Garcia-Arenzana, J. M., & Aldazabal, P., (2002). Probiotic supplementation reduces the risk of bacterial translocation in experimental short bowel syndrome. *Journal of Pediatric Surgery, 37*, 699–702.

Ellendersen, L. S. N., Granato, D., Guergoletto, K. B., & Wosiacki, G., (2012). Development and sensory profile of a probiotic beverage from apple fermented with *Lactobacillus casei*. *Engineering in Life Sciences, 12*, 475–485.

Elli, M., Zink, R., Rytz, A., Reniero, R., & Morelli, L., (2000). Iron requirement of *Lactobacillus* spp. in completely chemically defined growth media. *Journal of Applied Microbiology, 88*, 695–703.

Escamilla-Hurtado, M. L., Olguı'n-Lora, P., & Prado-Barragan, L. A., (1993). Lactic acid fermentation of sour corn porridge of the Tzotzil ethnic group. *Revista Española de Ciencia y Tecnología de Alimentos [Spanish Journal of Food Science and Technology], 33*, 555–565.

Espírito-Santo, A. P., Perego, P., Converti, A., & Oliveira, M. N., (2011). Influence of food matrices on probiotic viability – A review focusing on the fruity bases. *Trends in Food Science and Technology, 22*, 377–385.

FAO/WHO, (2001). *Health and Nutritional Properties of Probiotics in Food Including Powder Milk with Live Lactic Acid Bacteria.* Food and Agriculture Organization of the United Nations and World Health Organization Expert Consultation Report, Cordoba, Argentina.

Farnworth, E. R., (2005). The beneficial effects of fermented foods-potential probiotics around the world. *Journal of Nutraceuticals, Functional and Medical Foods, 4*, 3, 4.

Feng, M., Chen, X., Li, C., Nurgul, R., & Dong, M., (2012). Isolation and identification of an exopolysaccharide-producing lactic acid bacterium strain from Chinese paocai and biosorption of Pb (II) by its exopolysaccharide. *Journal of Food Science, 77*, 111–117.

Fijan, S., (2014). Microorganisms with claimed probiotic properties: An overview of recent literature. *International Journal of Environmental Research and Public Health, 11*, 4745–4767.

Fonteles, T. V., Costa, M. G. M., Jesus De, A. L. T., & Rodrigues, S., (2011). Optimization of the fermentation of cataloupe juice by *Lactobacillus casei NRRL B-442. Food and Bioprocess Technology*. doi: 10.1007/s11947-011-0600-0.

Fonteles, T. V., Costa, M. G. M., Jesus De, A. L. T., & Rodrigues, S., (2012). Optimization of the fermentation of cantaloupe juice by *Lactobacillus casei NRRL B-442. Food and Bioprocess Technology, 5*, 2819–2826.

Food Safety and Standards Authority of India. (2017). *F. No. 1-4/Nutraceuticals/FSSAI-2003, Ministry of Health Welfare and Family*. New Delhi. URL: http://old.fssai.gov.in/Portals/0/pdf/Direction_Operationalisation_HS_SMP_NF_Nutra_24_11_2016.pdf (accessed on 27 September 2021).

Gardiner, G. E., O'Sullivan, E., Kelly, J., Auty, M. A., Fitzgerald, G. F., Collins, J. K., Ross, R. P., & Stanton, C., (2000). Comparative survival rates of human-derived probiotic *Lactobacillus paracasei* and *L. salivarius* strains during heat treatment and spray drying. *Applied and Environmental Microbiology, 66*, 2605–2612.

Gaudreau, H., Champagne, C. P., Remondetto, G. E., Bazinet, L., & Subirade, M., (2013). Effect of catechins on the growth of oxygen-sensitive probiotic bacteria. *Food Research International, 53*, 751–757.

Global Market Insights Inc., (2014). *Probiotics Market to Exceed $65bn by 2024*. URL: https://globenewswire.com/news-release/2017/10/10/1143574/0/en/Probiotics-Market-to-exceed-65bn-by-2024-Global-Market-Insights-Inc.html (accessed on 27 September 2021).

Global Market Insights Inc, (2017). *Probiotics Market to Exceed $65bn by 2024*. URL: https://globenewswire.com/news-release/2017/10/10/1143574/0/en/Probiotics-Market-to-exceed-65bn-by-2024-Global-Market-Insights-Inc.html (accessed on 27 September 2021).

Gotcheva, V., Pandiella, S. S., Angelov, A., Roshkova, Z. G., & Webb, C., (2000). Microflora identification of the Bulgarian cereal-based fermented beverage boza. *Process Biochemistry, 36*, 127–130.

Granato, D., Branco, G. F., Nazzaro, F., Cruz, A. G., & Faria, J. A. F., (2010). Functional food and nondairy probiotic food development: Trends, concepts, and products. *Comprehensive Reviews in Food Science and Food Safety, 9*, 292–302.

Grand View Research, Inc. (2016). *Probiotics Market Analysis by Application (Probiotic Foods and Beverages (Dairy Products, Non-Dairy Products, Cereals, Baked Food, Fermented Meat Products, Dry Food), Probiotic Dietary Supplements (Food Supplements,*

Nutritional Supplements, Specialty Nutrients, Infant Formula), Animal Feed Probiotics), by End-Use (Human Probiotics, Animal Probiotics) and Segment Forecast to 2024. Market research report. Report ID: 978-1-68038-093-4. URL: https://www.grandviewresearch.com/industry-analysis/probiotics-market (accessed on 27 September 2021).

Gu, Q., Zhang, C., Song, D., Li, P., & Zhu, X., (2015). Enhancing vitamin B12 content in soy-yogurt by *Lactobacillus reuteri*. *International Journal of Food Microbiology, 206*, 56–59.

Güven, S., & Aksoy, M., (2009). Some modifications in hardaliye production. II. In: *Traditional Foods Symposium Abstract Book* (pp. 675–678). Van.

Heller, K. J., (2001). Probiotic bacteria in fermented foods: Product characteristics and starter organisms. *The American Journal of Clinical Nutrition, 73*, 374–379.

Hemaiswarya, S., Raja, R., Ravikumar, R., & Carvalho, I. S., (2013). Mechanism of action of probiotics. *Brazilian Archives of Biology and Technology, 56*, 113–119.

Holzapfel, W. H., Haberer, P., Geisen, R., Bjo-Rkroth, J., & Schillinger, U., (2001). Taxonomy and important features of probiotic microorganisms in food and nutrition. *The American Journal of Clinical Nutrition, 73*, 365S–373S.

Hubalek, Z., (2003). Protectants used in the cryopreservation of microorganisms. *Cryobiology, 46*, 205–229.

Hussain, S. A., Patil, G. R., Reddi, S., Yadav, V., Pothuraju, R., Singh, R. R. B., & Kapila, S., (2017). Aloe vera (*Aloe barbadensis Miller*) supplemented probiotic lassi prevents shigella infiltration from epithelial barrier into systemic blood flow in mice model. *Microbial Pathogenesis, 102*, 143–147.

Indian Council of Medical Research, (2011). *ICMR-DBT Guidelines for Evaluation of Probiotics in Food*. Indian Council of Medical Research, New Delhi. URL: https://dbtindia.gov.in/sites/default/files/PROBIOTICS-GUIDELINES-PDF_0.pdf (accessed on 27 September 2021).

Jankovic, I., Sybesma, W., Phothirath, P., Ananta, E., & Mercenier, A., (2010). Application of probiotics in food products-challenges and new approaches. *Current Opinion in Biotechnology, 21*, 175–181.

Jay, J. M., Loessner, M. J., & Golden, D. A., (2005). *Modern Food Microbiology* (p. 790). Springer New York.

Jayabalan, R., Malbaša, R. V., Lončar, E. S., Vitas, J. S., & Sathishkumar, M., (2014). A review on kombucha tea—microbiology, composition, fermentation, beneficial effects, toxicity, and tea fungus. *Comprehensive Reviews in Food Science and Food Safety, 13*, 538–550.

Jayamanne, V. S., & Adams, M. R., (2004). Survival of probiotic bifidobacteria in buffalo curd and their effect on sensory properties. *International Journal of Food Science and Technology, 39*, 719–725.

Johnston, B. C., Supina, A. L., & Vohra, S., (2006). Probiotics for pediatric antibiotic-associated diarrhea: A meta-analysis of randomized placebo-controlled trials. *Canadian Medical Association Journal, 175*, 377–383.

Kalita, D., Mandal, M., & Mahanta, C. L., (2015). Characterization of litchi juice fermented by probiotic lactic acid bacteria. *International Journal of Food and Nutritional Sciences, 4*, 181–186.

Kalliomäki, M., Salminen, S., Arvilommi, H., Kero, P., Koskinen, P., & Isolauri, E., (2001). Probiotics in primary prevention of atopic disease: A randomized placebo-controlled trial. *The Lancet, 357*, 1076–1079.

Kausar, H., Saeed, S., Ahmad, M. M., & Salam, A., (2012). Studies on the development and storage stability of cucumber-melon functional drink. *Journal of Agricultural Research, 50*, 239–248.

Khurana, H. K., & Kanawjia, S. K., (2007). Recent trends in development of fermented milks. *Current Nutrition and Food Science, 3*, 91–108.

Kim, J., Muhammad, N., Jhun, B. H., & Yoo, J. W., (2016). Probiotic delivery systems: A brief overview. *Journal of Pharmaceutical Investigation, 46*(4), 377–386.

Koh, J. H., Kim, Y., & Oh, J. H., (2010). Chemical characterization of tomato juice fermented with bifidobacteria. *Journal of Food Science, 75*, 428–432.

Korbekandi, H., Mortazavian, A. M., & Iravani, S., (2011). Technology and stability of probiotic in fermented milks. In: Shah, N., Cruz, A. G., & Faria, J. A. F., (eds.), *Probiotic and prebiotic foods: Technology, Stability and Benefits to the Human Health* (pp. 131–169). Nova Science Publishers, New York.

Kourkoutas, Y., Kanellaki, M., & Koutinas, A. A., (2006). Apple pieces as immobilization support of various microorganisms. *LWT- Food Science and Technology, 39*, 980–986.

Kourkoutas, Y., Xolias, V., Kallis, M., Bezirtzoglou, E., & Kanellaki, M., (2005). *Lactobacillus casei* cell immobilization on fruit pieces for probiotic additive, fermented milk and lactic acid production. *Process Biochemistry, 40*, 411–416.

Kumar, S., Bansal, A., Chakrabarti, A., & Singhi, S., (2013). Evaluation of efficacy of probiotics in prevention of Candida colonization in a PICU—A randomized controlled trial. *Critical Care Medicine, 41*, 565–572.

Kun, S., Rezessy-Szabo, J. M., Nguyen, Q. D., & Hoschke, A., (2007). Changes of microbial population and some components in carrot juice during fermentation with selected *Bifidobacterium* strains. *Process Biochemistry, 43*, 816–821.

Lebeer, S., Vanderleyden, J., & De Keersmaecker, S. C., (2008). Genes and molecules of *Lactobacilli* supporting probiotic action. *Microbiology and Molecular Biology Reviews, 72*, 728–764.

Lee, Y. K., & Salminen, S., (2009). *Handbook of Probiotics and Prebiotics* (2nd edn.). John Wiley and Sons Inc., Hoboken.

Lei, V., & Jacobsen, M., (2004). Microbiological characterization and probiotic potential of koko and koko sour water, African spontaneously fermented millet porridge and drink. *Journal of Applied Microbiology, 96*, 384–397.

Li, J., Zhang, W., Wang, C., Yu, Q., Dai, R., & Pei, X., (2012). *Lactococcus lactis* expressing food-grade β-galactosidase alleviates lactose intolerance symptoms in post-weaning balb/c mice. *Applied Microbiology and Biotechnology, 96*, 1499–1506.

Luckow, T., & Delahunty, C., (2004). Which juice is healthier? A consumer study of probiotic non-dairy juice drinks. *Food Quality and Preference, 15*, 751–759.

Luckow, T., Sheehan, V., Fitzgerald, G., & Delahunty, C., (2006). Exposure, health information and flavor masking strategies for improving the sensory quality of probiotic juice. *Appetite, 47*, 315–325.

Makinen, K., Berger, B., Bel-Rhlid, R., & Ananta, E., (2012). Science and technology for the mastership of probiotic applications in food products. *Journal of Biotechnology, 162*, 356–365.

Malganji, S., Sohrabvandi, S., Jahadi, M., Nematollahi, A., & Sarmadi, B., (2015). Effect of refrigerated storage on sensory properties and viability of probiotic in grape drink. *Applied Food Biotechnology, 3*, 59–62.

Mangell, P., Nejdfors, P., Wang, M., Ahrne, S., Westrom, B., Thorlacius, H., & Jeppsson, B., (2002). *Lactobacillus plantarum* 299v inhibits *Escherichia coli* induced intestinal permeability. *Digestive Diseases and Sciences, 47*, 511–516.

Marhamatizadeh, M. H., Rezazadeh, S., Kazemeini, F., & Reza, M. K., (2012). The study of probiotic juice product conditions supplemented by culture of *Lactobacillus acidophilus* and *Bifidobacterium bifidum*. *Middle-East Journal of Scientific Research, 11*, 287–295.

Markowiak, P., & Slízewska, K., (2017). Effects of probiotics, prebiotics, and synbiotics on human health. *Nutrients, 9*, 1021. doi: 10.3390/nu9091021.

Martensson, O., Öste, R., & Holst, O., (2002). The effect of yoghurt culture on the survival of probiotic bacteria in oat-based, non-dairy products. *Food Research International, 35*, 775–784.

Martins, E. M. F., Ramos, A. M., Vanzela, E. S. L., Stringheta, P. C., Pinto De, C. L. O., & Martins, J. M., (2013). Products of vegetable origin: A new alternative for the consumption of probiotic bacteria. *Food Research International, 51*, 764–770.

Masuda, T., Kimura, M., Okada, S., & Yasui, H., (2010). *Pediococcus pentosaceus* Sn26 inhibits IgE production and the occurrence of ovalbumin-induced allergic diarrhea in mice. *Bioscience, Biotechnology, and Biochemistry, 74*, 329–335.

Mattila-Sandholm, T., Myllarinen, P. M., Crittenden, R., Mogensen, G., Fonden, R., & Saarela, M., (2002). Technological challenges for future probiotic foods. *International Dairy Journal, 12*, 173–182.

Mättö, J., Alakomi, H. L., Vaari, A., Virkajärvi, I., & Saarela, M., (2006). Influence of processing conditions on *Bifidobacterium animalis* subsp. *lactis* functionality with a special focus on acid tolerance and factors affecting it. *International Dairy Journal, 16*, 1029–1037.

Mättö, J., Malinen, E., Suihko, M. L., Alander, M., Palva, A., & Saarela, M., (2004). Genetic heterogeneity and technological properties of intestinal bifidobacteria. *Journal of Applied Microbiol*ogy, *97*, 459–470.

McMaster, L. D., Kokotta, S. A., Reidb, S. J., & Abratt, V. R., (2005). Use of traditional African fermented beverages as delivery vehicles for *Bifidobacterium* lactis DSM 10140. *International Journal of Food Microbiology, 102*, 231–237.

Meile, L., Ludwig, W., Rueger, U., Gut, C., Kaufmann, P., Dasen, G., Wenger, S., & Teuber, M., (1997). *Bifidobacterium lactis* sp. *nov*, a moderately oxygen tolerant species isolated from fermented milk. *System Applied Microbiology, 20*, 57–64.

Michetti, P., Dorta, G.,Wiesel, P. H., Brassart, D., Verdu, E., Herranz, M., Felley, C., Porta, N., Rouvet, M., Blum, A. L., & Corthésy-Theulaz, I., (1999). Effect of whey-based culture supernatant of *Lactobacillus acidophilus (johnsonii) La1* on *Helicobacter pylori* infection in humans. *Digestion, 60*, 203–209.

Mishra, S., & Mishra, H. N., (2012). Technological aspects of probiotic functional food development. *Nutrafoods, 11*, 117–130.

Mohammadi, R., Mortazavian, A. M., Khosrokhavar, R., & Cruz, A. G., (2011). Probiotic ice cream: Viability of probiotic bacteria and sensory properties. *Annals of Microbiology, 61*, 411–424.

Mohammed, S. I., Steenson, L. R., & Kirleis, A. W., (1991). Isolation and characterization of microoganisms associated with the traditional sorghum fermentation for production of Sudanese kisra. *Applied and Environmental Microbiology, 57*, 2529–2533.

Morelli, L., (2000). *In vitro* selection of probiotic lactobacilli: A critical appraisal. *Current Issues in Intestinal Microbiology, 1*, 59–67.

Mousavi, Z. E., Mousavi, S. M., Razavi, S. H., Emam-Djomeh, Z., & Kiani, H., (2011). Fermentation of pomegranate juice by probiotic lactic acid bactéria. *World Journal of Microbiology and Biotechnology, 27*, 123–128.

Muyanja, C. M. B. K., Narvhus, J. A., Treimo, J., & Langsrud, T., (2003). Isolation, characterization and identification of lactic acid bacteria from Bushera: A Ugandan traditional fermented beverage. *International Journal of Food Microbiology, 80*, 201–210.

Nase, L., Hatakka, K., & Savilahti, E., (2001). Effect of long-term consumption of *Lactobacillus* GG in milk on dental caries and caries risk in children. *Caries Research,* 35, 412–420.

Nicolesco, C. L., & Buruleanu, L. C., (2010). Correlation of some substrate parameters in growing *Lactobacillus acidophilus* on vegetable and fruit cocktail juices. *Bulletin UASVM Agriculture, 67*, 352–359.

Noorbakhsh, R., Yaghmaee, P., & Durance, T., (2013). Radiant energy under vacuum (REV) technology: A novel approach for producing probiotic enriched apple snacks. *Journal of Functional Foods, 5*, 1049–1056.

Nova, E., Warnberg, J., Gomez-Martinez, S., Diaz, L. E., Romeo, J., & Marcos, A., (2007). Immunodulatory effects of probiotics in different stages of life. *British Journal of Nutrition, 98*, S90–S95.

Oatley, J. T., Rarick, M. D., Ji, G. E., & Linz, J. E., (2000). Binding of aflatoxin B1 to bifidobacteria *in vitro*. *Journal of Food Protection, 63*, 1133–1136.

Oelschlaeger, T. A., (2010). Mechanisms of probiotic actions—A review. *International Journal of Medical Microbiology, 300*, 57–62.

Oh, S. Y., Lee, J. H., Jang, D. K., Heo, S. C., & Kim, H. J., (2005). Relationship of nutrients and food to colorectal cancer risk in Koreans. *Nutrition Research, 25*, 805–813.

Oh, Y. J., & Jung, D. S., (2015). Evaluation of probiotic properties of *Lactobacillus* and *Pediococcus* strains isolated from Omegisool, a traditionally fermented millet alcoholic beverage in Korea. *LWT Food Science and Technology, 63*, 437–444.

Ohland, C. L., & Macnaughton, W. K., (2010). Probiotic bacteria and intestinal epithelial barrier function. *American Journal of Physiology-Gastrointestinal and Liver Physiology, 298*, G807–819.

Oi, Y., & Kitabatake, N., (2003). Chemical composition of an East African traditional beverage, Togwa. *Journal of Agricultural and Food Chemistry, 51*, 7024–7028.

Olasupo, N. A., Olukoya, D. K., & Odunfa, S. A., (1997). Identification of *Lactobacillus* species associated with selected African fermented foods. *Zeitschrift fur Naturforschung C. Journal of Bioscience, 51*, 105–108.

Omemu, A. M., Oyewole, O. B., & Bankole, M. O., (2007). Significance of yeasts in the fermentation of maize for ogi production. *Food Microbiology, 24*, 571–576.

Onyango, C., Bley, T., Raddatz, H., & Henle, T., (2004). Flavor compounds in back slop fermented uji (an East African sour porridge). *European Food Research Technology, 218*, 579–583.

Onyango, C., Okoth, M. W., & Mbugua, S. K., (2003). The pasting behavior of lactic-fermented and dried uji (an East African sour porridge). *Journal of the Science of Food and Agriculture, 83*, 1412–1418.

Ouwehand, A. C., Lagström, H., Suomalainen, T., & Salminen, S., (2002). Effect of probiotics on constipation, fecal azoreductase activity and fecal mucin content in the elderly. *Annals of Nutrition and Metabolism, 46*, 159–162.

Ozdemir, S., Gocmen, D., & Kumral, A. Y. (2007). A traditional Turkish fermented cereal food: Tarhana. *Food Reviews International, 23*, 107–121.

Pakbin, B., Razavi, S. H., Mahmoudi, R., & Gajarbeygi, P., (2014). Producing Probiotic Peach Juice. *Biotechnology and Health Sciences, 1*, e24683. doi: 10.17795/bhs-24683.

Panghal, A., Janghub, S., Virkara, K., Gata, Y., Kumara, V., & Chhikaraa, N., (2018). Potential non-dairy probiotic products – a healthy approach. *Food Bioscience, 21*, 80–89.

Park, H. K., So, J. S., & Heo, T. R., (1995). Acid adaptation promotes survival of *Bifidobacterium breve* against environmental stress. *Food Biotechnology, 4*, 226–230.

Parvez, S., Malik, K. A., Ah Kang, S., & Kim, H. Y., (2006). Probiotics and their fermented food products are beneficial for health. *Journal of Applied Microbiology, 100*, 1171–1185.

Pereira, A. L. F., Maciel, T. C., & Rodrigues, S., (2011). Probiotic beverage from cashew apple juice fermented with *Lactobacillus casei*. *Food Research International, 44*, 1276–1283.

Pereira, D. I. A., & Gibson, G. R., (2002). Effects of consumption of probiotics and prebiotics on serum lipid levels in humans. *Critical Reviews in Biochemistry and Molecular Biology, 37*, 259–281.

Peres, C. M., Peres, C., Hernandez-Mendoza, A., & Malcata, F. X., (2012). Review on fermented plant materials as carriers and sources of potentially probiotic lactic acid bacteria -With an emphasis on table olives. *Trends in Food Science and Technology, 26*, 31–42.

Pimentel, T. C., Madrona, G. S., Garcia, S., & Prudencio, S. H., (2015). Probiotic viability, physicochemical characteristics and acceptability during refrigerated storage of clarified apple juice supplemented with *Lactobacillus paracasei* ssp. paracasei and oligofructose in different package type. *LWT – Food Science and Technology, 63*, 415–422.

Possemiers, S., Marzorati, M., Verstraete, W., & Wiele, V. T., (2010). Bacteria and chocolate: A successful combination for probiotic delivery. *International Journal of Food Microbiology, 141*, 97–103.

Prado, F. C., Parada, J. L., Inaba, J., Pandey, A., & Soccol, C. R., (2008). Trends in non-dairy probiotic beverages. *Food Research International, 41*, 111–123.

Prior, R. L., & Cao, G., (2000). Antioxidant phytochemicals in fruit and vegetables, diet and health implications. *HortScience, 35*, 588–592.

Puerari, C., Magalhaes-Guedes, K. T., & Schwan, R. F., (2015). Physicochemical and microbiological characterization of chicha, a rice-based fermented beverage produced by Umutina Brazilian Amerindians. *Food Microbiology, 46*, 210–217.

Ragaert, P., Verbeke, W., Devlieghere, F., & Debevere, J., (2004). Consumer perception and choice of minimally processed vegetables and packaged fruits. *Food Quality and Preference, 15*, 259–270.

Rakin, M., Vukasinovic, M., Siler-Marinkovic, S., & Maksimovic, M., (2007). Contribution of lactic acid fermentation to improved nutritive quality vegetable juices enriched with brewer's yeast autolysate. *Food Chemistry, 100*, 599–602.

Ranadheera, R. D. C. S., Baines, S. K., & Adams, M. C., (2010). Importance of food in probiotic efficacy. *Food Research International, 43*, 1–7.

Reddy, L. V., Min, J. H., & Wee, Y. J., (2015). Production of probiotic mango juice by fermentation of lactic acid bacteria. *Microbiology and Biotechnology Letters, 43*, 120–125.

Reid, G., (2001). Probiotic agents to protect the urogenital tract against infection. *The American Journal of Clinical Nutrition, 73*, 437S–443S.

Reid, G., (2008). Probiotics and prebiotics – progress and challenges. *International Dairy Journal, 18*, 969–975.

Reiff, C., & Kelly, D., (2010). Inflammatory bowel disease, gut bacteria and probiotic therapy. *International Journal of Medical Microbiology, 300*, 25–33.

Reina, L. D., Fleming, H. P., & Júnior, F. B., (2002). Bacterial contamination of cucumber fruit through adhesion. *Journal of Food Protection, 65*, 1881–1887.

Rickman, J. C., Barrett, D. M., & Bruhn, C. M., (2007). Nutritional comparison of fresh, frozen and canned fruits and vegetables. Part 1. Vitamins C and B and phenolic compounds. *Journal of the Science of Food and Agriculture, 87*, 930–944.

Riebroy, S., Banjakul, S., Visessanguan, W., & Tanaka, M., (2007). Effect of iced storage of bigeye snapper (Priacanthus tayenus) on the chemical composition, properties and acceptability of som-fug, a fermented Thai fish mince. *Food Chemistry, 102*, 270–280.

Roberts, J. S., & Kidd, D. R., (2005). Lactic acid fermentation of onions. *Lebensmittel Wissenschaft und Technologie [Food Science and Technology], 38*, 2185–2190.

Rodgers, S., (2007). Incorporation of probiotics in food service products: An exploratory study. *Journal of Foodservice, 18*, 108–118.

Ross, R. P., Desmond, C., Fitzgerald, G. F., & Stanton, C., (2005). Overcoming the technological hurdles in the development of probiotic foods. *Journal of Applied Microbiology, 98*, 1410–1417.

Rößle, C., Brunton, N., Gormley, R. T., Ross, P. R., & Butler, F., (2010). Development of potentially symbiotic fresh-cut apple slices. *Journal of Functional Foods, 2*, 245–254.

Rycroft, C. E., Jones, M. R., Gibson, G. R., & Rastall, R. A., (2001). A comparative *in vitro* evaluation of the fermentation properties of prebiotic oligosaccharides. *Journal of Applied Microbiology, 91*, 878–887.

Saarela, M., Mogensen, G., Fondén, R., Mätto, J., & Sandholm, T. M., (2000). Probiotic bacteria: Safety, functional and technological properties. *Journal of Biotechnology, 84*, 197–215.

Salminen, S., Bouley, C., Boutron-Ruault, M. C., Cummings, J. H., Franck, A., Gibson, G. R., Isolauri, E., et al., (1998). Functional food science and gastrointestinal physiology and function. *British Journal of Nutrition, 80*, S147–S171.

Sanders, M. E., Tompkins, T., Heimbach, J. T., & Kolida, S., (2005). Weight of evidence needed to substantiate a health effect for probiotics and prebiotics. *European Journal of Nutrition, 44*, 303–310.

Sandholm, T. M., Myllarinena, P., Crittenden, R., Mogensen, G., Fonden, R., & Saarela, M., (2002). Technological challenges for future probiotic foods. *International Dairy Journal, 12*, 173–182.

Santivarangkna, C., Kulozik, U., & Foerst, P., (2006). Effect of carbohydrates on the survival of *Lactobacillus helveticus* during vacuum drying. *Letters in Applied Microbiology, 42*, 271–276.

Sanz, M. L., Cote, G. L., Gibson, G. R., & Rastall, R. A., (2006). Influence of glycosidic linkages and molecular weight on the fermentation of maltose-based oligosaccharides by human gut bacteria. *Journal of Agricultural and Food Chemistry, 54*, 9779–9784.

Sapers, G. M., (2001). Efficacy of washing and sanitizing methods. *Food Technology Biotechnology, 39*, 305–311.

Saw, L. K., Chen, S., Wong, S. H., Tan, S. A., & Goh, K. K. T., (2011). Fermentation of tropical fruit juices by lactic acid bacteria. In: *12th ASEAN Food Conference* (pp. 80–87). Kangkok, Thailand.

Schatzmayr, G., Zehner, F., Taubel, M., Schatzmayr, D., Klimitsch, A., Loibner, A. P., & Binder, E. M., (2006). Microbiologicals for deactivating mycotoxins. *Molecular Nutrition and Food Research, 50*, 543–551.

Shah, N. P., (2007). Functional cultures and health benefits. *International Dairy Journal, 17*, 1262–1277.

Sharma, S., Arora, M., & Baldi, A., (2013). *Probiotics in India: Current Status and Future Prospects*. PharmAspire drug therapy/medical updates. URL: https://www.researchgate.net/profile/Dr_Ashish_Baldi/publication/257537041_ Probiotics_in_India_Current_status_and_future_prospects/links/0c960525661ad870f0000000/Probiotics-in-India-Current-status-and-future-prospects.pdf (accessed on 27 September 2021).

Sheehan, V. M., Ross, P., & Fitzgerald, G. F., (2007). Assessing the acid tolerance and the technological robustness of probiotic cultures for fortification in fruit juices. *Innovative Food Science and Emerging Technologies, 8*, 279–284.

Sheela, T., & Suganya, R. S., (2012). Studies on anti-diarrhoeal activity of symbiotic plums juice. *International Journal of Scientific and Research Publications, 2*, 1–5.

Soccol, C. R., Vandenberghe, L. P. S., Spier, M. R., Medeiros, A. B. P., Yamaguishi, C. T., Lindner, J. D., Pandey, A., & Thomaz-Soccol, V., (2010). The potential of probiotics: A review. *Food Technology and Biotechnology, 48*, 413–434.

Soni, S. K., Sandhu, D. K., Vilkhu, K. S., & Kamra, N., (1986). Microbiological studies on dosa fermentation. *Food Microbiology, 3*, 45–53.

Southon, S., (2000). Increased fruit and vegetable consumption within the EU: Potential health benefits. *Food Research International, 33*, 211–217.

Stanton, C., Fitzgerald, G., Ross, R. P., Desmond, C., Coakley, M., & Collins, J. K., (2003). Challenges facing development of probiotic containing functional foods. In: Farnworth, E. R., (ed.), *Handbook of Functional Fermented Foods* (pp. 29–40). CRC Press, Boca Raton.

Su, L. J., & Arab, L., (2006). Salad and raw vegetable consumption and nutritional status in the adult US population: Results from the third national health and nutrition examination survey. *Journal of the American Dietetic Association, 106*, 1394–1404.

Sun-Waterhouse, D., (2011). The development of fruit-based functional foods targeting the health and wellness market: A review. *International Journal of Food Science and Technology, 46*, 899–920.

Tamime, A. Y., & McNulty, D., (1999). Kishkda dried fermented milk/cereal mixture: 4. microbiological quality. *Le Lait, 79*, 449–456.

TechSci Research, (2014). *India Probiotic Market Forecast and Opportunities 2019*. URL: https://www.techsciresearch.com/news/261-india-probiotic-market-pegged-to-grow-at-around-20-by-2019.html (accessed on 27 September 2021).

Teixeira, P., Castro, H., & Kirby, R., (1994). Inducible thermotolerance in *Lactobacillus bulgaricus*. *Letters in Applied Microbiology, 18*, 218–221.

Thakur, M., & Sharma, R. K., (2017). Development of probiotic pomegranate beverage and its physico-chemical and microbial characterization. *International Journal of Pure and Applied Bioscience, 5*, 35–41.

Thakur, M., Satwadhar, P. N., & Deshpande, H. W., (2015). Exploiting fruits and vegetables for development of probiotic products. *Trends in Biosciences, 8*, 6039–6046.

Thakur, N., Savitri, Saris, P. E. J., & Bhalla, T. C., (2015). Microorganisms associated with amylolytic starters and traditional fermented alcoholic beverages of North Western Himalayas in India. *Food Bioscience, 11*, 92–96.

Thapa, N., Pal, J., & Tamang, J. P., (2004). Microbial diversity in Ngari, Hentak and Tungtap, fermented fish products of North-East India. *World Journal of Microbiology and Biotechnology, 20*, 599–607.

Tiwari, U., & Cummins, E., (2013). Factors influencing levels of phytochemicals in selected fruit and vegetables during pre- and post-harvest food processing operations. *Food Research International, 50*, 497–506.

Tou, E. H., Guyot, J. P., Mouquet-Rivier, C., Rochette, I., Counil, E., Traore, A. S., & Treche, S., (2006). Study through surveys and fermentation kinetics of the traditional processing of pearl millet (Pennisetum glaucum) into ben-saalga, a fermented gruel from Burkina Faso. *International Journal of Food Microbiology, 106*, 52–60.

Tripathi, M. K., & Giri, S. K., (2014). Probiotic functional foods: Survival of probiotics during processing and storage. *Journal of Functional Foods, 9*, 225–241.

Tsen, J. H., Lin, Y. P., & King, V. A. E., (2003). Banana puree fermentation by *Lactobacillus acidophilus* immobilized in Ca-alginate. *The Journal of General and Applied Microbiology, 49*, 357–361.

Tsen, J. H., Lin, Y. P., Huang, H. Y., & King, V. A. E., (2007). Accelerated storage testing of freeze-dried immobilized *Lactobacillus acidophilus*-fermented banana media. *Journal of Food Processing and Preservation, 31*, 688–701.

Vanajakshi, V., Vijayendra, S. V. N., Varadaraj, M. C., Venkateswaran, G., & Agrawal, R., (2015). Optimization of a probiotic beverage based on Moringa leaves and beetroot. *LWT – Food Science and Technology, 63*, 1268–1273.

Vijaya, K. B., Vijayendra, S. V. N., & Reddy, O. V. S., (2015). Trends in dairy and non-dairy probiotic products: A review. *Journal of Food Science and Technology, 52*, 6112–6124.

Vinderola, C. G., & Reinheimer, J. A., (2003). Lactic acid bacteria: A comparative "*in vitro*" study of probiotic characteristics and biological barrier resistance. *Food Research International, 36*, 895–904.

Vinderola, C. G., Costa, G. A., Regenhardt, S., & Reinheimer, J. A., (2002). Influence of compounds associated with fermented dairy products on the growth of lactic acid starter and probiotic bacteria. *International Dairy Journal, 12*, 579–589.

Weinberg, E. D., (1997). The *Lactobacillus* anomaly: Total iron abstinence. *Perspectives in Biology and Medicine, 40*, 578–583.

Yahia, E. M., Celis, M. E. M., & Svendsen, M., (2009). The contribution of fruit and vegetable consumption to human health. In: De La Rosa, L. A., Alvarez-Parrilla, E., & González-Aguilar, G. A., (eds.), *Fruit and Vegetable Phytochemicals: Chemistry, Nutritional Value, and Stability* (pp. 3–51). Wiley-Blackwell, Oxford.

Yan, P. M., Xue, W. T., Tan, S. S., Zhang, H., & Chang, X. H., (2008). Effect of inoculating lactic acid bacteria starter cultures on the nitrite concentration of fermenting Chinese paocai. *Food Control, 19*, 50–55.

Yerlikaya, O., (2014). Starter cultures used in probiotic dairy product preparation and popular probiotic dairy drinks. *Food Science and Technology (Campinas), 34*, 221–229.

Yoon, K. Y., Woodams, E. E., & Hang, Y. D., (2004). Probiotication of tomato juice by lactic acid bacteria. *Journal of Microbiology, 42*, 315–318.

Yoon, K. Y., Woodams, E. E., & Hang, Y. D., (2005). Fermentation of beet juice by beneficial lactic acid bacteria. *Lebensmittel Wissenschaft and Technologie [Food Science and Technology], 38*, 73–75.

Yoon, K. Y., Woodams, E. E., & Hang, Y. D., (2006). Production of probiotic cabbage juice by lactic acid bacteria. *Bioresource Technology, 97*, 1427–1430.

Zheng, X., Yu, Y., Xiao, G., Xu, Y., Wu, J., Tang, D., & Zhang, Y., (2014). Comparing product stability of probiotic beverages using litchi juice treated by high hydrostatic pressure and heat as substrates. *Innovative Food Science and Emerging Technologies, 23*, 61–67.

CHAPTER 5

Ice Cream as Probiotic Food and Its Potential Benefits in Human Health

TATIANA COLOMBO PIMENTEL,* MICHELE ROSSET, SUELLEN JENSEN KLOSOSKI, CARLOS EDUARDO BARÃO, VANESSA APARECIDA MARCOLINO, VANESSA GUIMARÃES ALVES OLHER, and ADRIANO GOMES DA CRUZ

*Corresponding author. E-mail: tatiana.pimentel@ifpr.edu.br

ABSTRACT

Ice cream is a product with much acceptability and attractive sensory attributes to the consumer, ensuing in high sales. This product can be an excellent carrier medium of probiotic microorganisms because of its ingredients and chemical components, including milk lipids, proteins, and lactose, along with other components. Additionally, it has the advantage of being stored at freezing temperatures, making it less susceptible to exposure to abusive temperatures. Ice creams also have relatively high pH values (5.5–6.5), during storage which favors the endurance of the microbial cultures; and results in greater acceptance by consumers, especially those who prefer low acid products. Conversely, the development of probiotic ice creams needs to overcome some of the technological intrinsic parameters linked with the processing steps. The objective of this chapter is to appraise the technological challenges associated with the production of ice creams containing probiotic microorganisms, presenting strategies for the efficient incorporation of probiotic cultures and their potential impact on the physicochemical properties, sensorial features, and storage stability. Furthermore, the health effects associated with the consumption of probiotic ice cream are discussed.

5.1 INTRODUCTION

The concern about consuming nutritious foods that can decrease the risk of illnesses during the life has increased exponentially. This new trend stems from the awareness and knowledge that food has a major impact on the health and quality of life of people (Thamer and Penna, 2006). Some people's diet has attracted attention because of the low incidence of some diseases. The Eskimos and the French, for example, have presented low rates of cardiovascular disease, a fact that may be correlated with the daily intake of fish rich in omega 3 and 6, and red wine, rich in phenolic substances (Anjo, 2004).

With the increasing search for foods that offer wellness, can regulate the body functions, and aid protection against diseases, a new conception of food, namely 'functional foods,' has emerged. In addition, basic nutritional functions, functional foods can produce physiological, metabolic, or beneficial effects on human health (Roberfroid, 2002). In addition, they should be safe and not contain medical restrictions. In most cases, the functional food is obtained by adding one or more chemical and/or biological component to the matrix. Milk matrices such as yogurts, fermented milk, cheeses, and ice creams stand out due to the preservation capacity of the components added in the processing and storage and due to the palatability offered by these matrices.

Ice cream is manufactured from a stabilized emulsion, also called pasteurized syrup, which, through a freezing process under continuous stirring (beat) and incorporation of air, produces a creamy, smooth, and pleasant substance to the palate. This emulsion comprises dairy products, fat, water, sugar, emulsifier, stabilizer, dye, and flavoring agents. The ice cream is also rich in the most of vitamins such as A, C, D, B_1, B_2, B_6, K, and E, phosphorus, calcium like minerals; hence, can be considered as a complete food from the nutritional point of view (Souza et al., 2010).

The manufacturing process of ice cream involves the liquid mix preparation through blending the preferred ingredients, homogenization, heat treatment (pasteurization), and aging. Subsequently, it is preceded to the whipping and freezing stages. Then after, in the frozen mix flavoring agents are incorporated and for desired shape placed in molds. Under static conditions, it is subject to freezing (hardening). During the first freezing step generally swept-surface freezers are used, whereas for the second freezing step, forced convection freezers like plate-type conduction freezers or air blast tunnels or rooms are employed (Deosarkar et al., 2016).

Probiotic cultures are viable microorganisms generally belongs to the *Lactobacillus* and *Bifidobacterium* genera. Few strains can replace the intestinal microbiota, which has been altered in some way by antibiotic, chemotherapeutic or any other stress (Tannock, 1998), thus promoting health, but not the cure of the diseases. The selected strains should be safe, stable, and metabolically active to ensure positive effects to the host (Havenaar et al., 1992). The daily intake of approximately 10^8–10^9 viable cells of these microorganisms is sufficient to provide benefits, therefore, the consumption of 100 g of a product with 10^6–10^7 cfu/g is sufficient (Lee and Salminen, 1995; Soukoulis et al., 2014).

Ice cream like products can be considered an outstanding vehicle for the insertion of probiotic microorganism in the human diet, because it has high pH (5.5–6.5), preventing lactic acid from interfering with the probiotic viability, and suitable composition, with fat, protein, and lactose, contributing to the bacterial survival (Cruz et al., 2009; Villalva et al., 2017). However, the low storage temperatures, the acidity development and the mechanical stress originated by the agitation step of the ice cream can cause damage to the probiotic cells, reducing their viability. Mechanical stress may favor the production and growth of ice crystals and increase the electrolyte concentration (resulting from the extraction of water), the oxygen uptake and the disintegration by physical process of cells during freezing, reducing the viability of the probiotic cultures. Furthermore, probiotic addition can alter the sensory characteristics of the finished products and, consequently, reduce the consumers acceptance of the products.

The objective of this chapter is to appraisal the technological challenges related to the probiotic ice creams production, presenting strategies for the efficient incorporation of probiotic cultures and their potential impact on the physicochemical properties, sensorial features, and storage stability. Furthermore, the health effects associated with the consumption of probiotic ice cream are discussed.

5.2 ICE CREAMS: GENERAL CHARACTERISTICS

Ice cream is a relatively delicious, easily digestible, healthy, and well-balanced food. Its nutritional quality is quite interesting, presenting high palatability, and for people with stomach and throat illnesses, who are unable of ingesting other foods is an ideal food product. Their characteristics stimulate the flow of digestive juices improving and accelerating digestion. Ice creams are products made primarily of milk and/or dairy products, consisting of a fat and protein emulsion, and that are subjected to freezing under conditions to guarantee the preservation of the product in the partially frozen or frozen state during storage, transportation, and delivery for consumption.

The composition of the ice cream is quite varied, usually exhibiting from 8 to 20% fat, 8 to 15% non-fat milk solids, 13 to 20% sugar and 0 to 0.7% emulsifier-stabilizer, but the composition depends on the region and the different markets. Ice cream is an outstanding source of energy, mainly owing to its high content of carbohydrates and fat. Milk proteins account for 34–36% of their non-fat solids, and ice cream contains high levels of minerals and vitamins, which depends primarily on the amount of milk solids used in the formulation (Souza et al., 2010).

The proteins present in the milk have excellent biological value since it has all essential amino acids, such as lysine and tryptophan and high digestibility. As milk is the chief ingredient used to manufacture ice cream, there is the transfer of these characteristics to the product. The ice cream has about 4 times more carbohydrates than milk, being mostly mono and disaccharides. In ice cream, lactose accounts for 20% of the carbohydrates. This sugar favors the development of *Lactobacillus acidophilus* in the intestines of consumers by raising the acidity in this environment, also favoring the absorption of calcium and phosphorus. Milk fat considerably contributes to the nutritional significance of the product, containing a significant amount of arachidonic acid and linoleic acid, type of essential fatty acids. It is also the energy source and carrier of the fat-soluble vitamins (A, D, E, and K).

Milk and dairy products are sources of essential minerals to the diet. The presence of lactose in ice cream favors the absorption of calcium due to its metabolism in the human body. The calcium contents in the ice cream are around 0.132 g/100 g and phosphorus of 0.105 g/100 g, respectively. Ice cream is a rich source of several vitamins that are essential for the maintenance of normal growth and health. Among the liposoluble vitamins, the ice cream possesses vitamin A, E, and has low content of vitamin D. Vitamin K is present in small amounts and it is destroyed in the manufacturing process, as in the pasteurization stage. Among the vitamins that are water-soluble, in ice cream, vitamin B_1 (thiamine) has a standard concentration of 0.48 mg/kg. For vitamin B_2 (riboflavin), ice cream is a reasonable source, providing on average 2.3 mg/kg. Vitamin B_6 (pyridoxine) is present in ice cream at a concentration of 0.0047 mg/kg, as well as vitamin B_{12} (cyanocobalamin). Vitamin C is most commonly found in fruit ice creams, with an average concentration of 3 mg/kg.

The world's largest consumer of ice cream is China, with 4.3 million liters in 2016, followed by the United States with 2.7 million liters (Mintel Annual Report, 2017). Per capita consumption is led by New Zealand (28.4 liter per year), followed by the United States of America (USA), Australia, and Finland *viz.,* 20.8, 18, and 14.2 liters/year, respectively (World Atlas, 2017). Currently, the markets that have gained prominence in the consumption of ice cream are of India and China. In India, ice cream has a low consumption, but in the last years there was recognition of this dessert and a strong growth was observed. China, on the other hand, has the highest growth rate since 2013 (Mintel Annual Report, 2017). The world ice cream market has revealed a gradual boost in recent years (Figure 5.1).

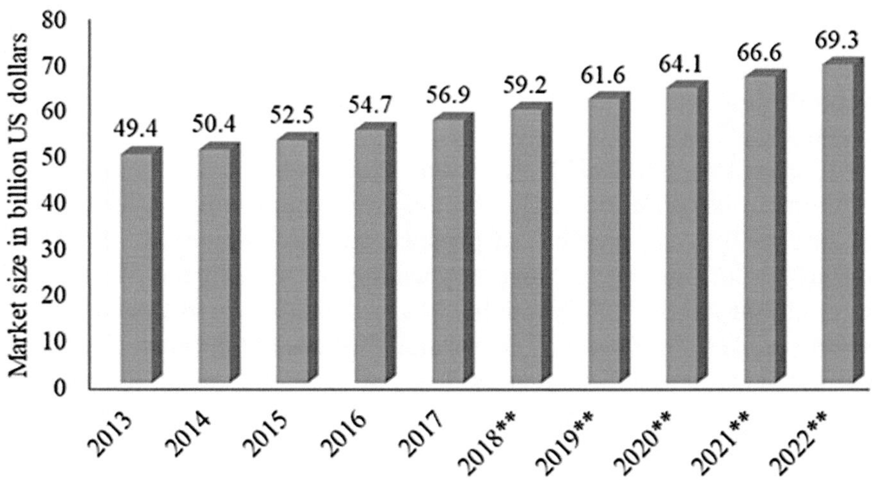

FIGURE 5.1 The current and predicted future global market of ice cream (Rolled Ice, 2016) [**Forecast].

Source: Adapted from: Rolled Ice (2016).

In established markets, there is difficulty of growth and one of the alternatives is to explore the product's diversification. Frozen yogurt and probiotic ice creams stand out as alternatives. According to Mintel, the emergence of new types of ice cream, with variations such as premium, gourmet, organic, and vegan, should grow up to 81% by 2020 (Sebrae, 2017). Consumers are concerned about the use of sugar in their diets, resulting in a decrease in ice cream consumption in the USA. However, the principles of healthy eating are ignored when the ice cream quality exceeds the expectations of the consumers, being observed behavior in the USA, United Kingdom, and Australia. Thus, the offer of functional ice cream may be a viable substitute for the maintenance and expansion in the market, mainly considering the consumers that look for healthy eating.

5.3 ICE CREAM: A CARRIER OF PROBIOTICS

Probiotics are defined as living microorganisms which when administered in adequate amounts can result in benefits in the health of the consumer (FAO/WHO, 2001). The regulation regarding probiotic cultures is complex and diverse, and there are differences between probiotics added to foods (or food supplement) with a specific health claim and probiotics which are marketed as a pharmaceutical component. The scientific criteria for judging health claims for probiotics differ greatly from nation to nation and generally are very rigid (Reid, 2015). The probiotic metabolic activity and viability should be conserved at all stages of food processing, since its production until consumer ingestion and survival in the gastrointestinal (GI) tract.

To exert favorable health effects on consumers, several authors suggest a minimum concentration of probiotics in products with values between 10^6 and 10^7 cfu per g or mL of product, but these values are dependent on factors such as the intended application (expected benefit), in which foods the probiotic will be applied, as well as the microorganism used (species and strain). The minimum amount of probiotics in foods is established to ensure that sufficient amounts of probiotics reach the intestine after consumption, because the numbers are reduced significantly during their passage through the digestive system. In other health benefits, the minimum counts ensure the observation of the health effect (Guarner and Schaafsma, 1998).

Ice creams are foods with a great potential as a carrier for probiotic cultures, presenting some advantages: (1) are ate by all ages people and in all parts of the world, (2) are nutritionally rich foods, with milk proteins, lipids, lactose, vitamins, and minerals in their composition, (3) are stored at freezing temperatures, making them less susceptible to exposure to abusive temperatures, (4) have high pH values (5.5–6.5), during storage which favors the endurance of the cultures; and results in greater acceptance by consumers, especially those who prefer low acid products, and (5) the more solids content in ice cream (like milk solids and fat) protect the probiotic cultures (product and at some stage in passage through the GI-tract) (Pimentel et al., 2017).

Over the last few years, several reports have suggested that for the delivery of probiotics ice cream serve as an outstanding ability as a medium in the human intestine if careful is taken during the processing stages and also the storage of the ice cream. Milk based frozen products provide the best surroundings for probiotic microorganisms to remain live for long-term manufacturing, distribution as well as storage. Furthermore, such product creates excellent conditions for growth of probiotics at large scale plus endurance at the storage stage. Regarding the ice cream characteristics, it has been found that the probiotic added ice cream generally has similar taste, texture, or composition characteristics to the conventional product. There are numerous ways to improve the organoleptic attributes of ice cream, such as the supplementation with oligofructose and inulin (prebiotics) to improve the stability of probiotics as well as the physicochemical and sensory attributes of the symbiotic product.

5.3.1 PROBIOTIC CULTURES EMPLOYED IN THE PRODUCTION OF ICE CREAM

The selection of suitable probiotic bacterial strain greatly depends on their potential to survive in the product (ice cream) conditions (elevated concentrations of sucrose, high levels of O_2, freezing temperature and storage temperature), acidic, and alkaline circumstances (to simulate gastric and intestinal environment, respectively). For efficient production of probiotic ice creams, some factors must be taken into account due to the processing steps, and it is essential to pick strains that are resistant to oxygen, since at the time of overrun where air is incorporation in the mixture is detrimental to the development of probiotic cultures, which are anaerobic or microaerophilic. The main microorganisms used as probiotic cultures in ice creams are *Lactobacillus* and *Bifidobacteria*. Table 5.1 lists some studies involving probiotic ice creams and the cultures used.

In general, in food processing, the addition of probiotic strains to food matrices should ensure their viability at high levels both during processing and storage, without altering their physical, chemical, and sensory characteristics. Therefore, the steps of incorporation of air, storage at low temperatures, how the probiotic is added to the product and the correct choice of the ingredients are important to the quality of the finished product. Additionally, the acidification of the final product, the types of acids produced, the oxygen content presented during storage, the antimicrobial compounds and the loss of milk nutrients may reduce the probiotic viability and, consequently, the probiotic properties of the product. In addition, product melting at the time of consumption may also impair probiotic cells (Jay et al., 2005).

TABLE 5.1 Probiotic Cultures Used in Probiotic Ice Cream Formulations

Type of Product	Probiotic Culture	References
Ice cream added prebiotic ingredients (inulin and oligofructose) and probiotic culture	*Bifidobacterium animalis* Bb-12 and *Lactobacillus acidophilus* La-5	Akalin and Eris (2008)
Ice cream with aspartame and sucrose and probiotic microorganisms	*L. rhamnosus, L. agilis,* and *L. acidophilus*	Bas Ygˇit et al. (2006)
Probiotic ice cream in simulated industrial and retail conditions	*L. johnsonii* La1	Alamprese et al. (2002)
	L. rhamnosus GC	Alamprese et al. (2005)
Probiotic ice cream	*Bifidobacterium bifidum* and *L. acidophilus*	Hekmat and McMahon (1992)
Probiotic ice cream	*L. acidophilus* and *L. rhamnosus*	Abghari et al. (2011)
Ice creams added phenolic compounds and probiotic culture	*L. casei Shirota*	Sagdic et al. (2012)
Probiotic ice cream containing inulin and concentrated whey	*L. acidophilus* NCDC 14	Pandiyan et al. (2012)
Ice cream with FOS and probiotic culture	*L. acidophilus (La-5)*	Ahamadi et al. (2012)
Ice cream with dietary fibers (orange, apple, oat, wheat, and bamboo) and probiotic cultures	*L. acidophilus* and *Bifidobacterium lactis*	Akalın et al. (2018)
Probiotic ice cream with different forms of probiotic addition	*L. acidophilus* (ATCC 4356)	Arslan et al. (2016)
Probiotic ice cream with and without hydrolysis of lactose	*L. acidophilus La-5*	Chiquetti et al. (2016)

5.4 TECHNOLOGICAL PROCESSING OF PROBIOTIC ICE CREAMS

The ingredients utilized in ice creams can be classified according to the quantities that are used in the production of the products. The primary ingredients are those used in higher quantities, like milk, cream (fat), sugar, and water, while the secondary ingredients are used in lower amounts, such as emulsifiers, stabilizers, and flavorings/dyes. Other ingredients can still be added to give specific characteristics to the ice cream, such as chocolates, fruits, nuts, and others. This type of product also contains a significant proportion of air (volume), but this is not considered as an ingredient. Because of its composition and production process, ice cream may be considered appropriate for the incorporation of probiotics.

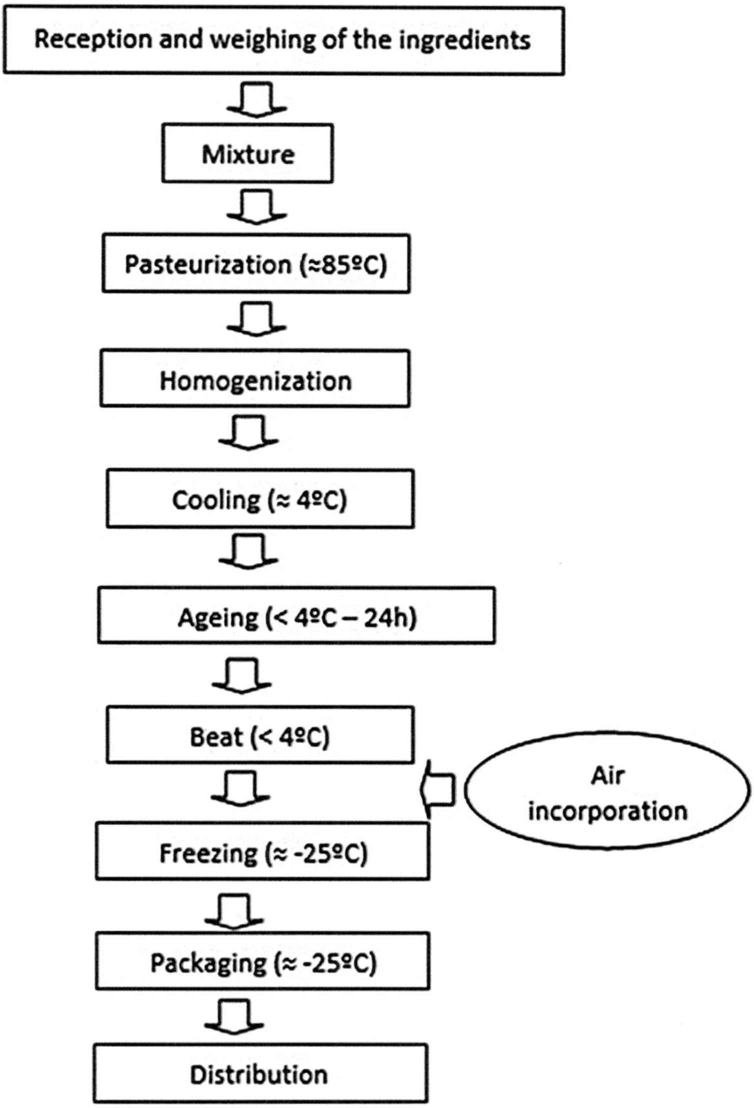

FIGURE 5.2 Technological processing of conventional ice creams.

The selection of good ingredients and proper handling are extremely important factors in the successful processing of any food, guaranteeing a clean, fresh flavor and adequate palatability. The different components used in the preparation of ice creams-milk products, sugar, stabilizer, emulsifier, fat, flavoring, and coloring-perform functions related to the product quality, such as body, texture, creaminess, color, aroma, and flavor. Stabilizers are used in small amounts (0.1 to 0.5%) in the ice cream mix, giving uniformity and softness to the texture of the product. The main functions of stabilizers in ice cream involves: (1) during storage, prevention as well as inhibition of the ice crystals and lactose, and recrystallization, occurs due to fluctuations in temperature, (2) to improve the beating properties, (3) to increase the viscosity of the syrup, (4) to help in the improvement of the consistency and texture of the final ice cream, (5) to improve the melting properties, (6) to prevent serum separation, (7) to facilitate the incorporation and distribution of air during manufacture of the ice cream, and (8) to promote better stability during storage. The emulsifiers are employed in product to: (1) promote homogeny during the beating, (2) lessen the beat time of the syrup, (3) manage the agglomeration and regrouping of the fat while freezing (stabilizes the fat emulsion), (4) assist in the air bubbles distribution, producing smooth body and typical creamy texture in ice cream, (5) reduce the negative effects caused by temperature fluctuation and, (6) increase the resistance to melting (Souza et al., 2010). The processing steps of conventional ice cream are revealed in Figure 5.2.

The steps are reception and weight of the raw ingredients (including milk, milk powder, emulsifiers, stabilizers, and sugar), mixture, pasteurization, homogenization, cooling at 4°C and aging for 24 hours at this temperature. The product of these stages is called mixture. This mixture is subsequently whipped and frozen to obtain the ice cream, which is packaged and kept at a freezing temperature during transportation, distribution, commercialization, and storage for consumption.

The preparation of the mixture is the former step in the processing of ice cream, by mixing, dispersing, and hydrating (dissolve) the ingredients in adequate amounts and in a specific order to obtain consistent and optimum mixing quality and utmost use of the components. Afterwards, the ice cream mix is warm up in the tank and agitated. Heating with continuous stirring must be cautiously controlled so that all the used ingredients can be dispersed and dissolved effectively. The fluid components, including cream, milk, water, sugar syrup, etc., are initially added plus before addition, the solid fats are also melted down. Then after, milk powder, emulsifiers, stabilizers, sugars, etc., like dry ingredients, are mixed. Prior to the addition to the mixing tank, stabilizing agents can be dry blended with an equal weight of sugar to facilitate dissolution, since stabilizers are the most difficult ingredients to dissolve. They are slowly added to the ice cream tank to guarantee the dispersion and also to avoid the lumps formation. The outlet of the mixing tank may contain a filter to eliminate the lumps if formed. At this stage, dyes and flavors are usually added unless they are heat-sensitive; in this case, they are introduced after the heat treatment or pasteurization (Clark, 2004).

For human consumption, the mixture is then pasteurized to eliminate or diminish the concentration of viable bacteria to a safe level. The binomial generally used is 80.5°C during 31 s, but other binomials (temperature × time) can be used, as soon as they have the same efficiency. The temperature and time used in the pasteurization of the ice cream mixture are greater as compared to milk, because the addition of sugar and fat difficult the heat transference and can protect the microorganisms. However, the temperature must not exceed 85°C in order to avoid milk proteins denaturation and the production of off-flavors (boiled milk taste). The pasteurization also produces the fusion of the emulsifiers and activates the stabilizers in colloidal solution, also improving the effect of the whey proteins. By denaturing the whey protein, the lipophilic part of the molecule inside the structure is broken. Under these situations, the whey proteins reduce the interfacial tension fat/water, thus acting as emulsifying agents. The water retention ability of the whey protein increases about three times its value due to pasteurization, which reaches numbers similar to those of casein. Furthermore, denaturation of proteins has an affirmative action on the ice cream quality, which results in to a creamier product with more uniform consistency and body-texture (Souza et al., 2010). Under high pressure, the hot mixture (>70°C) in the homogenizer is forced through a small valve in which the bigger-sized fat droplets are elongated and broken into very small droplets with fine emulsion, considerably escalating the surface area of fat.

After homogenization, the mixture is cooled to 4°C. Cooling inhibits the growth of microbes and prepares the mixture for aging. In the aging process, the cooled mixture is pumped to tanks that are intended to reduce the exposure of the ice cream mix to the atmosphere plus other probable contamination sources. The mixture is maintained at 0 to 4°C and is gently shaken periodically. The heat sensitive ingredients such as dyes, flavorings, and fruits may be added at this stage. Two important processes occur during aging: in the first step, on the surface of the fat droplets, the emulsifiers adsorb by replacing milk proteins up to a certain level and, in the latter part, the fat inside the droplets slowly starts to crystallize since nucleation should occur within each droplet individually. The aging stage is also important for the viscosity increase, better air incorporation and improved melting resistance. Ice creams that do not have the aging step frequently exhibit variable whipping abilities lower viscosity, and exhibit are very wet at extrusion from the dynamic freezer (Deosarkar et al., 2016).

After aging, the product goes to the freezing, which is the heart of the ice cream manufacturing method. At this stage, the mixture is transformed into ice cream by airing, freezing, and thumping, generating the air bubbles, ice crystals, and the matrix. At about 4°C, the ice cream mixture is pumped from the aging tank into the aeration tank where it is aerated and frozen. The air is injected into the tank through a filter system to make sure it is dry, clean, and free from contamination of microorganisms. Initially, the air makes large bubbles, but it is essential a dispersion of tiny air bubbles to get ice cream of high quality. In the ice cream manufacturing stage, the balance between protein, fat, and emulsifier is decisive as it controls the emulsion stability and thus the ease of aeration (Clarke, 2004). After freezing, the flavors, chocolates, dry fruits, etc., are mixed, and the finished product is immediately hygienically packed and put in a hardening room (minimum –30°C with plate-type or forced convection conduction freezers) in which most of the residual water freezes. Initially, rapid freezing is important to form as many crystal nuclei as possible so that their size stays minute. Similarly, subsequent hardening is also done rapidly to keep small-sized ice crystals (Deosarkar et al., 2016).

The packaging should protect the finished product from photooxidation, temperature fluctuations, odor transmittance and dehydration, and should also take into account distributional factors like thermal shock and package integrity. After packaging, the ice cream is kept for storage and distributed depending on the market share, scale of manufacture, consumer preferences and point of sale. Once it is manufactured, it must be maintained in the frozen state during storage, transportation, distribution, and selling (Deosarkar et al., 2016).

The difference in the processing steps between the probiotic and conventional ice cream is the incorporation of the probiotic culture. Figure 5.3 presents the step of probiotic addition. In a general view, the steps to produce an ice cream with probiotic bacteria are: reception and weight of the ingredients, mixture, pasteurization, homogenization, cooling at 4°C, aging for few hours, probiotic addition, beat, freezing, hardening, packaging, and lastly storage. Therefore, independently of the type of ice cream, the probiotic addition occurs after aging and before beat and freezing.

In non-fermented ice creams, after the aging, the mixture is warm up to 35–40°C and the lyophilized probiotic culture is added. This temperature

Villalva et al. (2017) formulated low-calorie peach-flavored ice cream with inulin and probiotic culture (*B. lactis* Bb-12), and evaluated the sensory, physic-chemical, and survival capacity of the probiotic culture in the product. The caloric value of the product was 70 Kcal and the moisture, ash, protein, and lipid contents were 76.47%, 1.22%, 6.37% and 0.15%, respectively. During the 21-day storage period, the *B. lactis* count was greater than 10^7 cfu/mL, i.e., higher than the limit established by the FDA (10^6 cfu per mL) to be established as a functional food. The results of the sensorial analysis showed that only 5% of the consumers did not like the product, that is, the ice cream presented good acceptability. Diniz et al. (2014) analyzed the sensorial characteristics of ice cream manufactured from goat's milk with the addition of *Bifidobacterium animalis* subsp. *lactis* BLC1 and compared with a control treatment. Incorporation of probiotics have no significant effect of the on the global appearance, aroma, and flavor acceptance, while an improved consistency was observed for the probiotic product. Cruxen et al. (2017) developed an ice cream based on butiá, native fruit of South America, added with *Bifidobacterium lactis* (Bl-04) and the product obtained high acceptability (6.8 to 7.8 on a nine-point hedonic scale) and purchase intention (80%) among the consumers. However, Di Criscio et al. (2010) mentioned that probiotic ice creams (*L. rhamnosus* GG and *L. casei*) were less accepted (taste and aroma) than the conventional products. The impact was observed to be strain specific, with the *L. casei* added formulation being better rated than the *L. rhamnosus* GG added formulation.

5.7 HEALTH BENEFITS OF PROBIOTIC ICE CREAMS

Besides the concern to keep intact the probiotic matrix during its processing, storage, and consumption, several studies have evaluated the clinical results obtained in the administration of probiotic foods to the human diet. Several beneficial effects by adding probiotic food to the diet are already proven and others need more research. Among the reported effects, it is highlighted the antimicrobial action, reduction of symptoms caused by lactose intolerance, antimutagenic, and anticarcinogenic action, antidiarrheal action, immune system stimulation and intestinal microbiota maintenance (Shah and Ravula, 2000; Shah et al., 1995, 2007; Chaikham and Rattanasena, 2017).

The species *B. lactis* Bb-12 and *L. rhamnosus* GG have been shown to suppress, *in vitro*, lymphocyte responses (Mattila-Sandholm and Kauppila, 1998). Positive effects were observed in children having atopic eczema caused by food allergy, and a clinical improvement in the symptoms was evidenced in children who were fed *B. lactis* Bb-12 and *L. rhamnosus* GG, in comparison to the placebo (Mattila-Sandholm and Kauppila, 1998; Alander and Mattila-Sandholm, 2000). The anticarcinogenic and antimutagenic potential of some bacteria may be associated with the degradation of pro-carcinogens in the gut and the inhibition of tumors through the immune response (Saarela et al., 2000).

Intake of probiotics can improve the immune response in several ways. Some probiotics are able to inhibit the adhesion and displacement of pathogenic microorganisms, for example, *Escherichia coli* and *Staphylococcus aureus*, sinking the risk of infectious diseases (Kechagia et al., 2013). During antibiotic treatment, the consumption of probiotic cultures may be beneficial to keep the intestinal microbiota healthy and results in the reduction of cases of antibiotic-associated diarrhea (Plummer et al., 2005). In addition, probiotics may be used in the treatment of hypercholesterolemia, ulcerative colitis (UC), irritable bowel disease, and Crohn's disease (CD). The health effects associated with probiotic ice creams consumption are presented in Table 5.3.

TABLE 5.3 Probiotic Ice Creams and Their Health Effects

Probiotic Strain	Health Effect	References
Propionibacterium jensenii 702, *Bifidobacterium animalis* subsp. Bb-12 and *Lactobacillus acidophilus* La-5	Possible anticarcinogenic property evidenced by the adherence to human adenocarcinoma cells (Caco-2)	Ranadheera et al. (2012)
Lactobacillus acidophilus La-5 and *Bifidobacterium lactis* Bb-12	Decrease of caries-causing agents in children aged 6–12 years, leading to the improvement of oral health as well as in the prevention of associated diseases	Ashwin et al. (2015)
Bifidobacterium lactis Bb-12 and *Lactobacillus acidophilus* La-5	Reduction of *Streptococci mutans* in children aged 6 to 12 years	Mahantesha et al. (2015)
Lactobacillus acidophilus La-5 and *Bifidobacterium lactis* Bb-12 ATCC27536	Reduction of *Streptococci mutans* in children aged 12 to 14 years	Singh, Damle, and Chawla (2011)
Bifidobacterium lactis Bb-12	In young adults' consumption of probiotic ice-cream containing bifidobacteria on daily basis may decrease the salivary levels of *S. mutans*	Çaglar et al. (2008)

TABLE 5.3 *(Continued)*

Probiotic Strain	Health Effect	References
Lactobacillus casei 431	Inhibition of the demineralization of dental enamel in a highly cariogenic environment	Nadelman et al., 2017
Lactobacillus casei 01 *and Lactobacillus acidophilus LA5*	In proximal and distal colon vessels, elevated synthesis of acetate, propionate, butyrate, and lactic acid was observed while the harmful levels of ammonia decreased	Chaikham and Rattanasena (2017)
	It led to stimulate the lactobacilli and bifidobacteria counts in colon, whereas other growth of harmful bacteria like fecal coliforms and clostridia declined gradually	

Ranadheera et al. (2012) assessed the GI survival of *Propionibacterium jensenii* 702, *Bifidobacterium animalis* subsp. Bb-12, and *Lactobacillus acidophilus* La-5 in diverse matrices, such as goat's whey, plain, and fruit-flavored yogurts plus ice cream samples. In this study it was observed that when exposed to acidic surroundings: like 2.0 pH and 3.0% of bile, the matrix significantly influenced the GI tolerance of the used probiotics *in vitro*. The ice cream increased the acid resistance and bile tolerance of the three probiotic cultures in comparison to the other matrices. In the study, the *in vitro* adhesion capacity of the few screened probiotic strains was evaluated. In all cases, a significant viable count (10^5–10^6 cfu/g) was able to adhere to human adenocarcinoma cells (Caco-2). *In vitro* assays using the propidium monoazide dye (PMA) and qPCR showed that the species *L. acidophilus* La-5 and *Bifidobacterium lactis* Bb-12 have excellent GI resistance with survival rates higher than 50% (Matias et al., 2016).

In current times, the insertion of probiotic foods into the human diet has been considered an effective strategy to reduce *Streptococci mutans*, which is the main causative agent of oral caries (**Çaglar** et al., 2008). Ashwin et al. (2015) suggested that the consumption of probiotic ice cream containing *L. acidophilus* La-5 and *Bifidobacterium lactis* Bb-12 may lead to diminish caries-causing agents in children aged 6–12 years, leading to the enhancement of oral health as well as in the prevention of associated diseases. In a separate study, Mahantesha et al. (2015), demonstrated the decrease of *Streptococci mutans* in children aged 6 to 12 years after ingestion of probiotic ice creams prepared with the addition of *L. acidophilus*

La-5 and *Bifidobacterium lactis* Bb-12 species. The combination of two or more probiotics has also been used in the development of new bioproducts capable of inhibiting the *S. mutans* salivary bacteria. Ice creams containing *L. acidophilus* La-5 and *Bifidobacterium lactis* Bb-12 ATCC27536 were able to satisfactorily lessen the levels of bacteria linked with oral cavity in children 12 to 14 years old (Singh, Damle, and Chawla, 2011). It has recently been reported that probiotic matrices based on sheep's milk (fermented milk and ice cream) are able to inhibit demineralization of dental enamel in a highly cariogenic environment (Nadelman et al., 2017).

Besides the various advantageous health effects, consumption of probiotic ice cream has not been shown to cause adverse effects. *In vivo* studies report that ingestion of *Bifidobacterium* increases the levels of this microorganism in the colon (Niness, 1999; Bouhnik, 1996, 1997) and consequently, increases intestinal health.

The development of symbiotic ice creams for therapeutic purposes has also been widely explored by the scientific academy. Symbiotic foods are products that have probiotics as well as prebiotics (non-digestible food components). *In vivo* investigations on the effects of prebiotics have shown that these components can improve the survival rate of microorganisms in the gastrointestinal tract (GIT) (Pandiyan et al., 2012) thus guaranteeing the probiotic food action.

According to Mcnamara (2016), the consumption of ice cream containing *Lactobacillus kasei* KE99 and *Bifidobacterium bifidum* does not cause negative effects in patients with age from 21 to 28 years. In this study, the authors evaluated the rate of flatulence, fecal enlargement, and differences in the Bristol scale, and all parameters evaluated did not show any variations during the study, indicating no side effects.

Finally, the benefits and physiological effects observed for probiotic ice creams are still challenging and require further investigations into the mechanism of action that produce the biological responses. However, the development of a comprehensive way in the field of probiotic ice creams can be considered a futuristic technological innovation.

5.8 SUMMARY AND CONCLUSION

Probiotics containing ice cream are advantageous to the consumer, as it is a food with health benefits and contains dairy raw material, minerals, and vitamins. This matrix can serve as a probiotic carrier, and when the processing steps are evaluated carefully, probiotic survival remains at adequate levels to give away the recommended health effects. Furthermore, it does not have a negative impact on overrun, melting, physicochemical characteristics and sensory acceptance. The stability of probiotic cultures can be enhanced in ice cream selecting more resistant strains, eliminating the dissolved oxygen, adjusting the product formulation and also by means of application of recent techniques like microencapsulation. More studies must be conducted to explore the health benefits of probiotic ice creams, but the ones that are available confirm their oral, intestinal, immunological, and anticancer benefits.

KEYWORDS

- **acid tolerance**
- ***Bifidobacterium***
- **cryoprotectant**
- **emulsion**
- **fermented ice cream**
- **functional foods**
- **gut microflora**
- **health effects**
- **homogenization**
- **immune system**
- **inulin**
- **lactic acid bacteria**
- ***Lactobacillus***
- **microbiota**
- **microencapsulation**
- **non-fermented ice cream**
- **oligofructose**
- **overrun**
- **oxygen toxicity**
- **package permeability**
- **pasteurization**
- **pH**
- **prebiotic fiber**

- **probiotic inoculum**
- **probiotic viability**
- **stabilizer**
- **symbiosis**
- **wellness**

REFERENCES

Abghari, A., Sheikh-Zeinoddin, M., & Soleimanian-Zad, S., (2011). Nonfermented ice cream as a carrier for *Lactobacillus acidophilus* and *Lactobacillus rhamnosus*. *International Journal of Food Science and Technology, 46*, 84–92.

Ahmadi, A., Milani, E., Madadlou, A., Mortazavi, S. A., Mokarram, R. R., & Salarbashi, D., (2014). Synbiotic yogurt-ice cream produced via incorporation of microencapsulated *Lactobacillus acidophilus* (la-5) and fructooligosaccharide. *J. Food Sci. Technol., 51*, 1568.

Akalin, A. S., & Erisir, D., (2008). Effects of inulin and oligofructose on the rheological characteristics and probiotic culture survival in low-fat probiotic ice cream. *Journal of Food Science, 73*(4), 184–188.

Akalın, A. S., Kesenkas, H., Dinkci, N., Unal, G., Ozer, E., & Kınık, O., (2018). Enrichment of probiotic ice cream with different dietary fibers: Structural characteristics and culture viability. *Journal of Dairy Science, 101*(1), 37–46.

Akalın, A. S., Kesenkas, H., Dinkci, N., Unal, G., Ozer, E., & Kınık, O., (2017). Enrichment of probiotic ice cream with different dietary fibers: Structural characteristics and culture viability, *Journal of Dairy Science, 101*, 1–10.

Akın, M. B., Akın, M. S., & Kırmacı, Z., (2007). Effects of inulin and sugar levels on the viability of yogurt and probiotic bacteria and the physical and sensory characteristics in probiotic ice cream. *Food Chemistry, 104*, 93–99.

Alamprese, C., Foschino, R., Rossi, M., Pompei, C., & Corti, S., (2005). Effects of *Lactobacillus rhamnosus* CG addition in ice-cream. *International Journal of Dairy Technology, 58*(4), 200–206.

Alamprese, C., Foschino, R., Rossi, M., Pompei, C., & Savani, L., (2002). Survival of *Lactobacillus johnsonii La 1* and influence of its addition in retail-manufactured ice cream produced with different sugar and fat concentrations. *International Dairy Journal, 12*, 201–208.

Alander, M., & Mattila-Sandholm, T., (2000). Functional foods for EU-health in 2000. In: *Workshop Probdemo Vtt FAIR CT96–1028* (4th edn., p. 198). Rovaniemi.

Ampatzoglou, A., Schurr, B., Deepika, G., Baipong, S., & Charalampopoulos, D., (2010). Influence of fermentation on the acid tolerance and freeze-drying survival of *Lactobacillus rhamnosus GG*. *Biochemical Engineering Journal, 52*, 65–70.

Anjo, D. L. C., (2004). Alimentos funcionais em angiologia e cirurgia vascular. *Journal Vascular Brasileiro, 3*(2), 145–154.

Arslan, A. A., Gocer, E. M. C., Atamer, M. D. Z., Hinrichs, J., et al., (2016). Viability of *Lactobacillus acidophilus* ATCC 4356 incorporated into ice cream using three different methods. *Dairy Science & Technology, 96*(4), 477–487.

Ashwin, D., Vijayaprasad, K. E., Taranath, M., Ramagoni, N. K., Nara, A., & Sarpangala, M., (2015). Effect of probiotic-containing ice-cream on salivary mutans streptococci (SMS) levels in children of 6–12 years of age: A randomized controlled double-blind study with six-months follow up. *Journal of Clinical and Diagnostic Research, 9*(2), ZC06–ZC09.

Bas yg˘it, G., Kuleas an, H., & Karahan, A. G., (2006). Viability of human-derived probiotic lactobacilli in ice-cream with sucrose and aspartame. *Journal of Industrial Microbiology and Technology, 33*(9), 796–800.

Bouhnik, Y., Flourie, B., Andrieux, C., Bisetti, N., Briet, F., & Rambaud, J. C., (1996). Effects of *Bifidobacterium sp* fermented milk ingested with or without inulin colonic *Bifidobacteria* and enzymatic activities in healthy humans. *European Journal of Clinical Nutrition, 50*(4), 267–273.

Bouhnik, Y., Flourie, B., D'agay-Abensour, L., Pochart, P., Gramet, G., Durand, M., & Rambaud, J. C., (1997). Administration of transgalacto-oligofructoses increases fecal bifidobacteria and modifies colonic fermentation metabolism in healthy humans. *Journal of Nutrition, 127*(3), 444–448.

Burgain, J., Gaiani, C., Linder, M., & Scher, J., (2011). Encapsulation of probiotic living cells: From laboratory scale to industrial applications. *Journal Food Engineering, 104*, 467–483.

Çaglar, E., Kuscu, O. O., Kuvvetli, S. S., Cildir, S. K., Sandalli, N., & Twetman, S., (2008). Short-term effect of ice-cream containing *Bifidobacterium lactis* Bb-12 on the number of salivary mutans streptococci and lactobacilli. *Acta Odontologica Scandinavica, 66*(3), 154–158.

Capela, P., Hay, T. K. C., & Shah, N. P., (2006). Effect of cryoprotectants, prebiotics and microencapsulation on survival of probiotic organisms in yogurt and freeze-dried yoghurt. *Food Research International, 39*, 203–211.

Chaikham, P., & Rattanasena, P., (2017). Combined effects of low-fat ice cream supplemented with probiotics on colon microfloral communities and their metabolites during fermentation in a human gut reactor. *Food Bioscience, 17*, 35–41.

Chiquetti, R. L., Castro, E. M., Valério, G. D., Bernini, L. J., Suguimoto, H. H., Santana, E. H. W., Alegro, L. C. A., & Souza, C. H. B., (2016). Viability of the probiotic *Lactobacillus acidophilus* La-5 in ice cream: Effect of lactose hydrolysis and overrun. *International Food Research Journal, 23*(6), 2631–2637.

Clarke, C., (2004). *The Science of Ice Cream* (p. 187). Cambridge: The Royal Society of Chemistry.

Cruxen, C. E., Dos, S., Hoffmann, J. F., Zandoná, G. P., Fiorentini, Â. M., Rombaldi, C. V., & Chaves, F. C., (2017). Probiotic butiá (Butia odorata) ice cream: Development, characterization, stability of bioactive compounds, and viability of *Bifidobacterium* lactis during storage. *LWT – Food Science and Technology, 75*, 379–385.

Cruz, A. G., Antunes, A. E. C., Sousa, A. L. O. P., Faria, A. F. J., & Saad, S. M. I., (2009). Ice cream as a probiotic food carrier. *Food Research International, 42*(9), 12339.

Deosarkar, S. S., Khedkar, C. D., Kalyankar, S. D., & Sarode, A. R., (2016). Ice cream: Uses and method of manufacture. In: Caballero, B., Finglas, P. M., & Toldrá, F., (eds.), *Encyclopedia of Food and Health* (pp. 391–397).

Di Criscio, T., Fratianni, A., Mignogna, R., Cinquanta, L., Coppola, R., Sorrentino, E., & Panfili, G., (2010). Production of functional probiotic, prebiotic, and synbiotic ice creams. *Journal of Dairy Science, 93*, 4555–4564.

EL-Sayed, H. S., Salama, H. H., & EL-Sayed, S. M., (2014–15). Production of synbiotic ice cream. *International Journal Chemical Technology Research, 7*(1), 138–147.

Fatemeh, A., Bakr, S. A., & Salihin, B. A., (2016). Effects of the replacement of cow milk with vegetable milk on the count of probiotics and changes in sugar and amino acid contents in fermented ice creams. *LWT – Food Science and Technology, 70*, 261–270.

Ferraz, J. L., Cruz, A. G., Cadena, R. S., Freitas, M. Q., Pinto, U. M., Carvalho, C. C., Faria, J. A. F., & Bolini, H. M. A., (2012). Sensory acceptance and survival of probiotic bacteria in ice cream produced with different overrun levels. *Journal of Food Science, 77*(1), 24–28.

Food and Agriculture Organization of United Nations; World Health Organization – FAO/WHO., (2001). *Evaluation of Health and Nutritional Properties of Probiotics in Food Including Powder Milk with Live Lactic Acid Bacteria*. Report of a Joint FAO/WHO expert consultation, Córdoba, Argentina. URL: ftp://ftp.fao.org/es/esn/food/probioreport_en.pdf (accessed on 27 September 2021).

Gibson, G. R., & Wang, X., (1994). Bifidogenic properties of different types of fructooligosaccharides. *Food Microbiology, 11,* 491–498.

Goff, H. D., (2008). 65 years of ice-cream science. *International Dairy Journal, 18*(7), 754–758.

Guarner, F., & Sachaafsma, G. J., (1998). Probiotics. *International Journal of Food Microbiology, 39*(3), 237, 238.

Havenaar, R., Brink, B. T., & Huis In't, V. J. H. J., (1992). Selection of strains for probiotic use. In: Fuller, R., (ed.), *Probiotics* (pp. 209–224). The scientific basis.

Haynes, I. N., & Playne, M. J., (2002). Survival of probiotic cultures in low-fat ice-cream. *Australian Journal of Dairy Technology, 57*, 1.

Hekmat, S., & Mcmahon, D. J., (1992). Survival of *Lactobacillus acidophilus* and *Bifidobacterium bifidum* in ice cream for use as a probiotic food. *Journal of Dairy Science, 75,* 1415–1422.

Homayouni, A., Azizi, A., Ehsani, M. R., Yarmand, M. S., & Razavi, S. H., (2008). Effect of microencapsulation and resistant starch on the probiotic survival and sensory properties of symbiotic ice-cream. *Food Chemistry, 111*(1), 50–55.

Homayouni, A., Ehsani, M. R., Azizi, A., Razavi, S. H., & Yarmand, M. S., (2008). Growth and survival of some probiotics strains in simulated ice cream conditions. *Journal of Applied Sciences, 8*(2), 379–382.

Jay, J. M., Loessner, M. J., & Golden, D. A., (2005). *Modern Food Microbiology*. Springer, New York.

Kailapasathy, K., & Sultana, K., (2003). Survival and b-D-galactosidase activity of encapsulated and free *Lactobacillus acidophilus* and *Bifidobacterium lactis* in ice cream. *Australian Journal of Dairy Technology, 58*(3), 223–227.

Karthikeyan, N., Elango, A., Kumaresan, G., Gopalakrishnamurty, T. R., & Pandiyan, C., (2014). Augmentation of probiotic viability in ice cream using microencapsulation technique. *International Journal of Advanced Veterinary Science and Technology, 2*(1), 76–83.

Kechagia, M., Basoulis, D., Konstantopoulou, S., Dimitriadi, D., Gyftopoulou, K., Skarmoutsou, N., & Fakiri, E. M., (2013). Health benefits of probiotics: A review. *ISRN Nutrition, 5*, 1–7.

Leandro, E. S., De Araújo, E. A., Da conceição, L. L., De Moraes, C. A., & De carvalho, A. F., (2013). Survival of *Lactobacillus delbrueckii* UFV H2b20 in ice cream produced with different fat levels and after submission to stress acid and bile salts. *Journal of Functional Foods, 5*(1), 503–507.

Lee, Y. K., & Salminen, S., (1995). The coming of age of probiotics. *Trends in Food Science & Technology, 6*, 241–244.
Lorenz, J. G., (2009). Comparison of emulsification and spray drying methods in microencapsulation of *Lactobacillus acidophilus* (la-5) and application in ice cream. Dissertation in Science of Food, Federal University of Santa Catarina, Florianopoli, p. 79.
Low, R. H. P., Baba, A. S., & Aboulfazli, F., (2015). Effects of different levels of refined cane sugar and unrefined coconut palm sugar on the survivability of *Lactobacillus acidophilus* in probiotic ice cream and its sensory and antioxidant properties. *Food Science and Technology Research, 21*(6), 857–862.
Mahantesha, T., Reddy, K. M. P., Kumar, N. H. P., Nara, A., Ashwin, D., & Buddiga, V., (2015). Comparative study of probiotic ice cream and probiotic drink on salivary *Streptococcus mutans* levels in 6–12 years age group children. *Journal of International Oral Health : JIOH, 7*(9), 47–50.
Matias, N. S., Padilha, M., Bedani, R., & Saad, S. M. I., (2016). *In vitro* gastrointestinal resistance of *Lactobacillus acidophilus* La-5 and *Bifidobacterium animalis* Bb-12 in soy and/or milk-based synbiotic apple ice creams. *International Journal of Food Microbiology, 234*, 83–93.
Mattila-sandholm, T., & Kauppila, T., (1998). Functional food research in Europe. In: *Workshop Probdemo VTT FAIR CT96–1028* (3rd edn., p. 125). 8., 1998, Haikko Symposium 187. Haikko, Finland.
Mcnamara, E., (2016). *Synbiotic ice Cream as a Probiotic Carrier Tested in a Human Blind Crossover Trial* (p. 78). [thesis]. University of Nebraska – Lincoln.
Medici, M., Vinderola, C. G., & Andperdigón, G., (2004). Gut mucosal immune modulation by probiotic fresh cheese. *International Dairy Journal, 14*, 611–618.
Mintel, (2017). *Ice Cream – Global Annual Review – 2017*. URL: http://store.mintel.com/ice-cream-global-annual-review-2017 (accessed on 27 September 2021).
Mohammadi, R., Amir, M. M., Roya, K., & Gomes, A. C., (2011). Probiotic ice cream: Viability of probiotic bacteria and sensory properties. *Annals of Microbiology, 61*, 411–424.
Nadelman, P., Frazão, J. V., Vieira, T. I., Balthazar, C. F., Andrade, M. M., Alexandria, A. K., & Maia, L. C., (2017). The performance of probiotic fermented sheep milk and ice cream sheep milk in inhibiting enamel mineral loss. *Food Research International, 97*, 184–190.
Niness, K. R., (1999). Inulin and oligofructose: What are they? *Journal of Nutrition, 129*(7), 1402S–1406S.
Nousia, F. G., Androulakis, P. I., & Fletouris, D. J., (2011). Survival of *Lactobacillus acidophilus* LMGP-21381 in probiotic ice cream and its influence on sensory acceptability. *International Journal of Dairy Technology, 64*, 130–136.
Pandiyan, C., Annal, V. R., Kumaresan, G., Murugan, B., & Gopalakrishnamurthy, T. R., (2012). Development of synbiotic ice cream incorporating *Lactobacillus acidophilus* and *Saccharomyces boulardii*. *International Food Research Journal, 19*(3), 1233–1239.
Pimentel, T. C., & Garcia, S., & Prudencio, S. H., (2017). Produtos lácteos funcionais. In: Luiz, A. N., Adriano, G. D. C., & Luciano, S. B., (eds.), (Org.). *Production, Processing and Inspection of Milk and Derivatives* (1st edn., Vol. 1, pp. 205–226). Atheneu.
Plummer, S., Garaiova, I., Sarvotham, T., Cottrell, S., Le Scouiller, S., Weaver, M., Tang, J., et al., (2005). Effects of probiotics on the composition of the intestinal microbiota following antibiotic therapy. *International Journal of Antimicrobial Agents, 26*(1), 69–74.
Prasertsiriphan, S., & Kusump, S., (2015). Impact of selected cultures of probiotics on quality of vanilla low fat ice cream during storage. *Journal of Science and Technology Ubon Ratchathani University, 7*(3), 19–27.

Purnomo, H., (2007). Sugar components of coconut sugar in Indonesia. *ASEAN Food Journal, 7*, 200, 201.
Ranadheera, C. S., Evans, C. A., Adams, M. C., & Baines, S. K., (2012). *In vitro* analysis of gastrointestinal tolerance and intestinal cell adhesion of probiotics in goat's milk ice cream and yogurt. *Food Research International, 49*(2), 619–625.
Ranadheera, C. S., Evans, C. A., Adams, M. C., & Baines, S., (2013). Production of probiotic ice cream from goat ' s milk and effect of packaging materials on product quality. *Small Ruminant Research, 112*, 174–180.
Reid, G., (2015). The growth potential for dairy probiotics. *International Dairy Journal, 49*, 16–22.
Roberfroid, M. B., Van, L. J. A. E., & Gibson, G. R., (1998). The bifidogenic nature of chicory inulin and its hydrolysis products. *Journal Nutrition, 128*(1), 11–19.
Roberfroid, M., (2002). Functional food concept and its application to prebiotics. *Digestive and Liver Disease, 34*(2), 105–110.
Rolled Ice, (2016). *Ice Cream Market Report*. URL: http://icecreamrollfranchise.com/ice-cream-market-report/ (accessed on 27 September 2021).
Saarela, M., Mogensen, G., Fonde, R., Maltto, J., & Mattila-sandholm, T., (2000). Probiotic bacteria: safety, functional and technological properties. *Journal of Biotechnology, 84*, 197–215.
Sagdic, O., Ozturk, I., Cankurt, H., & Tornuk, F., (2012). Interaction between some phenolic compounds and probiotic bacterium in functional ice cream production. *Food Bioprocess Technology, 5*, 2964–2971.
Schmidt, K. A., Lundy, A., Reynolds, J., & Yee, L. N., (1993). Carbohydrate or protein based fat mimicker effects on ice milk proteins. *Journal Food Science, 58*, 761–793.
Sebrae, (2017). How to Stand Out in the Ice Cream Market. URL: http://www.sebrae.com.br/sites/PortalSebrae/artigos/como-se-destacar-no-mercado-de-sorvetes,a49d99a5a995b510VgnVCM1000004c00210aRCRD (accessed on 27 September 2021).
Senanayake, S. A., Fernando, S., Bamunuarachchi, A., & Arsekularatne, M., (2013). Application of *Lactobacillus acidophilus* (LA 5) strain in fruit-based ice cream. *Food Science & Nutrition, 1*(6), 428–431.
Shah, N. P., & Ravula, R. R., (2000). Microencapsulation of probiotic bacteria and their survival in frozen fermented dairy desserts. *Australian Journal of Dairy Technology, 55*, 139–144.
Shah, N. P., et al., (1995). Survival of *Lactobacillus acidophilus* and *Bifidobacterium bifidum* in commercial yogurt during refrigerated storage. *International Dairy Journal, 5*, 515–521.
Shah, N. P., et al., (2007). Functional cultures and health benefits. *International Dairy Journal, 17*(11), 1262–1277.
Silva, P. D. L., Bezerra, M. F., Santos, K. M. O., & Correia, R. T. P., (2015). Potentially probiotic ice cream from goat's milk: Characterization and cell viability during processing, storage and simulated gastrointestinal conditions. *LWT – Food Science and Technology, 62*, 452–457.
Singh, R., Damle, S. G., & Chawla, A., (2011). Salivary mutans streptococci and lactobacilli modulations in young children on consumption of probiotic ice-cream containing *Bifidobacterium lactis* Bb12 and *Lactobacillus acidophilus* La5. *Acta Odontologica Scandinavica, 69*(6), 389–394.

Son, J. H., Kim, K. H., Nam, Y. K., Park, J. K., & Kim, S. K., (2004). *Optimization of Cryoprotectants for Cryopreservation of Rat Hepatocyte* (Vol. 26, pp. 100–715). Department of Chemical Engineering, Dongguk University, Seoul, Korea.

Songtummin, S., & Leenanon, B., (2016). Survival of *Lactobacillus acidophilus* TISTR1338 and *Lactobacillus casei* TISTR390 in probiotic Gac ice cream. *International Food Research Journal, 23*(2), 790–796.

Soukoulis, C., & Tzia, C., (2008). Impact of the acidification process, hydrocolloids and protein fortifiers on the physical and sensory properties of frozen yogurt. *International Journal Dairy Technol*ogy, *61*, 170–177.

Soukoulis, C., Fisk, I. D., & Bohn, T., (2014). Ice cream as a vehicle for incorporating health-promoting ingredients: Conceptualization and overview of quality and storage stability. *Comprehensive Reviews in Food Science and Food Safety, 13*, 627–655.

Soukoulis, C., Lebesi, D., & Tzia, C., (2009). Enrichment of ice cream with dietary fibre: Effects on rheological properties, ice crystallization and glass transition phenomena. *Food Chemistry, 115*, 665–671.

Souza, J. C. B., Costa, M. R., De Rensis, C. M. V. B., & Sivieri, K., (2010). Ice cream: Composition, processing and viability of probiotic addition. *Food and Nutrition, 21*(1), 155–165.

Souza, J. C. B., Gergoletto, K. B., Garcia, S., & Sivieri, K., (2011). Feasibility of adding Lactobacillus casei (lc-1) protected with trehalose and acacia gum in ice cream. *Food and Nutrition, Araraquara, 22*(2), 231–237.

Statista-The Statistic Portal. (2017). *Size of the Global Ice Cream Market from 2013 to 2022 (in billion U.S. dollars)*. URL: https://www.statista.com/statistics/326315/global-ice-cream-market-size/ (accessed on 27 September 2021).

Streit, F., Delettre, J., Corrieu, G., & Beal, C., (2008). Acid adaptation of *Lactobacillus delbrueckii subsp. bulgaricus* induces physiological responses at membrane and cytosolic levels that improves cryotolerance. *Journal of Applied Microbiology, 105*, 1071–1080.

Tannock, G. W., (1998). Studies of the intestinal microflora: A prerequisite for the development of probiotics. *International Dairy Journal, 8*, 527–533.

Thamer, K. G., & Penna, A. L. B., (2006). Characterization of functional fermented dairy beverages by probiotics plus prebiotic. *Food Science Technology, 26*(3), 589–595.

Valsiljevic, T., & Shah, N. P., (2008). Probiotics-from Metchnikoff to bioactives. *International Dairy Journal, 18*(7), 714–728.

Wang, X., & Gibson, G. R., (1993). Effects of the *in vitro* fermentation of oligofructose ad inulin by bacteria growing in the human large intestine. *Journal of Applied Bacteriology, 75*, 373–380.

World Atlas, (2017). *Which Country Eats the Most Ice Cream*. URL: https://www.worldatlas.com/articles/the-top-ice-cream-consuming-countries-of-the-world.html (accessed on 27 September 2021).

Part III

Industrial Aspects of Microbial Food Technologies

CHAPTER 6

Lactulose: A High Food Value-Added Compound and Its Industrial Application in Food

MAJID NOOSHKAM* and ZAHRA ZAREIE

*Corresponding author. E-mail: Nooshkamma@gmail.com

ABSTRACT

Dairy industry is producing a large amount of liquid wastes, mainly whey and whey permeates. These by-products contain lactose, proteins, fats, and minerals, with high BOD and COD values, due to the lactose molecules. Therefore, lactose is undergoing some processes to produce lactose-derived compounds, such as lactosucrose, epilactose, tagatose, and lactulose. Lactulose is recognized as a prebiotic disaccharide with versatile applications in the food and medical sciences. It is mainly synthesized from lactose by several chemical isomerization methods. However, due to the use of noxious compounds and high production costs of this route, the food industry has been motivated to introduce new green and low-cost synthesis methods for lactose to lactulose conversion. In this way, enzymatic synthesis is currently gaining a lot of attention to produce lactulose from pure and also crude lactose sources, such as whey and whey permeates. Lactulose is used in medical science to treat constipation, hepatic encephalopathy, and other implications. In food industry, it is applied as a prebiotic and a functional ingredient. This book chapter therefore focuses on the different lactose to lactulose conversion methods. The purification techniques, the potential food/pharmaceutical applications, and bioactive-derived compounds of lactulose are also discussed in the chapter.

6.1 INTRODUCTION

Dairy industry generates dilute aqueous solutions containing lactose, proteins, fats, minerals, and non-proteinaceous compounds. Whey, a common coproduct and waste stream of cheese production and casein manufacture, is a highly biodegradable (~99%) and rich source of lactose (0.18–60 kg m^3), protein (1.4–33.5 kg m^3), and fat (0.08–10.58 kg m^3) (Bacenetti et al., 2018). It has been reported that both sweet and acid whey contain mainly water (94%) and lacrosse (4.5%), and the acid content of the latter is higher than the former (pH value of 4.3–5.1 *vs* 5.9–6.6). In addition, about 70% of the sweet whey is undergoing several processes to produce pure lactose and whey protein; whilst, acid whey has limited applications and discarded mainly as a waste, leading to water pollution and environmental damages (Lindsay et al., 2018). This by-product has significantly high biological and chemical oxygen demands (BOD and COD) of 40–60 and 50–80 g L^{-1}, respectively, and about 90% of the BOD and COD values is due to the presence of disaccharide lactose. About 50% of the produced whey is treated to separate lactose and protein for feed and food applications. However, the rest is discarded and causes serious economic and environmental problems. This motivates the dairy industry to find effective solutions to recover lactose and proteins in conjugation with producing several high-added value products from whey, such as single-cell protein, butanol, acetone, ethanol, organic acids, biopolymers, lactulose, lactobionic acid, lactitol, hydrogen, and biodegradable plastics (Díez-Antolínez et al., 2018).

The aim of this book chapter is to focus on the conversion of whey lactose to several high-added value compounds, especially lactulose. As well, the production, purification, and food/pharmaceutical applications of lactulose are discussed. Lactulose-based bioactive compounds are also highlighted in this chapter.

6.2 LACTOSE DERIVATIVES

Food- and pharmaceutically grade lactose is frequently applied in many products, such as confectionaries, bread, infant formula, animal feed, as a good fermentation source in culture media, and as a medicinal carrier molecule. Furthermore, it is used as a crude compound to produce several lactose-derivatives including rare sugars, oligosaccharides, ethanol, and sugar alcohols in laboratory and industrial scales (Figure 6.1). It is worthwhile to note that some of the lactose-based derivatives have unique characteristics and are synthesized industrially and vastly applied in the food and pharmaceutical sciences (Seki and Saito, 2012), which are summarized hereunder.

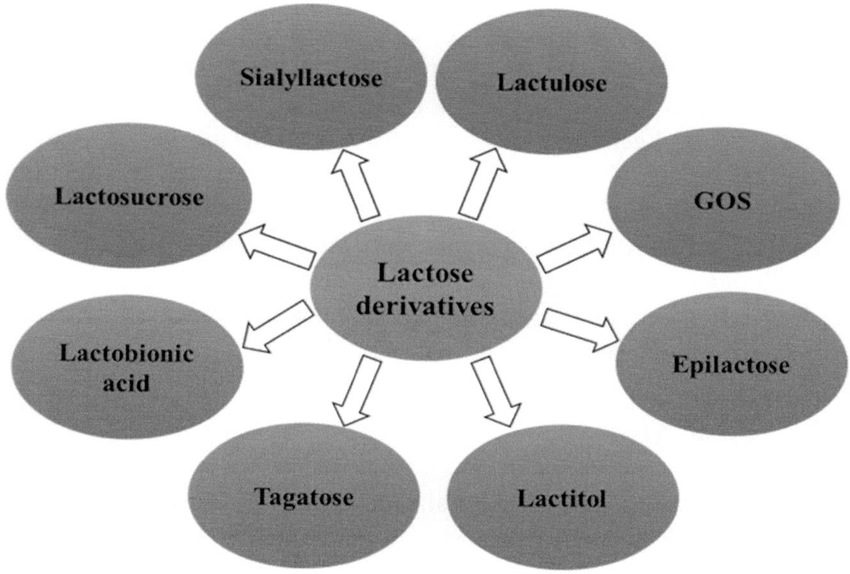

FIGURE 6.1 Lactose-based high added-value derivatives.

6.2.1 *LACTOSUCROSE*

Trisaccharide lactosucrose (O-β-D-galactopyranosyl-(1-4)-O-α-D-glucopyranosyl-(1-2)-β-D-fructofuranoside) is synthesized from disaccharides lactose and sucrose through an enzymatic transglycosylation reaction (Duarte et al., 2017). Lactosucrose undergoes a little hydrolysis in digestional tract (prebiotic sugar) and increases the number of *Bifidobacterium* species remarkably in fecal microflora (bifidogenic agent). In addition, the balance of intestinal microenvironment can be maintained by lactosucrose and this bioactive prebiotic ingredient renders other health-promoting effects, such as anticancer activity, improving calcium absorption, and anti-allergic activity (Li et al., 2015).

6.2.2 *LACTOBIONIC ACID*

Lactobionic acid (4-O-β-D-galactopyranosyl-D-gluconic acid) with molecular weight of 358.30 Da, chemical formula of $C_{12}H_{22}O_{12}$, and pKa of 3.8, is considered as an aldonic acid product from lactose oxidation. It is a high added-value lactose derivative with versatile functions in food and medical sectors because of its chelating, antioxidant, and humectant features. In lactobionic acid, a gluconic acid molecule is attached to a galactose moiety through an ether-like linkage. It is highly hygroscopic molecules with the ability to form gels with 14% water content, mainly due to its high content of hydroxyl groups. Lactobionic acid and its mineral salts like sodium, calcium, and potassium lactobionate are commercially manufactured in small scales for medical, industrial, and research applications (Gutiérrez et al., 2012). Recently, lactobionic acid has gained a lot of attention as a biologically active molecule and a superb platform to synthesize biodegradable and biocompatible drug delivery systems and biomaterials. As well, it has an important role in treating hepatic disorders (Alonso et al., 2013).

6.2.3 EPILACTOSE

Disaccharide epilactose (4-O-β-galactopyranosy-D-mannose; $C_{12}H_{22}O_{11}$; 342 g mol^{-1}) comprises of galactose and mannose, and is an epimerization product of lactose (Seki and Saito, 2012). Heat-treated bovine milk contains extremely low content of epilactose and its chemical synthesis is difficult. It is a non-digestible and prebiotic carbohydrate with potential to improve intestinal mineral absorption, post gastrectomy osteopenia, and anemia *via* increasing ferrum and calcium absorption rates (Mu et al., 2013).

6.2.4 TAGATOSE

Tagatose is a D-fructose stereoisomer and D-galactose isomer with generally recognized as safe (GRAS) status for food and beverage usages (Espinosa and Fogelfeld, 2010). It is found naturally in some dairy products and fruits and industrially synthesized from lactose by a combination of enzymatic and chemical reactions (Fujimaru et al., 2012). Tagatose has ~90% of sucrose sweetness, low calorie value (<1.5 kcal g^{-1}, corresponding to 0.33 of sucrose calorie), and minimal glycemic index (Livesey and Brown, 1996; Fujimaru et al., 2012). Therefore, tagatose is classified as a prebiotic sugar and has been successfully applied for curing type 2 diabetes (Lu et al., 2008). It is worth to note that tagatose shows similar physical characteristics, such as browning, appearance, and bulking features to sucrose molecule (Fujimaru et al., 2012).

6.2.5 LACTITOL

Lactitol ((+)-4-O-β-D-galactopyranosyl-D-sorbitol/glucitol; $C_{12}H_{24}O_{11}$; 344 g mol^{-1}) is polyalcohol sugar and lactulose analog. Its glucose aldehyde group is reduced to OH group (lactose hydrogenation) and therefore it does not contribute to the Maillard reaction and has a remarkable high heat stability (Seki and Saito, 2012). Lactitol presents low calorie value and has similar taste to sucrose sugar, and therefore it is used as a sweetener, food additive, and osmotic laxative to increase peristaltic movement and treat constipation and hepatic encephalopathy (Faruqui and Joshi, 2012; Seki and Saito, 2012; Miller et al., 2014).

6.2.6 GALACTOOLIGOSACCHARIDES (GOS)

Galactooligosaccharides (GOS; Gal-$(Gal)_n$-Glu) are produced from lactose through a reverse glycosyl transfer reaction catalyzed by β-galactosidase with similar structure to human milk oligosaccharides. Bioactive GOS are frequently used as prebiotic agents in infant formula (Olano and Corzo, 2009).

6.2.7 SIALYLLACTOSE

Sialic acid is an *N*- and *O*-substituted derivative, which is found in human and bovine milk as sialyllactose (N-acetyl neuraminic acid; $C_{23}H_{39}NO_{19}$; 633 g mol^{-1}) and is combined with lactose (Ten Bruggencate et al., 2014). Although, sialyllactose is highly expensive compound, it is currently isolated from cheese whey by ultrafiltration process and applied in infant formula to increase the ability of defense system against infection (Shimatani et al., 1993).

6.2.8 LACTULOSE

6.2.8.1 CHEMISTRY

Lactulose (4-O-β-D-galactopyranosyl-D-fructose) is a synthetic and ketose disaccharide, consisting of one molecule galactose and one molecule fructose attached together with β-1,4-glycosidic linkage (Figure 6.2) (Bisinella et al., 2017). It does not exist naturally and is synthesized firstly in 1930 in a lactose solution under alkaline conditions through a well-known reaction, Lobry de Bruyn-Alberda van Ekenstein (LA) rearrangement (Wang et al., 2013).

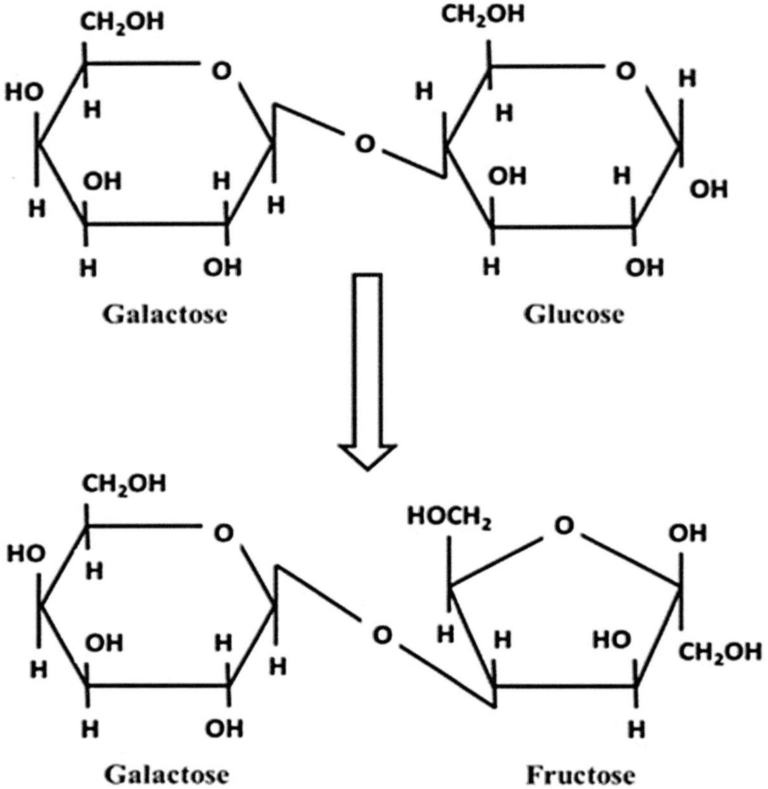

FIGURE 6.2 Lactose to lactulose isomerization.

It has a similar empirical formula ($C_{12}H_{22}O_6$) and molecular weight (342.3 g mol^{-1}) to lactose. Lactulose is ether-insoluble, partly soluble in methanol, and water-soluble. It was also demonstrated that lactulose has higher alcohol solubility (25.4 *vs.* 1.5 g L^{-1} in methanol at 30°C), water solubility (2060 *vs.* 180 g L^{-1} at 20°C), and sweetness in comparison to lactose (0.6–0.8 *vs.* 0.17–0.2), and its superb heat-stability under acidic conditions leads the sugar to be used as a prebiotic food additive in acidic foods, particularly fruit juices (Nooshkam et al., 2018). Lactulose shows versatile properties such as, improving color and flavor of the final product, stabilizing effect, texturizing behavior, antiplaque ability, good water solubility, low energy value of 2.0 kcal g^{-1}, viscosity close to saccharose solution, and high sweetness (60–80% compared to sucrose), which make the lactulose a potent sucrose alternative in food products (Schumann, 2002). It is found in two anhydrate and trihydrate white crystalline forms, and the former is a commercially available form of lactulose (Seki and Saito, 2012). Furthermore, the lactulose synthesis methods resulted in different lactulose isomeric forms in the following order: β-furanose> β-pyranose>α-furanose. However, α-pyranose isomeric form has not been detected in lactulose solutions after isomerization reaction (Aït-Aissa and Aïder, 2014).

It has been reported that lactulose syrup is a clear, odorless, and yellowish liquid with 80% solid content containing 66% lactulose and other sugars like galactose, fructose, epilactose, and tagatose (Nooshkam and Madadlou, 2016b). It is noteworthy that lactulose has melting point of 168.5–170°C, heat of solution of –11.68 j g^{-1}, specific rotation of –51.5°, and glass transition temperature of –47°C (Nooshkam et al., 2018).

6.2.8.2 LACTULOSE PRODUCTION

Lactulose is an isomerization product from lactose and is synthesized through chemical, enzymatic, and electro-activation methods, which are discussed in subsections.

6.2.8.2.1 Chemical Methods

Chemical lactulose production is mainly based on the LA rearrangement *via* the isomerization of glucose moiety of lactose to fructose molecule (Figure 6.3). Firstly, the deprotonation of glucose molecule at carbon number 2 (C_2) leads to the formation of $C_1=C_2$ followed by degradation of $C_1=O$ bond to leave the O atom in a reactive state. Secondly, the adjacent H atom (-OH from C_2) is attracted by the O atom, due to its high negativity, to stabilize C_1 and leave the O atom of C_2 in a reactive state. The formed intermediate compound is 1,2-enediol, and finally $C_1=C_2$ rearranging to C_1–C_2 results in $C_2=O$ double bond, i.e., fructose formation (Sitanggang et al., 2016). Several catalysts have been used for lactose to lactulose isomerization, which should have following properties (Aider and de Halleux, 2007):

- Providing the highest lactulose yield and the lowest reaction secondary products;
- No significant toxic effect on environment and human;
- Having low-cost and be easily accessible;
- It can be removed easily from the isomerized reaction by common demineralization methods;
- Giving repetitive isomerization results.

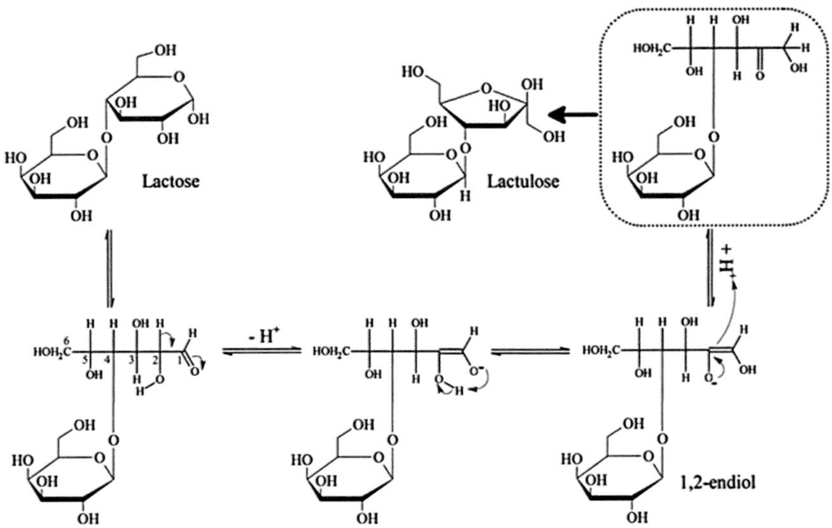

FIGURE 6.3 Lobry de Bruyn-Alberda van Ekenstein (LA) arrangement of lactose to lactulose. *Source*: Reprinted with permission from Sitanggang et al., 2016. © Springer Nature.

1. **Soluble/Homogeneous Catalysts:** Lactulose was firstly produced in a lactose solution at 30°C for 36 h in the presence of calcium hydroxide as alkaline catalyst, and about 30% of lactose was isomerized to lactulose (Montgomery and Hudson, 1930). Different chemical catalysts such as NaOH, $Ca(OH)_2$, KOH, magnesium oxide, tertiary amines, and sodium and potassium carbonates were then used for lactulose synthesis. These catalysts, based on the catalyst type, resulted in an isomerization yield ranged between ~11% (Hashemi and Ashtiani, 2010; Nooshkam and Madadlou, 2016b) and ~30% (Zokaee et al., 2002) at 70–100°C. The highest isomerization efficiency (around 20–33%) is obtained within a few hours and then lactulose content decreases after prolong heating period (Schuster-Wolff-Bühring et al., 2010).

   It was claimed that lactose to lactulose isomerization reaction is comprised of the three stages. In the first stage, the highest lactulose formation and minimum lactulose degradation to by-products are observed. The second stage is recognized by the growth of secondary degradation products and pH decrement in conjugation with lactulose formation. The third stage corresponds to the stabilization of solution color development at sufficiently high temperature of isomerization reaction (72±2°C) and pH decrement reaches to neutral condition (Aider and de Halleux, 2007). In this way, the maximum lactulose yield can be maintained by stopping both lactose isomerization reaction and lactulose degradation *via* cooling and reducing the pH of isomerized solution (Schuster-Wolff-Bühring et al., 2010).

It can be deduced that due to the complicity, high catalyst concentration, solubility of catalysts in isomerized solution, and poor specificity of the chemical lactulose synthesis, unwanted by-products such as tagatose, isosaccharic acid, galactose, epilactose, and colored components, are generated in large amounts. In this context, some cumbersome and expensive purification procedures are required to discard catalysts and other side-products to purify lactulose solution (Panesar and Kumari, 2011).

2. **Complex Catalysts:** Lactulose degradation into secondary products during isomerization reaction has motivated the food and pharmaceutical sections to use alternative alkaline catalysts to obtain the highest lactulose content besides minimum side-products. Aluminates and borates are complex reagents with potential to form stable and insoluble complexes with lactulose under basic pHs. The reaction equilibrium is therefore shifted to synthesize more lactulose (70–80%) along with minimizing lactulose degradation. However, this significantly high efficiency is achieved in the presence of large amounts of borate and sodium aluminate in reacting medium. Furthermore, the elimination of the catalysts from the mixture is difficult and requires tedious and sophisticated downstream processes to be reused for further isomerization cycles (Hicks and Parrish, 1980; Zokaee et al., 2002; Sitanggang et al., 2016).
3. **Natural Catalysts:** Calcium carbonate-based heterogeneous catalysts along with alkaline-treated sepiolites and limestone were introduced as easily accessible catalysts to lactose to lactulose isomerization, with potential to remarkably reduce energy consumption and waste generation (Nooshkam and Madadlou, 2016a; Paseephol et al., 2008; Montilla et al., 2005; Villamiel et al., 2002). These natural catalysts are commonly used in powdered forms, and therefore they can be easily separated from the isomerized solution by centrifugation. Additionally, a low amount of the catalysts is needed to perform isomerization reaction and, in turn, lower lactulose degradation by-products can be formed during the process. Lactulose yield of approximately 17–25% is obtained using natural catalysts in milk ultrafiltration and concentration permeates, which is nearly close to the lactulose efficiency acquired by homogenous catalysts (Nooshkam and Madadlou, 2016a; Nooshkam et al., 2018).

6.2.8.2.2 Electro-Activation Method

In recent years, the electro-activation lactulose production based on the electrocatalytic procedure has been introduced as a very low energy-intensive and reagentless isomerization procedure, which the high alkalinity required for the process is self-generated in the reaction medium. However, the lactose to lactulose conversion efficiency is lower compared to complex reagents (Sitanggang et al., 2016). The electrolysis reactor is consisted of two electrodes inserted in an electrolytic solution, having the ability to conduct ions upon applying an external electric field (Figure 6.4) (Aider and Gimenez-Vidal, 2012). As it is seen, the reactor has also three anodic, central, and cathodic compartments, containing lactose (in cathodic part) and electrolyte (in central and anodic parts) solutions. In addition, the central compartment is coupled to other parts by anion and cation exchange membranes and the isomerization process is completed in the cathodic compartment (Aït-Aissa and Aïder, 2014). It is worthwhile to note that water electrolysis performs at the cathodic interface to generate alkaline medium as proton acceptors for lactose to lactulose conversion (Aider and Gimenez-Vidal, 2012).

6.2.8.2.3 Enzymatic Method

Enzymatic-based lactulose production is receiving more attention regarding to the synthesis feasibility and environmental considerations in comparison to chemical lactose to lactulose isomerization (Schuster-Wolff-Bühring et al., 2010). Additionally, the pure substrates are not required for lactulose synthesis through enzymatic route, and whey/whey permeates can be applied as lactose sources. A well, the mild reaction conditions and good selectivity result in a manifestly lower downstream operation costs (Guerrero and Wilson, 2016). β-galactosidase and glycosidase are commonly employed for lactulose production, and β-galactosidase is a well-documented catalyst for *trans*-galactosylation reaction (Panesar and Kumari, 2011). In enzymatic-mediated lactulose synthesis and through a rapid *trans*-galactosylation mechanism, β-galactosidase leads to lactose hydrolysis to glucose and galactose in order to form a galactosyl-β-galactosidase intermediate complex. In the final stage, the galactosyl moiety is transferred to one fructose molecule, as an acceptor, and yields bioactive lactulose (Figure 6.5) (Wang et al., 2013).

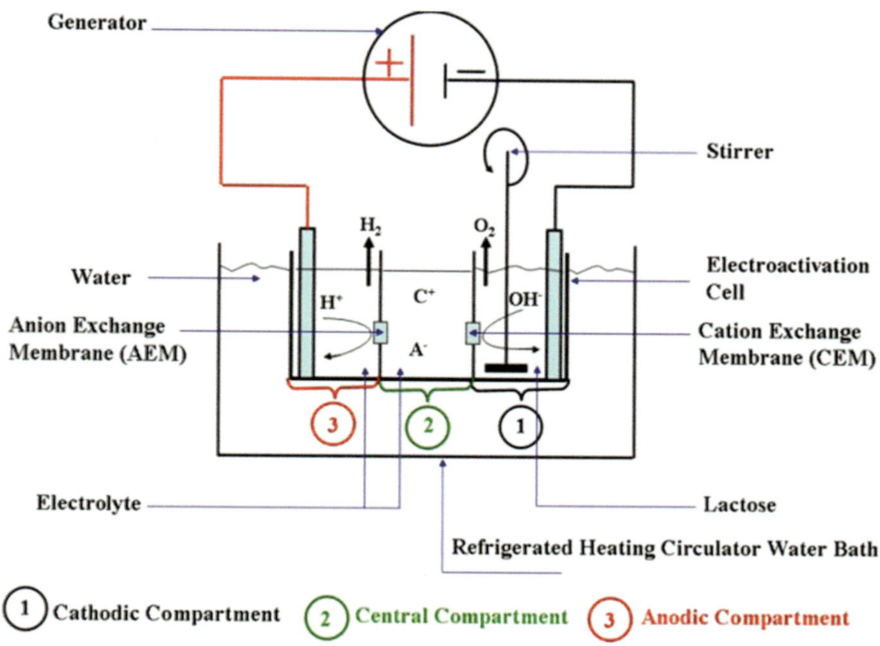

FIGURE 6.4 Schematic illustration of electro-activation reactor for lactose to lactulose isomerization. A^- and C^+ are anion and cation, respectively.

Source: Reprinted with permission from Aissa et al., 2013. © Elsevier.

1. **Free Enzymes:** Lactulose synthesis through enzymatic *trans*-galactosylation by *Aspergillus oryzae*' β-galactosidase and *Pryococcus furiosus*' β-glycosidase leads to lactulose productivity of 30% and 44% relative to lactose, respectively (Mayer et al., 2004). Lee et al. (2004) tested the ability of β-galactosidase from *Escherichia coli*, *Kluyveromyces fragilis*, *Kluyveromyces lactis*, and *A. oryzae* to synthesize lactulose and the results demonstrated that *K. lactis* originated β-galactosidase produced the highest lactulose yield. It is noteworthy that lactose to lactulose bioconversion yield is always lower than that of chemical isomerization methods and high levels of fructose is required to trigger the lactulose synthesis, leading to the high content of unreacted fructose in the reaction mixture (Gosling et al., 2010; Guerrero et al., 2015). In this way, glucose isomerase was coupled with glucose to fructose isomerization to lower fructose concentration in the isomerized solution. However, this synthesis route provided low lactulose yield and was not commercially attractive (Song et al., 2013a). Thus,

several attempts have been made to increase lactulose efficiency. For instance, the use of borate in combination of thermostable cellobiose 2-epimerase from *Caldicellulosiruptor saccharolyticus* resulted in a lactulose yield of up to 51%, which was close to the isomerization yields of chemical complex catalysts. The sophisticated purification processes for the elimination of borate and enzyme inactivation, however, increased the processing costs (Kim et al., 2013).

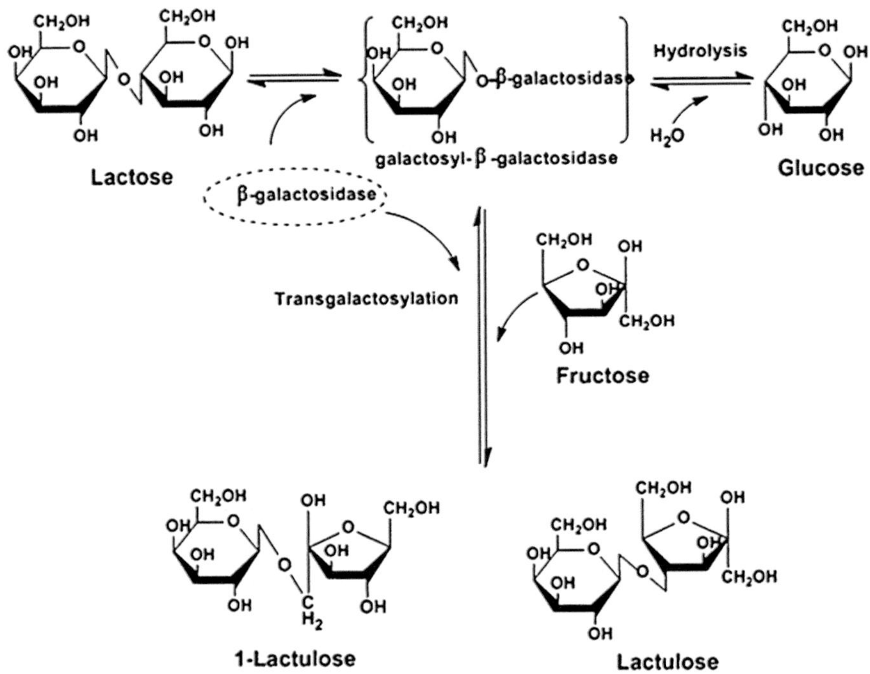

FIGURE 6.5 β-galactosidase-catalyzed *trans*-galactosylation reaction in the presence of fructose as an acceptor to produce lactulose and 1-lactulose.

Source: Reprinted with permission from Wang et al., 2013. © Springer Nature.

2. **Immobilized Enzymes:** Immobilization of β-galactosidase provides a continuous condition to isomerize lactose to lactulose. By circulating galactose and fructose solutions through columns with immobilized β-galactosidase, Ajisaka et al. (1988) reported a lactulose yield of 11.3% after 24 h isomerization reaction. In another study, glucose isomerase and β-galactosidase were

immobilized and the reaction was performed in two organic and aqueous media to obtain the highest isomerization yield. This dual-enzymatic mode resulted in a lactulose production efficiency of 19%, due likely to the decreased water activity of isomerization medium by the solvent, i.e., cyclohexane-buffer (Hua et al., 2010). It was reported that the presence of fructose, galactose, and glucose can inhibit the lactose hydrolysis and *trans*-galactosylation reactions during lactulose synthesis, and galactose and glucose showed greater inhibitory effects on the latter reaction than the former (Song et al., 2013b). In this way, the use of continuous reactor containing β-galactosidase significantly reduced the inhibitory effects and provided a high lactose to lactulose isomerization yield of 19.1 g L^{-1} (Song et al., 2013a).

3. **Whole-Cell Biocatalysts:** The production cost of β-galactosidase is the main barrier to commercialize enzymatic-based lactulose synthesis and therefore the food industry is focused on developing cost-effective procedures for this purpose. Intracellular β-galactosidase, a whole-cell biocatalyst, is used to reduce β-galactosidase production cost due to the lack of complex purification steps. The whole cells have a potential to be applied as biocatalysts for lactulose production. However, the cell envelope has a permeability barrier for products and substrates, causing remarkably low reaction rates in whole cells, particularly those obtained from yeasts. In this context, the permeabilized yeast cells can be used to prepare whole-cell biocatalysts and lower the permeability barrier (Panesar and Kumari, 2011). It was demonstrated that the permeabilization of yeast cells by ethanol could be a convenient, easy, economical, and safe process for the enzymatic-based biotransformation in conjugation with product formation (Panesar et al., 2007a, b; Panesar, 2008; Lee et al., 2004). The permeabilized cells are able to increase the lactulose yield manifestly (1.3–2.1 fold) in comparison to the untreated cells (Lee et al., 2004).

6.2.8.3 LACTULOSE PURIFICATION

The isomerized solution is not pure and usually contains lactose, galactose, glucose, epilactose, and other lactose/lactulose degradation products like tagatose, formic acid, and isosaccharic acid. There are several simple ways to prevent lactose and lactulose degradations during isomerization, such as decreasing pH and temperature along with cooling the reaction mixture. Afterwards, the isomerized solution undergoes some purification processes to separate lactulose from the isomerized solution. It is worth to mention that the lactulose purity is influenced by both chemical and physical treatments and therefore lactulose recovery during downstream processes is an important reaction (Panesar and Kumari, 2011). It was observed that the use of catalyst boric acid plus triethylamine precipitates the alkaline hydroxide ions and prevents the formation of degradation secondary products. Low-pressure distillation or resins can remove the triethylamine, while the addition of absolute methanol and evaporation under low pressure are used to remove boric acid as methyl borate (Hicks and Parrish, 1980). In addition, the acidification of the isomerized solution can lead to the precipitation of borate as boric acid and sodium aluminate as gel-like aluminum hydroxide (Carobbi and Innocenti, 1990; Kozempel et al., 1995). The heterogeneous catalysts such as eggshell powder are easily separated by centrifugation and the remained solution is decolorized *via* activated carbon without affecting lactulose content (Montilla et al., 2005). Due to the higher alcohol solubility of lactulose than lactose, different alcoholic solutions have been successfully employed to crystallize lactose and purify the isomerized solution (Nooshkam and Madadlou, 2016a, b). Pressurized liquid extraction has also been used to purify lactulose, leading to the lactulose recovery and purity of 84.4% and >90%, respectively (Ruiz-Matute et al., 2007). Furthermore, cation-exchange resins could be employed to separate lactose, lactulose, and galactose from the reaction mixture (Dendene et al., 1995).

6.2.8.4 LACTULOSE APPLICATIONS

6.2.8.4.1 Pharmaceutical Science

Lactulose is an effective drug in the pharmaceutical industry to treat constipation, hepatic encephalopathy, tumor prevention, complications of liver disease, and inflammatory bowel disease. It also improves mineral absorption, maintains blood glucose and insulin levels, and has an anti-endotoxin effect (Table 6.1) (Nooshkam and Madadlou, 2016a; Pranami et al., 2017).

TABLE 6.1 Food and Medical/Pharmaceutical Applications of Lactulose

Potential Applications of Prebiotic Lactulose	
In Food Science	**In Medical and Pharmaceutical Sciences**
• Bifidogenic agent	• Therapy of constipation
• Enhancing the survivability of probiotics	• Therapy of hepatic encephalopathy
• Stabilizing effect	• Therapy of inflammatory bowel disease
• Texturizing effect	• Anti-tumor effect
• Improving sensory and browning properties	• Anti-carcinogenic effect
• High water solubility	• Anti-diabetic effect (lowering blood glucose and insulin levels)
• 60–80% sweetness compared to sucrose	• Anti-gallstone formation
• Providing similar viscosity to sucrose solution	• Anti-endotoxin effect
• Low energy value (2.0 kcal g^{-1})	• Anti-plaque effect
• A sucrose substitute	• Prevention of infection
• An effective sweetener for diabetics	• Hypocholesterolemic effect
• A potent sucrose substitute in soft drinks, baby formulas, and dairy, bakery, and confectionery products	• Enhancing mineral absorption
• A fortifier in formulated food products for the elderly	• Lowering caecal pH
• High heat stability under acidic conditions; fortifying fruit juices	• Decreasing resident time of toxins
• Maintaining the characteristics of infant formulas during prolong storage	• Improving osmotic pressure and peristalsis

1. **Constipation:** Lactulose has been administrated since 1960 as a drug of choice to treat chronic constipation in all age groups (Aït-Aissa and Aïder, 2014). Its β-glycosidic linkage is not cleaved by digestive enzymes of mammalian and therefore the sugar is passed through the stomach and small intestine without being absorbed or degraded. It is then metabolized by *Bifidobacterium* and *Lactobacillus* species in the colon to generate low molecular weight organic acids like acetic, lactic, and formic acid along with CO_2. The produced acids lower the caecal pH, increase intestinal osmolarity, enhance colonic mobility, draw the fluid into the bowel, soften the feces, evacuate the stool

simply, reduce the intestinal transit time (Panesar and Kumari, 2011). Lactulose is therefore used as a laxative drug and its action is dependent on the ingested dose, body weight, health condition, dietary habits, age, and gender (Schuster-Wolff-Bühring et al., 2010).

2. **Hepatic Encephalopathy:** It is caused by rising contents of blood ammonia, which the ammonia functions as a neurotoxin in the human brain. It is demonstrated that the nitrogen of food proteins is consumed by intestinal putrefactive bacteria to form ammonia (Seki and Saito, 2012; Pranami et al., 2017). In this context, the protein intake was rigorously reduced in conjugation with antibiotic consumption to treat the symptoms of hepatic encephalopathy and lower the nitrogenous metabolites derived from bacterial degradation or digestion in the colon and subsequently in blood serum. It was further confirmed that lactulose digestion can cure neuropsychiatric signs of hepatic encephalopathy through (a) ammonia protonation to ammonium ions by acidifying the caecal content, which the ammonium ions remain in the colon and subsequently evacuate through defecation, and (b) increasing the number of carbohydrate-metabolizing bacteria and lowering the activity of proteolytic microorganisms contributing to the ammonia formation (Schuster-Wolff-Bühring et al., 2010).
3. **Inflammatory Bowel Disease (Crohn's, Ulcerative Colitis):** It has been proved that bacteria and their endotoxins play a key role in the pathogenesis of inflammatory bowel disease. Some therapeutic drugs like lactulose can affect both bacterial growth and endotoxin formation, and therefore they can be used for treating this disease. Lactulose digestion in the colon generates organic acids with potential to drop fecal pH, thereby providing conditions to increase the activity of *Lactobacillus* and *Bifidobacterium* strains and to inhibit the growth of Bacteroides, coliforms, *Shigella*, and *Salmonella* (Liao et al., 1994). Additionally, it was indicated that bacterial translocation and subsequent sepsis are mainly accompanied by obstructive jaundice, and lactulose administration could inhibit systemic endotoxemia and in turn inflammatory response in obstructive jaundice (Koutelidakis et al., 2003). As well, lactulose fermentation in the colon produces a remarkable level of endogenous hydrogen as a potent antioxidant and protective agent for DSS-triggered colitis, thus lowering oxidative stress and boosting the disease symptoms in humans (Chen et al., 2011).
4. **Anti-Tumor Effect and Immunology:** Bifidobacteria have the ability to prevent mammary and liver carcinogenesis, and possess

anti-tumor and immunologic effects specifically and non-specifically. It has been recently reported that lactulose has DNA-protective effect in human flora-associated rats, which were exposed to dimethylhydrazine. Due to its convenient metabolism by *Bifidobacterium* species, lactulose can therefore induce both anti-tumor and immunologic effects of *Bifidobacteria* (Schumann, 2002).

5. **Anti-Carcinogenic Effect:** Biochemical processes in mucosa, lumen, and adjacent tissues in the colon lead to the formation of cancer in the large intestine. The microflora and its metabolism products in the colon can influence the colon carcinogenesis. In this way, probiotics, and prebiotics are able to lower the risk of cancer formation in the large intestine (Panesar and Kumari, 2011).
6. **Anti-Endotoxin Effect:** The anti-endotoxin ability of lactulose has been clinically approved in patients with obstructive jaundice. In addition, the endotoxin-dependent complications like renal dysfunction were approximately completely inhibited by lactulose administration during 2–3 days before operation (Pain et al., 1991; Özçelik et al., 1997). Lactulose also presents anti-endotoxin effect in metabolic diseases such as, hypercholesterolemia, diabetic Mellitus, exocrine pancreatic dysfunction, and hepatorenal syndrome (Kramer, 1988; Liao and Florén, 1993; Yelich et al., 1992; Tabatabaie et al., 1997; Mack et al., 1992).
7. **Salmonella Carriers:** Lactulose, up to 60 g per day, has been successfully employed as a superb prebiotic sugar in non-typhoid *Salmonella* carrier, through lowering colonic pH and subsequently inhibiting the growth of *Salmonella* (Schumann, 2002).
8. **Urinary Tract and Respiratory Tract Infections:** It is observed that lactulose intake results in a remarked decrease in fecal *Shigella* and *Salmonella* species and, in turn, lowers the prevalence of respiratory tract and urinary tract infections in patients, mainly due to its significant effect on the colonic bacteria and reestablishing the balance of intestinal ecosystem. Furthermore, lactulose administration prevents the creation of secondary bile acids contributing to the colon cancers (Liao et al., 1994).
9. **Blood Glucose and Insulin:** The glucose decrement ability of lactulose (about 40%) was reported in the literature (Hosaka et al., 1972; Bianchi et al., 1994). In addition, endotoxin can decrease the production of pancreatic insulin and therefore the administration of substances with endotoxin-lowering effect like lactulose can possess anti-diabetic effect (Cornell, 1985).

10. **Hypocholesterolemic Effect:** Probiotic bacteria have documented hypocholesterolemic effect and prebiotics, due to enhancing the growth of probiotics, have gained a lot of attention in cholesterol studies (Ooi and Liong, 2010), and hypocholesterolemic effect of inulin (Causey et al., 2000), long-chained fructan (Mortensen et al., 2002), and lactulose (Conte et al., 1977) has been reported in the literature.

    Prebiotics like inulin and fructooligosaccharides are recognized as indigestible, soluble, fermentable, and viscous compounds, contributing to the hypocholesterolemia through: (i) lowering cholesterol absorption, which is accompanied by the increased cholesterol evacuation in feces; and (ii) the generation of low molecular weight fatty acids from prebiotics fermented by intestinal bacterial microflora. Indigestible prebiotic compounds have the potential to enhance the viscosity of gastrointestinal (GI) tract in conjugation with increasing the thickness of the unstirred layer of the small intestine. These inhibit the cholesterol uptake and may have resulted in a greater catabolism of cholesterol in the liver (hypocholesterolemic effect). It has been reported that the hypocholesterolemic effect of indigestible prebiotics is mainly ascribed to the creation of short-chain fatty acids such as acetate, butyrate, and propionic. Butyrate is able to prevent the synthesis of liver cholesterol and function as an energy source for colon epithelial cells; whilst, propionate inhibits fatty acid production in the liver and lowers the secretion rate of triacylglycerol. As well, propionate controls hepatic cholesterol synthesis and decreases the synthesis rate of cholesterol, thereby lowering the levels of plasma cholesterol (Ooi and Liong, 2010).
11. **Lithogenic Index:** Serum cholesterol and bile' lithogenic index were significantly decreased upon lactulose and lignin administrations. It was hypothesis in 1979 that lactulose showed a preventive effect towards colorectal carcinoma, and a manifestly reduced lithogenic index in the presence of lactulose was reported in 1981. This indicates the lactulose ability to reduce the risk of cholesterol gallstone formation. The protective effect of lactulose is attributed to the intestinal transit time, increased the evacuation of neutral sterols (as an index of decreased cholesterol absorption), decreased toxins and secondary bile acids, and remarkable protective role of butyrate (Schumann, 2002).
12. **Mineral Absorption:** Prebiotics such as lactulose improve the mineral absorption by increasing the permeability of intestinal

mucosa and enhancing mineral solubility by lowering the colonic pH (Schuster-Wolff-Bühring et al., 2010). Lactulose is able to promote the absorption and retention of some minerals such as, Fe, Ca, Mg, Cu, and Zn in the body. Additionally, it was shown that Ca and Mg absorptions were increased in adult men as a function of lactulose administration (Aït-Aissa and Aïder, 2014).

13. **Diagnostic Applications:** The diagnosis of colonic disorders *via* H_2*-breath test* (*breath hydrogen test*) is another important lactulose application in the pharmaceutical industry. In this essay, a single dose of lactulose is administrated and then the exhaled hydrogen is measured during the following hours. Hydrogen formation is due to the lactulose metabolism by bacteria that exist in the intestine. In this context, the hydrogen formation/exhalation time and concatenation determine the microbial colonization and colonic transit time (Simrén and Stotzer, 2006). The "small intestine bacterial overgrowth" and "lactase deficiency" are determined by the lactulose *breath test* (Schumann, 2002), which in the former, abnormal bacterial growth in the small intestine resulted in a faster lactulose metabolism and hydrogen exhalation in comparison to the normal bacterial metabolism in the colon of healthy people (Nucera et al., 2005).
14. **Colon Targeted Delivery System (CODES™):** Due to its indigestible property in the upper part of the digestional tract, lactulose has been successfully used in CODES™ to deliver several potent drugs and nutraceuticals. A CODES™ is consisted of an outer enteric-soluble material layer, an inner acid-soluble material layer, and a core containing drug/nutraceutical and lactulose (Figure 6.6) (Yang et al., 2003). The drug located in the core of CODES™ is not released in the stomach because of the enteric coating layer. This protective layer is then dissolved in the small intestine; however, the presence of inner acid-soluble coating layer does not allow the drug to be released from the system core. The GI fluids are able to penetrate into CODES™ core and dissolve lactulose during transition of the system through the small intestine. In the colon, lactulose is leached through the acid-soluble material layer and fermented by *Enterobacteria* to generate organic acids with potential to dissolve the acid-soluble coating layer and the drug is finally released from the CODES™ (Katsuma et al., 2002). Therefore, lactulose acts as a trigger agent to release the drug in the large intestine and to activate the CODES™ (Dehghan et al., 2013). It was reported that lactulose is employed

in the CODES™ to entrap and protect the encapsulant from harsh environmental conditions. The releasing profile of the encapsulant in CODES™ is influenced by the coating polymeric layers along with the composition of colonic microflora and fermentation rate of lactulose (Nooshkam et al., 2018).

6.2.8.4.2 Food Science

In recent years, the use of non-digestible oligosaccharides as health-enhancing compounds has been steadily increased in the food industry to develop functional foods. Lactulose, a superb prebiotic, possesses excellent functional features to the food products and, therefore, the food sector has been motivated to use lactulose as a prebiotic sugar and as a functional additive (Guerrero and Wilson, 2016).

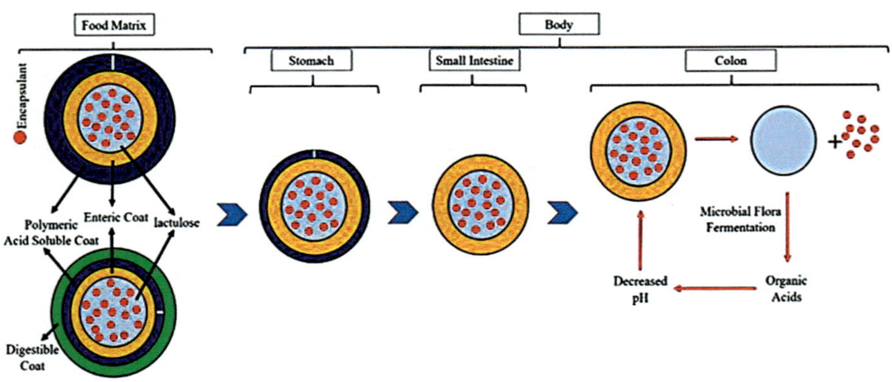

FIGURE 6.6 Schematic illustration of lactulose-based CODES™.
Source: Reprinted with permission from Nooshkam et al., 2018. © Elsevier.

1. **Prebiotic Activity:** Lactulose was considered in 1957 as a bioactive compound with bifidogenic activity and this explains why lactulose was firstly recognized as a therapeutic drug in place of a food ingredient (Schumann, 2002). A prebiotic compound should have the following criteria: (i) it is passed through the upper digestional tract without being fermented or absorbed; (ii) is preferentially fermented by colonic microflora and increases the growth of health-promoting bacteria; (iii) is shifting the microbial balance of the colon to a healthier composition; and (iv) is resulting in an enhanced health status (Schuster-Wolff-Bühring et al., 2010). As mentioned earlier,

lactulose is not hydrolyzed in the stomach and small intestine, and renders its therapeutic role in the colon to mainly treat chronic constipation and other implications. Although, its high doses lead to diarrhea and increase in intestinal gas formation, low doses of the lactulose are prescribed for pregnant women and infants (Macfarlane et al., 2008; Schuster-Wolff-Bühring et al., 2010).

It has been demonstrated that a daily intake of 4–10 g (adult) lactulose exerts prebiotic action (Mizota et al., 2002; Venema et al., 2003). In addition, lactulose administration at two doses of 10 g per day resulted in a remarked alteration in bacteria gut flora without any GI problems (Ballongue et al., 1997). In general, lactulose applications in food are mainly attributed to its prebiotic condition. It increases the mineral absorption at 0.04–0.2 g kg^{-1} of body weight and doses between 0.07 and 0.2 g kg^{-1} of body weight led to an increase in the growth of probiotic bacteria, thereby lactulose is applied in food supplements, functional foods, and nutraceuticals as an excellent health promoter compound (Panesar and Kumari, 2011; Schuster-Wolff-Bühring et al., 2010). Lactulose has been also employed as a potent prebiotic in animal feed, to increase the proliferation rate of *Lactobacillus* and *Bifidobacterium* species, to lower intestinal pathogens, and to enhance the intestinal motility. As well, it was used as an animal growth promoter (Aït-Aissa and Aïder, 2014; Schumann, 2002).

2. **Food Additive:** Lactulose confers multiple health-promoting effects to consumers and, therefore, its incorporation into food products and, in turn, in a daily regular diet provides a healthy and equilibrated nutrition (Olano and Corzo, 2009). It is frequently used as a prebiotic in various dairy products and milk formulas. It was shown that the addition of 0.5% lactulose in infant formula induces the healthy intestinal *Bifidobacteria* and higher incorporation level (1.0%) resulted in a mild laxative effect (Nagendra et al., 1995a). Furthermore, lactulose addition to infant formula maintained the characteristics of the final product for prolonged storage period (Nagendra et al., 1995b).

The functional properties of yogurt can be improved when lactulose is applied as a prebiotic and health promoter agent in this product. For example, lactulose-sweetened yogurt acted as a mild laxative in healthy human volunteers and increased the number of fecal formations. As well, the consumption of yogurt containing lactulose resulted in a faster intestinal transit time in comparison to glucose solution (Porkka et al., 1988).

In another study, the childhood constipation was effectively treated upon consuming lactulose-enriched yogurt, which was comparable to the yogurt contained inulin, resistant starch, fiber soy, and *trans*-GOS (Kokke et al., 2008). In addition, the incubation period of bio-yogurt making containing *Bifidobacterium bifidum* and *Lactobacillus acidophilus* was reduced in the presence of lactulose, and the cell counts of *B. bifidum* during refrigerated storage was significantly increased, as well. This demonstrates that lactulose supplementation could keep the viability of probiotic cells in yogurt during cold storage and above their minimum therapeutic effects, i.e., ~10^7 cfu g^{-1} (Özer et al., 2005). The ability prebiotic compounds to protect probiotic bacteria from bile acids and cold storage in yogurt and fermented milks could be attributed to (a) the fact that prebiotics act as an alternative or extra energy source for probiotics, thereby they can respond better to cold storage and bile acid stresses, (b) the increased growth of probiotics once the gut pH reached to an optimal value, influencing the bile acid and their physiological characteristics, and (c) the high levels of prebiotics, which may reduce the solubility of bile acids and likely their toxic effects (Nooshkam et al., 2018).

Interestingly, the incorporation of lactulose (0.5% w/v) into soymilk led to the probiotics *B. longum* and *B. animalis* to convert isoflavone glycosides (inactive phytochemical components) to bioactive isoflavone aglycones with estrogenic effects (Pham and Shah, 2008). Lactulose has also remarkably high stability under thermal-acidic conditions and, therefore, it is applied to fortify acidic foods like fruit juices (Seki and Saito, 2012). It has gained a manifest attention in the food industry as a potent sweetener for diabetics and as a sucrose replacer in baby formulas, yogurt, dairy products, soft drinks, bakery products, confectionaries, and formulated foods for old people (Panesar and Kumari, 2011; Schumann, 2002). Likewise, it was reported that about 10% of sucrose or glucose can be replaced by lactulose in pectin gels prepared by apple puree and different sugars. In this regard, it was able to increase the free moisture content (MC) and lower the effective viscosity of the final pectin gel (Dorohovich et al., 2016).

6.2.8.5 LACTULOSE-BASED BIOACTIVE COMPOUNDS

6.2.8.5.1 Galactooligosaccharides (GOS)

Human milk is known as the first oligosaccharide source with the potential to act as prebioticsa to promote the growth of beneficial bacteria located in the colon of breast-fed infants (Boehm and Moro, 2008). The accessibility of human milk is limited, and therefore it is mandatory to find another alternative source of prebiotics to supplement infant formulas for providing a beneficial prebiotic effect on the microflora of infants' digestional tract. In this context, lactulose is resistant to hydrolysis in the upper part of the gastrointestinal tract (GIT) and may open new horizon in the production of lactulose-derived oligosaccharides *via* an enzyme-mediated *tarns*-galactosylation reaction (Martínez-Villaluenga et al., 2007). Lactulose acts as galactosyl donor and acceptor to produce tri- or higher GOS and it was shown that β-galactosidase of *Kluyveromyes lactis* can hydrolyze lactulose to form lactulose-based trisaccharides (Martínez-Villaluenga et al., 2007). Lactose-derived oligosaccharides contain repeating units of galactose and a lactose moiety at their reducing end. Thus, these compounds are prone to isomerization reaction and the reducing aldose at the end-group is converted to ketose, leading to the generation of corresponding lactulose-derived GOS (Martínez-Villaluenga et al., 2008).

6.2.8.5.2 Lactulose-Protein Conjugates

The Maillard reaction is occurred between compounds containing free amino and carbonyl groups, i.e., proteins/peptides/amino acids and reducing saccharides, during heating and prolonged storage period (Nooshkam et al., 2019). It improves organoleptic properties and antioxidative activity of food products, the functional features of proteins, and forms bioactive melanoidins with antimicrobial, antioxidant, and antihypertensive activities (Nooshkam et al., 2018).

Lactulose has been successfully used in the Maillard glycation to create bioactive conjugates with novel functionalities. The sodium caseinate-lactose/lactulose Maillard conjugates showed no or less mutagenic effect towards *Salmonella typhimurium* compared to the conjugates obtained from monosaccharides glucose, fructose, galactose, and tagatose (Brands et al., 2000). In another research, hydrolyzed caseinomacropeptide was conjugated to lactulose, lactulose-derived GOS, and lactose-derived GOS. The bifidogenic activity and the concertation of lactic acid and short-chain fatty acids of the fermented conjugates were similar to that of non-conjugated carbohydrates, indicating health-promoting effects of the prebiotic-based Maillard conjugates (Hernandez-Hernandez et al., 2011). In addition, lactose, and lactulose conjugation to whey protein isolate (WPI) and its hydrolysate fabricated conjugates with improved solubility, foaming capacity, emulsifying activity, and antioxidant effect. As well, the lactulose-whey hydrolysate conjugate was completely degraded in *in vitro* gastric digestion (Nooshkam and Madadlou, 2016a, b).

6.2.8.5.3 Lactulose-Fatty Acid Conjugates

The OH groups of sugars can be converted to ester forms *via* the esterification reaction. Lipases are frequently employed to synthesize sugar esters due to regiospecifying the esterification reaction and using different natural substrates. Lactulose palmitates were successfully produced though lipase-mediated esterification process, which have the functional properties of lactulose and commonly used palmitic acid in food materials (Bernal et al., 2014, 2015). It is believed that sugar esters are biodegradable, non-toxic, odorless, and amphiphilic compounds with the potential to be applied as emulsifiers, antimicrobial compounds, and potent additives in food products and pharmaceutical industry (Bernal et al., 2015).

6.3 CONCLUSIONS

The incorporation of biologically active prebiotics such as lactulose into food products and subsequently in daily diet is recognized as a new approach towards a balanced and healthy nutrition. Lactulose is a synthetic and ketose disaccharide obtained from lactose through many lactose to lactulose isomerization methods like chemical, electro-activation, and enzymatic ones. Electro-isomerization and enzymatic-based lactulose production usually have lower degradation secondary products and therefore the purification and downstream operation costs are significantly lower in comparison to chemical methods. Lactulose has many beneficial effects in the pharmaceutical sector as a therapeutic drug to treat constipation, hepatic encephalopathy, inflammatory bowel disease, and other GI disorders. It is used in the food industry as a prebiotic and functional ingredient to improve the organoleptic properties of food products. In addition, due to its high sweetness and low energy value, lactulose is a potent sucrose substitute for diabetics and a bioactive ingredient in soft drinks, baby formulas, and dairy products. Lactulose is also undergoing some reactions such as, the Maillard conjugation, esterification reaction, and isomerization process to form lactulose-based bioactive compounds with novel functionalities.

KEYWORDS

- **anti-tumor effect**
- **chemical isomerization**
- **dairy products**
- **encephalopathy**
- **enzymatic synthesis**
- **esterification**
- **functional ingredient**
- **galactooligosaccharides**
- **health-promoting effect**
- **lactosucrose**
- **lactulose**
- **Maillard reaction**
- **natural catalysts**
- **pharmaceutical**
- **trans-galactosylation**

REFERENCES

Aider, M., & De Halleux, D., (2007). Isomerization of lactose and lactulose production. *Trends in Food Science & Technology, 18*(7), 356–364.

Aider, M., & Gimenez-Vidal, M., (2012). Lactulose synthesis by electro-isomerization of lactose: Effect of lactose concentration and electric current density. *Innovative Food Science & Emerging Technologies, 16*, 163–170.

Aissa, A. A., & Aïder, M., (2013). Ion exchange membranes controlled electro-catalytic synthesis of lactulose from lactose under refrigerated conditions. *Innovative Food Science & Emerging Technologies, 20*, 299–309.

Aït-Aissa, A., & Aïder, M., (2014). Lactulose: Production and use in functional food, medical and pharmaceutical applications. Practical and critical review. *International Journal of Food Science & Technology, 49*(5), 1245–1253.

Ajisaka, K., Fujimoto, H., & Nishida, H., (1988). Enzymic synthesis of disaccharides by use of the reversed hydrolysis activity of β-D-galactosidases. *Carbohydrate Research, 180*(1), 35–42.

Alonso, S., Rendueles, M., & Díaz, M., (2013). Bio-production of lactobionic acid: Current status, applications and future prospects. *Biotechnology Advances, 31*(8), 1275–1291.

Bacenetti, J., Bava, L., Schievano, A., & Zucali, M., (2018). Whey protein concentrate (WPC) production: Environmental impact assessment. *Journal of Food Engineering, 224*, 139–147.

Ballongue, J., Schumann, C., & Quignon, P., (1997). Effects of lactulose and lactitol on colonic microflora and enzymatic activity. *Scandinavian Journal of Gastroenterology, 32*(sup222), 41–44.

Bernal, C., Illanes, A., & Wilson, L., (2014). Heterofunctional hydrophilic-hydrophobic porous silica as support for multipoint covalent immobilization of lipases: Application to lactulose palmitate synthesis. *Langmuir, 30*(12), 3557–3566.

Bernal, C., Illanes, A., & Wilson, L., (2015). Improvement of efficiency in the enzymatic synthesis of lactulose palmitate. *Journal of Agricultural and Food Chemistry, 63*(14), 3716–3724.

Bianchi, G. P., De, M. M., Bugianesi, E., Abbiati, R., Fabbri, A., & Marchesini, G., (1994). Lowering effects of a preparation containing fibres and lactulose on glucose and insulin levels in obesity. *The Italian Journal of Gastroenterology, 26*(4), 174–178.

Bisinella, R. Z. B., Ribeiro, J. C. B., De Oliveira, C. S., Colman, T. A. D., Schnitzler, E., & Masson, M. L., (2017). Some instrumental methods applied in food chemistry to characterise lactulose and lactobionic acid. *Food Chemistry, 220*, 295–298.

Boehm, G., & Moro, G., (2008). Structural and functional aspects of prebiotics used in infant nutrition. *The Journal of Nutrition, 138*(9), 1818S–1828S.

Brands, C. M., Alink, G. M., Van, B. M. A., & Jongen, W. M., (2000). Mutagenicity of heated sugar– casein systems: Effect of the Maillard reaction. *Journal of Agricultural and Food Chemistry, 48*(6), 2271–2275.

Carobbi, R., & Innocenti, F., (1990). *U.S. Patent No. 4,957,564*. Washington, DC: U.S. Patent and Trademark Office.

Causey, J. L., Feirtag, J. M., Gallaher, D. D., Tungland, B. C., & Slavin, J. L., (2000). Effects of dietary inulin on serum lipids, blood glucose and the gastrointestinal environment in hypercholesterolemic men. *Nutrition Research, 20*(2), 191–201.

Chen, X., Zuo, Q., Hai, Y., & Sun, X. J., (2011). Lactulose: An indirect antioxidant ameliorating inflammatory bowel disease by increasing hydrogen production. *Medical Hypotheses, 76*(3), 325–327.

Conte, D., Bardella, M. T., Bernardi, R., Pasetti, G., & Bianchi, P. A., (1977). Hypocholesterolemic effect of lactulose in man: A preliminary report. *Panminerva Medica, 19*(1), 1–4.

Cornell, R. P., (1985). Endogenous gut-derived bacterial endotoxin tonically primes pancreatic secretion of insulin in normal rats. *Diabetes, 34*(12), 1253–1259.

Dehghan, M. H. G., Gupta, V. R. M., Asif, S. M., Darwis, Y., Rizwan, M., & Mundada, V. P., (2013). Assessment of isomalt for colon-specific delivery and its comparison with lactulose. *AAPS PharmSciTech, 14*(1), 53–59.

Dendene, K., Guihard, L., Balannec, B., & Bariou, B., (1995). Study of the separation of lactose, lactulose and galactose by liquid chromatography using cationic ion-exchange resin columns. *Chromatographia, 41*(5, 6), 561–567.

Díez-Antolínez, R., Hijosa-Valsero, M., Paniagua-García, A. I., Garita-Cambronero, J., & Gómez, X., (2018). Yeast screening and cell immobilization on inert supports for ethanol production from cheese whey permeate with high lactose loads. *PloS One, 13*(12), e0210002.

Dorohovich, A., Dorohovich, V., & Kambulova, J., (2016). The study of the rheological properties of pectin gels with mono-and disaccharides. *EUREKA: Life Sciences*, (4), 14–19.

Duarte, L. S., Da Natividade, S. J., Lorenzoni, A. S. G., Rodrigues, R. C., Rodrigues, E., & Hertz, P. F., (2017). A new bioprocess for the production of prebiotic lactosucrose by an immobilized β-galactosidase. *Process Biochemistry, 55*, 96–103.

Espinosa, I., & Fogelfeld, L., (2010). Tagatose: From a sweetener to a new diabetic medication?. *Expert Opinion on Investigational Drugs, 19*(2), 285–294.

Faruqui, A. A., & Joshi, C., (2012). Lactitol: A review of its use in the treatment of constipation. *Int. J. Recent Adv. Pharm. Res., 2*(1), 1–5.

Fujimaru, T., Park, J. H., & Lim, J., (2012). Sensory characteristics and relative sweetness of tagatose and other sweeteners. *Journal of Food Science, 77*(9), S323–S328.

Gosling, A., Stevens, G. W., Barber, A. R., Kentish, S. E., & Gras, S. L., (2010). Recent advances refining galactooligosaccharide production from lactose. *Food Chemistry, 121*(2), 307–318.

Guerrero, C., & Wilson, L., (2016). Enzymatic production of lactulose. In: Illanes, A., (ed.), *Lactose-Derived Prebiotics: A Process Perspective* (p. 312). Academic Press, London, UK.

Guerrero, C., Vera, C., Araya, E., Conejeros, R., & Illanes, A., (2015). Repeated-batch operation for the synthesis of lactulose with β-galactosidase immobilized by aggregation and crosslinking. *Bioresource Technology, 190*, 122–131.

Gutiérrez, L. F., Hamoudi, S., & Belkacemi, K., (2012). Lactobionic acid: A high value-added lactose derivative for food and pharmaceutical applications. *International Dairy Journal, 26*(2), 103–111.

Hashemi, S. A., & Ashtiani, (2010). The isomerization kinetics of lactose to lactulose in the presence of sodium hydroxide at constant and variable pH. *Food and Bioproducts Processing, 88*(2, 3), 181–187.

Hernandez-Hernandez, O., Sanz, M. L., Kolida, S., Rastall, R. A., & Moreno, F. J., (2011). *In vitro* fermentation by human gut bacteria of proteolytically digested caseinomacropeptide nonenzymatically glycosylated with prebiotic carbohydrates. *Journal of Agricultural and Food Chemistry, 59*(22), 11949–11955.

Hicks, K. B., & Parrish, F. W., (1980). A new method for the preparation of lactulose from lactose. *Carbohydrate Research, 82*(2), 393–397.

Hosaka, H., Tokunagaj, K., Sakumoto, I., Inaba, M., Yokoi, Y., & Abei, T., (1972). The influence of lactulose on intestinal absorption. *Journal of Gastroenterology, 7*(3), 316–317.

Hua, X., Yang, R., Zhang, W., Fei, Y., Jin, Z., & Jiang, B. O., (2010). Dual-enzymatic synthesis of lactulose in organic-aqueous two-phase media. *Food Research International, 43*(3), 716–722.

Katsuma, M., Watanabe, S., Kawai, H., Takemura, S., Masuda, Y., & Fukui, M., (2002). Studies on lactulose formulations for colon-specific drug delivery. *International Journal of Pharmaceutics, 249*(1, 2), 33–43.

Kim, Y. S., Kim, J. E., & Oh, D. K., (2013). Borate enhances the production of lactulose from lactose by cellobiose 2-epimerase from caldicellulosiruptor saccharolyticus. *Bioresource Technology, 128*, 809–812.

Kokke, F. T., Scholtens, P. A., Alles, M. S., Decates, T. S., Fiselier, T. J., Tolboom, J. J., Kimpen, J. L. L., & Benninga, M. A., (2008). A dietary fiber mixture versus lactulose in the treatment of childhood constipation: A double-blind randomized controlled trial. *Journal of Pediatric Gastroenterology and Nutrition, 47*(5), 592–597.

Koutelidakis, I., Papaziogas, B., Giamarellos-Bourboulis, E. J., Makris, J., Pavlidis, T., Giamarellou, H., & Papaziogas, T., (2003). Systemic endotoxaemia following obstructive jaundice: The role of lactulose. *Journal of Surgical Research, 113*(2), 243–247.

Kozempel, M. F., Kurantz, M. J., Craig, Jr. J. C., & Hicks, K. B., (1995). Development of a continuous lactulose process: Separation and purification. *Biotechnology Progress, 11*(5), 592–595.

Kramer, H. J., (1988). Therapie des hepatorenalen Syndroms. *DMW-Deutsche Medizinische Wochenschrift, 113*(14), 561–564.

Lee, Y. J., Kim, C. S., & Oh, D. K., (2004). Lactulose production by β-galactosidase in permeabilized cells of kluyveromyces lactis. *Applied Microbiology and Biotechnology, 64*(6), 787–793.

Li, W., Wang, K., Sun, Y., Ye, H., Hu, B., & Zeng, X., (2015). Lactosucrose and its analogues derived from lactose and sucrose: Influence of structure on human intestinal microbiota *in vitro*. *Journal of Functional Foods, 17*, 73–82.

Liao, W., & Florén, C. H., (1993). Endotoxin, cytokines, and hyperlipidemia. *Scandinavian Journal of Gastroenterology, 28*(2), 97–103.

Liao, W., (1994). Lactulose—a potential drug for the treatment of inflammatory bowel disease. *Medical Hypotheses, 43*(4), 234–238.

Lindsay, M. J., Walker, T. W., Dumesic, J. A., Rankin, S. A., & Huber, G. W., (2018). Production of monosaccharides and whey protein from acid whey waste streams in the dairy industry. *Green Chemistry, 20*(8), 1824–1834.

Livesey, G., & Brown, J. C., (1996). D-tagatose is a bulk sweetener with zero energy determined in rats. *The Journal of Nutrition, 126*(6), 1601–1609.

Lu, Y., Levin, G. V., & Donner, T. W., (2008). Tagatose, a new antidiabetic and obesity control drug. *Diabetes, Obesity and Metabolism, 10*(2), 109–134.

Macfarlane, G. T., Steed, H., & Macfarlane, S., (2008). Bacterial metabolism and health-related effects of galacto-oligosaccharides and other prebiotics. *Journal of Applied Microbiology, 104*(2), 305–344.

Mack, D. R., Flick, J. A., Durie, P. R., Rosenstein, B. J., Ellis, L. E., & Perman, J. A., (1992). Correlation of intestinal lactulose permeability with exocrine pancreatic dysfunction. *The Journal of Pediatrics, 120*(5), 696–701.

Martínez-Villaluenga, C., Cardelle-Cobas, A., Corzo, N., Olano, A., & Villamiel, M., (2008). Optimization of conditions for galactooligosaccharide synthesis during lactose hydrolysis by β-galactosidase from *Kluyveromyces lactis* (Lactozym 3000 L HP G). *Food Chemistry, 107*(1), 258–264.

Martínez-Villaluenga, C., Cardelle-Cobas, A., Olano, A., Corzo, N., Villamiel, M., & Jimeno, M. L., (2007). Enzymatic synthesis and identification of two trisaccharides produced from lactulose by transgalactosylation. *Journal of Agricultural and Food Chemistry, 56*(2), 557–563.

Mayer, J., Conrad, J., Klaiber, I., Lutz-Wahl, S., Beifuss, U., & Fischer, L., (2004). Enzymatic production and complete nuclear magnetic resonance assignment of the sugar lactulose. *Journal of Agricultural and Food Chemistry, 52*(23), 6983–6990.

Miller, L. E., Tennilä, J., & Ouwehand, A. C., (2014). Efficacy and tolerance of lactitol supplementation for adult constipation: A systematic review and meta-analysis. *Clinical and Experimental Gastroenterology, 7*, 241.

Mizota, T., Mori, T., Yaeshima, T., Yanagida, T., Iwatsuki, K., Ishibashi, N., Tamura, Y., & Fukuwatari, Y., (2002). Effects of low dosages of lactulose on the intestinal function of healthy adults. *Milchwissenschaft, 57*(6), 312–315.

Montgomery, E. M., & Hudson, C. S., (1930). Relations between rotatory power and structure in the sugar group. XXVII. Synthesis of a new disaccharide ketose (lactulose) from lactose 1. *Journal of the American Chemical Society, 52*(5), 2101–2106.

Montilla, A., Del, C. M. D., Sanz, M. L., & Olano, A., (2005). Eggshell as catalyst of lactose isomerization to lactulose. *Food Chemistry, 90*(4), 883–890.

Mortensen, A., Poulsen, M., & Frandsen, H., (2002). Effect of a long-chained fructan raftiline HP on blood lipids and spontaneous atherosclerosis in low density receptor knockout mice. *Nutrition Research, 22*(4), 473–480.

Mu, W., Li, Q., Fan, C., Zhou, C., & Jiang, B., (2013). Recent advances on physiological functions and biotechnological production of epilactose. *Applied Microbiology and Biotechnology, 97*(5), 1821–1827.

Nagendra, R., Baskaran, M. V., & Rao, S. V., (1995b). Shelf-life of spray-dried infant formula supplemented with lactulose. *Journal of Food Processing and Preservation, 19*(4), 303–315.

Nagendra, R., Viswanatha, S., Kumar, S. A., Murthy, B. K., & Rao, S. V., (1995a). Effect of feeding milk formula containing lactulose to infants on faecal bifidobacterial flora. *Nutrition Research, 15*(1), 15–24.

Nooshkam, M., & Madadlou, A., (2016a). Maillard conjugation of lactulose with potentially bioactive peptides. *Food Chemistry, 192*, 831–836.

Nooshkam, M., & Madadlou, A., (2016b). Microwave-assisted isomerization of lactose to lactulose and Maillard conjugation of lactulose and lactose with whey proteins and peptides. *Food Chemistry, 200*, 1–9.

Nooshkam, M., Babazadeh, A., & Jooyandeh, H., (2018). Lactulose: Properties, techno-functional food applications, and food-grade delivery system. *Trends in Food Science & Technology, 80*, 23–34.

Nooshkam, M., Varidi, M., & Bashash, M., (2019). The Maillard reaction products as food-born antioxidant and anti-browning agents in model and real food systems. *Food Chemistry, 275*, 644–660.

Nucera, G., Gabrielli, M., Lupascu, A., Lauritano, E. C., Santoliquido, A., Cremonini, F., Cammarota, G., Tondi, P., Pola, P., Gasbarrini, G., & Gasbarrini, A., (2005). Abnormal breath tests to lactose, fructose and sorbitol in irritable bowel syndrome may be explained by small intestinal bacterial overgrowth. *Alimentary Pharmacology & Therapeutics, 21*(11), 1391–1395.

Olano, A., & Corzo, N., (2009). Lactulose as a food ingredient. *Journal of the Science of Food and Agriculture, 89*(12), 1987–1990.

Ooi, L. G., & Liong, M. T., (2010). Cholesterol-lowering effects of probiotics and prebiotics: A review of *in vivo* and *in vitro* findings. *International Journal of Molecular Sciences, 11*(6), 2499–2522.

Özçelik, F., Pekmezci, S., Altinli, E., Eroğlu, C., Göksel, S., & Göksoy, E., (1997). Lactulose to prevent bacterial translocation in biliary obstruction. *Digestive Surgery, 14*(4), 267–271.

Özer, D., Akin, S., & Özer, B., (2005). Effect of inulin and lactulose on survival of *Lactobacillus acidophilus* LA-5 and *Bifidobacterium* bifidum BB-02 in acidophilus-Bifidus yoghurt. *Food Science and Technology International, 11*(1), 19–24.

Pain, J. A., Cahill, C. J., Gilbert, J. M., Johnson, C. D., Trapnell, J. E., & Bailey, M. E., (1991). Prevention of postoperative renal dysfunction in patients with obstructive jaundice: A multicentre study of bile salts and lactulose. *British Journal of Surgery, 78*(4), 467–469.

Panesar, P. S., & Kumari, S., (2011). Lactulose: Production, purification and potential applications. *Biotechnology Advances, 29*(6), 940–948.

Panesar, P. S., (2008). Application of response surface methodology in the permeabilization of yeast cells for lactose hydrolysis. *Biochemical Engineering Journal, 39*(1), 91–96.

Panesar, R., Panesar, P. S., Singh, R. S., & Bera, M. B., (2007a). Applicability of alginate entrapped yeast cells for the production of lactose-hydrolyzed milk. *Journal of Food Process Engineering, 30*(4), 472–484.

Panesar, R., Panesar, P. S., Singh, R. S., Kennedy, J. F., & Bera, M. B., (2007b). Production of lactose-hydrolyzed milk using ethanol permeabilized yeast cells. *Food Chemistry, 101*(2), 786–790.

Paseephol, T., Small, D. M., & Sherkat, F., (2013b). Lactulose production from milk concentration permeate using calcium carbonate-based catalysts. *Food Chemistry, 111*(2), 283–290.

Pham, T. T., & Shah, N. P., (2008). Effects of lactulose supplementation on the growth of bifidobacteria and biotransformation of isoflavone glycosides to isoflavone aglycones in soymilk. *Journal of Agricultural and Food Chemistry, 56*(12), 4703–4709.

Porkka, L., Salminen, E., & Salminen, S., (1988). The effects of lactulose-sweetened yoghurt on the rate of gastric emptying and intestinal transit in healthy human volunteers. *Magazine for Nutritional Science, 27*(3), 150–154.

Pranami, D., Sharma, R., & Pathak, H., (2017). Lactulose: A prebiotic, laxative and detoxifying agent. *Drugs & Therapy Perspectives, 33*(5), 228–233.

Ruiz-Matute, A. I., Sanz, M. L., Corzo, N., Martín-Álvarez, P. J., Ibáñez, E., Martínez-Castro, I., & Olano, A., (2007). Purification of lactulose from mixtures with lactose using pressurized liquid extraction with ethanol – water at different temperatures. *Journal of Agricultural and Food Chemistry, 55*(9), 3346–3350.

Schumann, C., (2002). Medical, nutritional, and technological properties of lactulose. An update. *European Journal of Nutrition, 41*(1), i17–i25.

Schuster-Wolff-Bühring, R., Fischer, L., & Hinrichs, J., (2010). Production and physiological action of the disaccharide lactulose. *International Dairy Journal, 20*(11), 731–741.

Seki, N., & Saito, H., (2012). Lactose as a source for lactulose and other functional lactose derivatives. *International Dairy Journal, 22*(2), 110–115.

Shimatani, M., Uchida, Y., Matsuno, I., Oyoshi, M., & Ishiyama, Y., (1993). *U.S. Patent No. 5,270,462*. Washington, DC: U.S. Patent and Trademark Office.

Simrén, M., & Stotzer, P. O., (2006). Use and abuse of hydrogen breath tests. *Gut, 55*(3), 297–303.

Sitanggang, A. B., Drews, A., & Kraume, M., (2016). Recent advances in prebiotic lactulose production. *World Journal of Microbiology and Biotechnology, 32*(9), 154.

Song, Y. S., Lee, H. U., Park, C., & Kim, S. W., (2013a). Optimization of lactulose synthesis from whey lactose by immobilized β-galactosidase and glucose isomerase. *Carbohydrate Research, 369*, 1–5.

Song, Y. S., Lee, H. U., Park, C., & Kim, S. W., (2013b). Batch and continuous synthesis of lactulose from whey lactose by immobilized β-galactosidase. *Food Chemistry, 136*(2), 689–694.

Tabatabaie, T., Kotake, Y., Wallis, G., Jacob, J. M., & Floyd, R. A., (2014b). Spin trapping agent phenyl N-tert-butyl nitrone protects against the onset of drug-induced insulin-dependent diabetes mellitus. *FEBS Letters, 407*(2), 148–152.

Ten, B. S. J., Bovee-Oudenhoven, I. M., Feitsma, A. L., Van, H. E., & Schoterman, M. H., (2014). Functional role and mechanisms of sialyllactose and other sialylated milk oligosaccharides. *Nutrition Reviews, 72*(6), 377–389.

Venema, K., Van, N. M. H., Van, D. H. E. G., Pool, W., & Van, D. V. J. M., (2003). The effect of lactulose on the composition of the intestinal microbiota and short-chain fatty acid production in human volunteers and a computer-controlled model of the proximal large intestine. *Microbial Ecology in Health and Disease, 15*(2, 3), 94–105.

Villamiel, M., Corzo, N., Foda, M. I., Montes, F., & Olano, A., (2002). Lactulose formation catalyzed by alkaline-substituted sepiolites in milk permeate. *Food Chemistry, 76*(1), 7–11.

Wang, H., Yang, R., Hua, X., Zhao, W., & Zhang, W., (2013). Enzymatic production of lactulose and 1-lactulose: Current state and perspectives. *Applied Microbiology and Biotechnology, 97*(14), 6167–6180.

Yang, L., Watanabe, S., Li, J., Chu, J. S., Katsuma, M., Yokohama, S., & Fix, J. A., (2003). Effect of colonic lactulose availability on the timing of drug release onset *in vivo* from a unique colon-specific drug delivery system (CODES™). *Pharmaceutical Research, 20*(3), 429–434.

Yelich, M. R., Schieber, C. K., Umporowicz, D. M., & Filkins, J. P., (1992). Polymyxin-B suppresses endotoxin-induced insulin hypersecretion in pancreatic islets. *Circulatory Shock, 38*(2), 85–90.

Zokaee, F., Kaghazchi, T., Zare, A., & Soleimani, M., (2002). Isomerization of lactose to lactulose—study and comparison of three catalytic systems. *Process Biochemistry, 37*(6), 629–635.

CHAPTER 7

Innovation in Technology for Processing of Sheep and Goat Milk and Dairy Products: An Overview

ANTONIO JOSÉ TRUJILLO* and BIBIANA JUAN

*Corresponding author. E-mail: Toni.Trujillo@uab.es

ABSTRACT

The need for enhancing microbial food safety, shelf life and quality, without altering the nutritional, functional, technological, and sensory characteristics of foods, has produced an increasing interest in low-temperature innovative processes for food preservation. Among these technologies, high pressure has been identified as a particularly promising technology for processing milk and dairy products. The aim of this chapter is to present the potential application of two innovative technologies based on high pressures (high hydrostatic pressure (HHP) and ultra-high-pressure homogenization) with respect to the classical heat treatments applied in goat and sheep milk and dairy products (cheese and yogurt), and discuss their effects on milk components and technological properties.

7.1 INTRODUCTION

Goat and sheep milks are of particular interest due to their specific composition, which has led them to be considered as raw materials for manufacturing high-quality foods. Dairy products from goat and sheep milks can offer a substitute to cow milk products due to their peculiar taste, texture, typicity, and their natural and healthy image. Goat and sheep milks are mainly processed into cheeses and fermented milks. Although the production of other products such as fluid milk (goat milk), butter, cream, ice cream, dry whole milk, etc., are also important in many countries. But limited publications are available on the production of these products (Pandya, and Ghodke, 2007; Raynal-Ljutovac et al., 2008; Ribeiro and Ribeiro, 2010).

The production of the different varieties of cheese and fermented milks involves a generally similar protocol, various steps of which could be modified to give a product with the desired characteristics. For hygienic, but also for technological and quality reasons, most cheese and yogurt making involves the heat treatment of milk; however, in industrial productions other conventional technological operations including standardization, homogenization, membrane concentration, etc., and new technologies can be considered.

Due to the increased demand of consumers for tasty, nutritious, natural, and easy-to-handle food products, the food industry is thoroughly investigating new techniques to replace the traditional preservation processes. There is a large list of innovative technologies at various stages of implementation from the laboratory to the industrial scale. These processes include new thermal treatments (microwaves, ohmic heating,), chemical treatments or non-thermal treatments. The most investigated new preservation technologies are the non-thermal inactivation such as high-pressure processing (HPP) and pulsed electric fields (PEFs), new packaging systems like modified atmosphere packaging (MAP) and active packaging, natural antimicrobial compounds and biopreservation (Fryer and Versteeg, 2008; Risch, 2009; García et al., 2010). Whether these technologies have a commercial future or not depends, amongst several other factors, on the state of the technology, the application, the company, and the country.

Studies on non-thermal technologies, particularly high hydrostatic pressure (HHP) and ultra-high-pressure homogenization (UHPH) are attracting interest from the dairy industry due to their ability to ensure safety without significantly affecting the nutritional composition and sensory characteristics of dairy products.

The aim of the chapter entitled "Innovation in technology for processing sheep and goat milk and dairy products: an overview" is to presents an overview of the main heat treatments applied to goat and sheep milk and their effects on cheese and yogurt manufacture, and the potential application of two innovative technologies based on high pressure *viz.* HHP and UHPH in milk and derived products from these species.

7.2 HEAT TREATMENT

Pasteurization is the traditional treatment to ensure potential pathogen destruction and spoilage microorganism's reduction in milk for cheese or yogurt making. Most cheeses are made from pasteurized milk rather than raw milk for hygienic purposes. Although in certain Mediterranean countries aged cheese is made from high quality milk that is not fully heat-treated. However, other heat treatments apart from conventional pasteurization could be used in cheese making, such as thermization or high temperature pasteurization.

Milk thermization aims at reducing the number of microorganisms (especially psychrotrophic bacteria) by a factor of 3 or 4 logs, minimizing the heat effects on the raw milk microbiota and on the functionality of milk proteins and salts. Typical temperature-time combinations commonly applied are 68°C for 10 s or 65°C for 20 s.

Conventional pasteurization systems rely on heating products to about 63–65°C for about 30 min in batch heating units, followed by rapid cooling; this process is referred to as low-temperature long-time (LTLT) pasteurization. This process is still used by small-scale milk processors. However, when higher volumes are required to be treated, continuous-flow plate or tubular heat exchanger-based pasteurizers, in which the product is heated to 72–74°C for at least 15 s (process designated high-temperature short-time called HTST pasteurization), are largely used.

More severe pasteurization treatment (80–95°C) are typically applied in yogurt milk, and in cheese milk in order to retain both casein and whey proteins in the curd.

There are various types of fluid goat milk around the world (Ribeiro and Ribeiro, 2010). Goat treated by ultra-high temperature (UHT) milk has the large market; however, fresh pasteurized goat's milk is also market by many companies in some countries. Pasteurized sheep's and UHT milk have a more distal distribution.

7.2.1 EFFECTS ON MAJOR CONSTITUENTS OF MILK

It is well known that heat treatments produce physicochemical changes in the milk constituents, including whey protein denaturation and whey protein interactions with casein micelles, reaction between lactose and protein (Maillard reactions), changes in casein micelle structure, transfer of soluble calcium and phosphate to colloidal phase, changes in fat globule membrane and in enzyme activities and decrease in pH (Singh and Waungana, 2001).

In an overview, Raynal-Ljutovac et al. (2007) revised the modifications and destabilizations that occur in goat and sheep milk after different heat treatments. The effects of heat treatments on whey protein denaturation in goat and sheep milks have been studied by different authors (Law, 1995; Montilla et al., 1995; Felipe et al., 1997; Law et al., 1998). With heat treatment, the order of irreversible denaturation of caprine and ovine whey proteins, based on loss of solubility at pH 4.6, is immunoglobulins (Ig) > serum albumin/lactoferinn (SA/Lf) > β-lactoglobulin (β-Lg) > α-lactalbumin (α-La), although there are differences in the rates of denaturation depending on the heat treatment applied. Caprine and ovine whey proteins seem to be less denatured than in cow milk at 70°C, but the opposite occurs at 80 and 90°C (Law, 1995). Under mild pasteurization conditions of milk (70°C for 5 min), approximately 7–10% of caprine whey proteins are denatured. However, when more severe pasteurization processes are applied, more whey proteins are denatured; a combination of 85°C for 5 min produces about 60% of α-La and 90% of each β-Lg, Igs, and SA denaturation (Law et al., 1998). Denatured β-Lg has been shown to associate with κ-casein involving a disulfide bond in goat milk during heat treatment (Henry et al., 2002). According to Raynal and Remeuf (1998) thermal treatments up to 90°C for 10 min have little impact on micelles of bovine casein if the interactions of casein with whey proteins are excluded. However, the size of caprine and ovine casein micelles starts to increase at 85°C, reaching an increase from 25% in goat milk to 75% in sheep milk.

Generally, during heat treatment a transfer of calcium and phosphate from soluble to colloidal phase occurs due to the solubility loss of these minerals, although changes could be reversible upon cooling at room temperature depending on the intensity of the heat treatment. The effect of heat treatment on the mineral balance of goat and sheep milk has been studied by De la Fuente et al. (1999) and by Raynal and Remeuf (1998) showing that pasteurization (85°C for 20 s) has no effect on the main soluble minerals but exposure to more severe heating (85°C for 10 and 30 min) reduces it (~20%). Heating (85°C for 30 min) results in a significant decrease in pH.

7.2.2 EFFECTS ON TECHNOLOGICAL PROPERTIES OF MILK

7.2.2.1 RENNET COAGULATION PROPERTIES

Several studies have shown the effects of heat treatments on cheese making properties of goat and sheep milks including different parameters related to the rennet coagulation properties, such as curd firmness (CF), curd syneresis, rate of curd firming (RCF), rennet clotting time (RCT), and potential cheese yield (Montilla et al., 1995; Balcones et al., 1996; Calvo and Balcones, 1998, 2000; Espinoza and Calvo, 1998; Remeuf and Raynal, 2001; Calvo, 2002).

The results published indicate that heating has a less pronounced effect on RCT and RCF in goat and sheep milks than in cow milk (Montilla et al., 1995; Raynal and Remeuf, 1998). The last authors studied the effect of heating at 75–90°C for 30 s to 10 min on renneting properties of goat milk. The RCT given by Formagraph increased with heating time and temperature; the increase of RCT reached a maximum between 1.3 and 1.5 times the RCT of fresh milk for heating intensity higher than 80°C for 1 min. At 75°C, the RCF did not increase markedly, but at higher temperatures the RCF values decreased (3.5 times lower than raw milk at 90°C) showing that coagulation is less rapid. Finally, firmness of rennet curds was almost unaffected by heating, showing that heated goat milk has the same cross-linking capacity as raw milk. However, Montilla et al. (1995); and Calvo (2002) observed that RCT was not affected by the different heat treatments assayed (65–85°C for 5–35 min), and only a slight decrease in RCF values was observed when goat and sheep milks were heated. Balcones et al. (1996) studied the effect of temperature (65–85°C), heating time (0–35 min), pH (6.2–6.8) and concentration of $CaCl_2$ (0–10 mM) on the rennet clotting characteristics of sheep milk. Heating at 85°C for 35 min, although β-Lg was totally denatured but induced slight modifications in the RCT. As the concentration of $CaCl_2$ and pH increased, the RCT decreased, and CF increased with decreased pH. The concentration of Ca^{2+} was the most important factor influencing the rennet clotting characteristics of sheep milk in the context of Balcones and coworker's research.

The heat-induced rise in RCT could be described by the denaturation of whey protein and bonding with κ-Cn, which is likely to slow the enzymatic stage of coagulation through the repulsion or steric blockage of the action of rennet enzymes (Van Hooydonk et al., 1987). However, RCF, and CF are not affected by heating, suggesting that enzymatic phase is delayed but not limiting for subsequent aggregation and curd shrinkage processes. According to Raynal and Remeuf (1998), and in relation to renneting process, the quantity of κ-Cn to be hydrolyzed for promoting micelle aggregation could be smaller in goat and sheep milks, due to their lower colloidal stability, if it is compared with cow milk. Thus, whey protein denaturation might not be a limiting factor for the secondary phase of coagulation in goat milk, as occurs in cow milk.

The effects of milk (cow, goat, and sheep), fat, curd incubation temperature (25, 30, and 35°C), heat treatment of milk (70°C for 5 or 30 min), and milk pH on the initial drained volume, syneresis rate, and drainage protein loss were studied by Calvo and Balcones (2000). Heat treatment of milk caused a significant decrease in the drainage rate of curds made from heated milks, and the influence of the intensity of the heat treatment was also detected. Significant differences between species (ovine > bovine > caprine) were found in curd from samples heated at 70°C for 30 min. Bovine and ovine samples showed similar drainage rates in the samples heated at 70°C for 5 min, whereas the caprine samples showed a significantly lower value. Differences in syneresis rate between ovine and caprine curds could be related to the heat-induced whey protein denaturation in sheep (~40%) and goat (20%) milks at 70°C for 30 min. The loss of proteins in whey was higher in cow milk than in goat and sheep milks, perhaps because of the higher denaturation of whey proteins in cow milk during heating. Differences between the two heat treatments were also significant for goat and sheep milk samples.

Calvo and Balcones (1998) analyzed the effect of the intensity of the heat treatment on the cheese yield of cow, sheep, and goat cheese milk submitted to different heat treatments (75°C for 5–30 min, 80°C for 30 min and 90°C for 5–30 min). Significant differences in cheese yield were only observed at 75°C for different times, with goat milk obtaining the highest cheese yield, but no significant differences were detected in goat and sheep milk samples submitted to higher heat treatments.

7.2.2.2 ACID COAGULATION PROPERTIES

Milk heat treatment is one of the most significant procedures that affect the texture and consistency of acid gels, such as yogurt (Lucey and Singh, 1998). As previously shown, heat treatment at a temperature >70°C causes whey protein denaturation, and it also promotes their interactions with casein micelles. It is likely that during acidification of heated milk denatured whey proteins become susceptible to aggregation, enhancing protein-protein interactions between denatured whey proteins in the serum and with denatured whey proteins related with micelles, increasing the number and strength of bonds between protein particles, and hence improving the gel formation (Lucey et al., 1997). For this reason, a common heat treatment applied to milk before the production of yogurt reaches temperatures between 85 and 95°C for a few minutes. However, when yogurt is made from severely heated milk (i.e., 90°C for 30 min) the acid gel presents texture defects (i.e., grainy) and it is more susceptible to whey separation (Lucey and Singh, 1998).

The development of caseins and whey proteins in ovine milk has undergone to various thermal treatments (63°C/30 min; 73°C/15 min; 85°C/10 min and 96°C/5 min) (El-Zahar et al., 2003), which was followed during fermentation of yogurts and during their storage up to 14 days. During the production and storage of yogurt, the contents of free amino acids and the susceptibility of α_s-, β-casein and whey proteins to proteolytic degradation by the starter increased in yogurt samples prepared from ovine milk subjected to higher heat treatment, although heat treatment of milk did not affect lactic acid production by the starter.

Zamberlin et al. (2010) examined the impact of sheep milk thermal treatment at 60°C/5 min and 90°C/5 min on the set yogurt composition and sensory characteristics. Results show that the yogurt produced by thermal treatment at the low temperature possessed acceptable sensory qualities and viable starter culture bacteria during the storage period; these authors recommend the usage of low temperature heat treatment of ovine milk in order to preserve a high amount of inherent functional and nutritional characteristics of ovine milk. However, in comparison with cow or goat milks, sheep milk contains significantly higher concentrations of total proteins and fat and consequently better yogurt making characteristics. However, the production of goat milk yogurt with a consistency similar to cow milk yogurt is difficult, mainly due to the low level of α_{s1}-casein and seasonal changes in chemical composition of the milk (Guo, 2003). Methods to improve texture characteristics of goat milk yogurt include product reformulation by the addition of powdered milk solids, polymerized whey proteins, water binding stabilizers such as pectin, gelatin, and carrageenan, or the use of microbial transglutaminase and processing techniques (i.e., evaporation) (Farnsworth et al., 2003; Posecion et al., 2005; Li and Guo, 2006).

Acid coagulation is also used to promote coagulation in the production of goat milk cheeses. In this product, milk coagulation is obtained by the predominant action of lactic mesophylic starter, in a process which is characterized by slow coagulation kinetics with a length of 16–24 h and a final drainage pH of ~4.40. Masle and Morgan (2001) studied the acidification capacity of goat milk, investigating the influence of different heat treatments applied to milk (thermization at 65°C/30 s; pasteurization at 72°C/30s and high pasteurization at 90°C/2 min). The results showed an increase of lactic starter activity at the beginning of the acidification period to increase the intensity of heat treatment. According to the authors heat treatment could produce the destruction of inhibitors (agglutinins, lactoperoxydase) and the release of certain growth factors, which could favor starter activity.

7.2.3 EFFECTS ON CHEESE RIPENING

The main biochemical event in cheese during ripening concerns the nature and extent of proteolysis having a direct and indirect role in the formation of texture and flavor of cheeses, although other biochemical reactions such as lipolysis, metabolism of residual lactose, lactate, and citrate, and the formation of volatile compounds are also important in the development of flavor compounds (Grappin and Beuvier, 1997).

Numerous studies involving different cheese varieties, including those from goat and sheep milks, have been published in these last two decades in which raw (RA) milk cheeses have been compared with pasteurized (PA) milk (for review see Beuvier and Buchin, 2004). A decrease in the number of indigenous microbiota and possibly the changes produced by pasteurization on indigenous milk enzymes may explain why PA milk cheese lacks the full aroma of traditional cheese made from RA milk (Grappin and Beuvier, 1997). According to Beuvier et al. (1997) the addition of indigenous microbiota, obtained in a retentate by microfiltration, to pasteurized milk is able to restore most of the characteristics of Swiss-type cheese made from RA milk.

Milk pasteurization produces a significant decrease of the amount of small peptides and free amino acids and different HPLC peptide profiles in goat and sheep milk cheeses, but there is not a clear trend in the primary proteolysis, being the breakdown of α- and β-caseins highly dependent on the cheese variety studied (Gómez et al., 1999; Ordóñez et al., 1999; Mendia et al., 2000; Xanthopoulos et al., 2000; Albenzio et al., 2001; Moatsou et al., 2001; Trujillo et al., 2002a; Gaya et al., 2005). Related to lipolysis, the total amount of free fatty acids appears to be lower in PA than in RA cheeses, which is attributed to the inactivation of indigenous milk lipase by pasteurization (Gaya et al., 1990; Sousa and Malcata, 1997; Chávarri et al., 2000; Albenzio et al., 2001; Buffa et al., 2001a; Morgan et al., 2001). Furthermore, RA cheeses generally have higher levels of volatile compounds than PA milk cheeses, with exception in certain volatile chemical families (Ortigosa et al., 2001; Fernández-García et al., 2002; Gómez-Ruiz et al., 2002). RA milk cheese is usually more flavored and ripened faster than pasteurized milk cheese, mainly due to the natural biodiversity of microorganisms present in raw milk (Mendia et al., 1999; Ordóñez et al., 1999; Ortigosa et al., 2001). Finally, differences in cheese texture related to the pasteurization treatment applied on milk seem to be not evident (Mendia et al., 1999; Buffa et al., 2001b).

There is a great variety of Mediterranean traditional cheeses made from goat and sheep raw milk with highly differentiated sensory properties compared to those made from PA milk; these differences are attributed mainly to the milk, the processing conditions, and the indigenous microbiota, as previously mentioned. The observation that RA milk cheeses develop a more intense flavor than those from PA milk, and the growing interest in preserving the characteristics of traditional RA milk cheeses, has stimulated the study of lactic acid bacteria (LAB) and non-starter LAB isolated from artisanal RA milk cheeses for use as autochthonous and adjunct cultures, for the industrial manufacture of pasteurized milk cheeses (Gómez et al., 1999; Madrau et al., 2006; Asteri et al., 2009; Nieto-Arribas et al., 2009).

Literature on cheese production from high heated milk goat and sheep milks is very scarce. Faccia et al. (2007) studied the effect of the milk type (goat and sheep) and time of the ripening on the biochemical changes in craft cheese (known as Cacioricotta) which is manufactured from overheated milk (90°C). Cheeses from overheated milk presented a significant higher primary proteolysis, but low ripening indices in ripened cheeses because of poor secondary proteolysis, with lipolysis proving to be the most important biochemical phenomenon of ripening in this type of cheese.

7.3 HIGH HYDROSTATIC PRESSURE (HHP)

High hydrostatic pressure (HHP) has been used in recent years in the food industry as an alternative to a wide range of food processing technologies, especially to thermal processing. In addition, more energy is saved in comparison to thermal stabilization techniques, because once the desired pressure is reached, it can be maintained without the need for further energy input. It uses pressure of up to 1,000 MPa, although the most conventional treatments applied to food are 400–600 MPa. HHP is transmitted isostatically and instantaneously, and thus the process is independent of the shape or size of the food, which can often be problematic in thermal processing. Compared to thermal processing, pressure has fewer detrimental effects on food and therefore the products preserve most of their natural characteristics such as color, flavor, and nutritional and functional properties (Cheftel, 1991; Smelt, 1998; Patterson et al., 2007).

7.3.1 EFFECTS ON MILK MICROBIOLOGY

The interest in milk treatment in HHP has increased largely because of the possibility in which, spoiling, and pathogenic microorganisms could be reduced, without any significant effect on flavor and nutritional.

Microorganisms have a high resistance to food stress depending on conditions for the HHP process (pressure, temperature, time, cycles, etc.), the food components, the characteristics of the microorganism and its physiological state. There is much published research on the changes induced by pressure treatment of microbial cells, including alterations in the cell membrane, cell morphology, effects on proteins, including enzymes, and effects on genetic mechanisms of microorganisms (Smelt, 1998). In the stationary phase, exponentially increasing cells are more susceptible to pressure than cells. The bacterial spores have always more resistance than the vegetative cells and can survive at 1,000 MPa pressure. However, bacterial spores can often be stimulated by pressure ranging from 50–300 MPa to germinate. Then, germinated spores may be killed through treatments with heat or mild pressure. Microorganisms that are gram-positive are more pressure resistant than microorganisms that are gram-negative. In both gram-positive and gram-negative microorganisms however, a significant variation in pressure resistance in strains of the same species has been shown. For inactivation of Gram-positive and Gram-negative microorganisms typically requires treatments of 500–600 MPa and 300–400 MPa pressure respectively for 10 min at 25°C. The vegetative forms of molds and yeasts are the most sensitive to pressure (Smelt, 1998).

Over the last decade, several studies have been conducted with cow's milk, as well as a few sheep's and goat's milks, in which HHP has been used to inactivate pathogenic and spoilage microorganisms (naturally present or inoculated). These studies have generally revealed that milk pressurized at 400 to 600 MPa for 10 to 60 min presents a comparable microbiological quality to the standard heat pasteurization treatment (72°C during 15 s), but not comparable to sterilized milk due to the high resistance of bacterial spores to the high pressures (Mussa and Ramaswamy, 1997; Buffa et al., 2001b). Buffa et al. (2001b) studied the bacteriological quality during ripening of raw, pasteurized (72°C, 15 s) and pressure-treated (PR) (500 MPa, 20°C, 15 min) goat milk assessed by enumeration of total bacteria, *Enterobacteriaceae*, enterococci, lactobacilli, lactococci, *Micrococcaceae,* and psychrotrophic bacteria. The high-pressure treatment was applied as effective as pasteurization to reduce the bacterial population of milk.

In order to determine the sensitivity of pathogenic and spoilage microorganisms in milk, a number of studies have been carried out on the effects of HHP on inoculated target microorganisms in sheep milk. For example: *Escherichia coli* CECT 405 (considered a good indicator of direct or indirect contamination of fecal origin), *Lactobacillus helveticus* CECT 414 (a non-pathogen microorganism but representative of lactic microbiota), *Listeria innocua* CECT 910 (indicator of human-pathogen *L. monocytogenes*), *Pseudomonas fluorescens* CECT 378 (indicator of *Pseudomonas* spp., major components of the spoilage microbiota of refrigerated milk), and *Staphylococcus aureus* CECT 534 (major components of the spoilage microbiota of mastitic milks) (Gervilla et al., 1997a, b, 1999a, b).

Gervilla et al. (1999a) reported an inactivation above 6 log cfu/ml for *E. coli* and *P. fluorescens* inoculated in ovine milk (6% fat) when the milk was treated at different pressures (300–500 MPa) at 50°C, although the level of inactivation reduced with the temperature. On the contrary, *Staphylococcus aureus* proved to be extremely resistant to pressure, and cell reductions above 7 log cfu/ml were only achieved when treatments of 500 MPa at 50°C for 15 min were applied. Gervilla et al. (1999b) described for *L. helveticus* that the pressure treatments were more effective at low (2 and 10°C) and moderately high (50°C) temperatures than at room temperature (25°C). In general, HP inactivation was higher on microorganism *P. fluorescens* and followed in an order by other microorganisms *E. coli, L. innocua*, *L. helveticus*, *S. aureus*. Temperature combined with HP on microorganisms had a different effect depending on the microorganism; *L. helveticus, L. innocua,* and *P. fluorescens* exhibited high resistivity to HP at room temperature (25°C) compared to low temperature (4°C), whereas *E. coli* and *S. aureus* revealed low resistivity to HP at room temperature compared to low temperature.

7.3.2 EFFECTS ON MAJOR CONSTITUENTS OF MILK

Treatment with HHP affects many components of milk, especially, the proteins and mineral equilibrium, and changes are induced in the functional properties of milk (Trujillo et al., 1997, 2002b; Huppertz et al., 2002; López-Fandiño, 2006). Almost all the information available is about bovine milk and few studies have been conducted on the effect of HHP on the physicochemical and technological characteristics of sheep and goat milk (López-Fandiño et al., 1998; López-Fandiño and Olano, 1998; Buffa et al., 2001c; Huppertz et al., 2006a, b). HHP treatment induces the disruption of casein micelles and casein dissociation from the micelle, the increase in non-micellar casein levels, and may be responsible for two key mechanisms: solubilization of micellar calcium phosphate and interruption of intramicellular hydrophobic and electrostatic interactions (Needs et al., 2000). HHP treatments >100 MPa considerably increases the non-micellar caseins, with a maximum increase at 250–300 MPa (López-Fandiño et al., 1998). The dissociation rate of individual caseins has been reported to be in the order κ- > β- > $\alpha_{\sigma 1}$- > $\alpha_{\sigma 2}$- in goat and sheep milks (López-Fandiño et al., 1998; Law et al., 1998).

β-Lg is denatured in milk by a pressure of over 100 MPa (estimated solubility loss). However, it appears that α-La and SA are fully resistant to pressures up to 400–500 MPa. The majority of denatured β-lg in HP-treated milk is bound to a small proportion of casein micelles which remain non-sedimentable in the form of whey protein additives or associated with smaller casein particles during an ultracentrifugal procedure (Felipe et al., 1997). HHP denaturation of whey protein differs from milk of different species; HHP denaturation of sheep and goat β-Lg at 250 MPa occurs at a faster rate than that of cow β-Lg in the order as ovine > caprine > cow (Felipe et al., 1997; López-Fandiño and Olano, 1998; Huppertz et al., 2006b). According to the findings of Huppertz et al. (2006b), treatment at 100–400 MPa for 30 min at 20°C has little effect on the level of native α-La in ovine milk, but treatment at 600 MPa reduces it by ~70%; the level of native β-Lg is not affected by treatment at 100 MPa, but is reduced by >90% after treatment at 400 or 600 MPa. Moatsou et al. (2008) observed denaturation of ovine α-La after a HHP treatment of 650 MPa and synergistic effects between this treatment and temperature; at 55°C the residual native α-La was 21% of that in raw milk, whereas at 40°C, the percentage was 55%.

HHP treatment increases the hydration of casein micelles, which is explained by the denatured β-Lg association with casein micelles. This raises net negative charge of micelles to increase micellar solvation and to lower casein micelles induced by the HHP, thereby increasing micellar hydration (Gaucheron et al., 1997).

Different authors have shown that the HHP-induced disintegration of the casein micelles in goat and sheep milk is accompanied by an increase in the levels of diffusible calcium and phosphate from colloidal phase (Law et al., 1998; López-Fadiño et al., 1998; De la Fuente et al., 1999; Huppertz et al., 2006a).

In relation to milk fat globule, the distribution of milk fat globules in sheep's milk has been shown to be mildly altered by HHP up to 500 MPa; pressure therapy at 25 and 50°C has tended to raise the number of tiny globules in the range 1–2 μm, whereas pressure treatment at 4°C tends to be the reverse (Gervilla et al., 2001). However, there was no damage to the milk fat globule membrane as the absence of lipolysis increased at 4°C during milk storage.

7.3.3 EFFECTS ON TECHNOLOGICAL PROPERTIES OF MILK AND DAIRY PRODUCTS

7.3.3.1 CHEESE MAKING PROPERTIES

The coagulation properties of goat milk (RCT, RCF, and CF) are not significantly affected by pressures < 200 MPa, but RCT and CF increase and RCF decreases after treatment at 300–400 MPa (López-Fandiño and Olano, 1998). Partial confirmatory results have been found by Trujillo et al. (1999a); and Buffa et al. (2001d) studying the rennet coagulation properties of pasteurized (72°C for 15 s) and HHP-treated (500 MPa for 15 min at room temperature) goat milk, assessed by using a Gelograph®-NT probe. The technological treatments caused a significant increase in RCT (HHP> pasteurization) compared to raw milk. Pasteurization did not affect RCF and CF; however, both parameters were higher in HHP-treated milk than in pasteurized and raw milks.

In the case of sheep milk, RCT reduces the pressure moderately to 100 MPa and then increases significantly to 200 and 300 MPa, again reducing to 400 MPa with untreated milk-like values., whereas CF is not considerably affected by the pressure treatment (López-Fandiño and Olano, 1998). The RCT lengthening observed in goat and sheep milks treated by HHP is due to the whey proteins denaturation and association with casein micelles, which sterically hinder the action of rennet on the altered micelles, increasing RCT. HHP-produced micelle casein disintegration and whey protein denaturation would explain the increase of CF in curds by increasing the surface area and consequently the potential bonds available for micelle aggregation.

Pressure treatment (400 MPa for 30 min) of goat and sheep milks increased wet curd yield, estimated by centrifugation, by 16.7 and 15.6%, respectively (López-Fandiño and Olano, 1998). Yield of fresh cheese made from caprine milk treated at 500 MPa was increased by 5% compared to pasteurized milk (Trujillo et al., 1999b). The yield increase in HHP-treated milk is mainly due to the incorporation of denatured whey proteins and, in a major extent, to an increase of water retention by the protein matrix of curd. These results are especially interesting for improving the poor cheese making properties of goat milk. It is well known that goat milk curds are fragile with low cohesion and with a great tendency to produce fine particles in cheese whey, decreasing cheese yield.

7.3.3.2 CHEESE FROM HHP-TREATED MILK AND RIPENING CHARACTERISTICS

As previously stated, HHP treatment decreases the number of microbes in milk so that microbiological safety and milk quality can be increased to produce high-quality “raw” cheeses. Trujillo et al. (1999a); and Buffa et al. (2001b) reported a comparable microbiological quality of goat pasteurized milk (72°C for 15 s) and HHP-treated milk (500 MPa for 15 min at 20°C). Results of experimental cheeses made from RA, PA, and PR milks suggest that goat cheese made from PR milk had similar microbiological properties to PA milk cheeses; there are no differences in numbers of starter bacteria during ripening, *Micrococcaceae*, and enterococci remain at a secondary level although differences can be observed between cheeses at the end of ripening, and the number of *Enterobacteriaceae* decreases during ripening, but faster in PR milk cheese than in PA and RA milk cheeses (Buffa et al., 2001b). However, PR milk cheeses present compositional differences compared to PA milk cheeses; fresh cheese made from HHP-treated (500 MPa) caprine milk has a higher pH throughout ripening, higher salt, and moisture content (MC) and a lower fat content than cheese made from PA milk. The lower fat content of PR milk cheese may be due to altered properties of the rennet-induced gel, and the higher salt content to differences in the MC, since the amount of salt absorbed by cheese during brine-salting is proportional to its MC (Trujillo et al., 1999a).

During ripening of PR milk goat cheeses, glycolysis, lipolysis, proteolysis, organic acid formation and texture characteristics are affected as compared to cheeses prepared from RA or PA milk. The fermentation of lactose in lactate (L- or D-) is generally higher in PA milk cheeses than PR milk cheeses. Possibly the starter activity and its ability to ferment residual lactose and reduce pH, and the isomerization of L-lactate in D-lactate in PR milk cheeses are influenced by their high salt levels (Trujillo et al., 1999a). PR milk cheeses (500 MPa) demonstrate the same level of lipolysis as RA milk cheeses, while PA milk cheeses have a lower level of lipolysis, with PR-milk cheeses being the highest in aroma and taste. This behavior can be clarified by the indigenous milk lipase’s heat-sensitive but resistant properties (Trujillo et al., 1999b; Buffa et al., 2001a).

Both pasteurization (72°C, 15 s) and pressurization (500 MPa, 15 min, 20°C) of goat milk affect primary and secondary proteolysis of cheeses (Trujillo et al., 2002). Electrophoretic analyzes show that PA milk cheeses have a more important hydrolysis of α_{s1}-CN than RA milk cheeses at the first stages of ripening (15 days), while PR milk cheeses have a level between those seen in PA and RA milk cheeses. Degradation of b-CN is hydrolyzed at essentially similar rates in PR and RA milk cheeses, but lower than in PA milk cheese. Secondary proteolysis, which gives the proportion of peptides and free amino acids by nitrogen fractionation or by HPLC profiles, shows variations in cheeses, PR-milk cheeses have substantially greater levels of free amino acid than those of RA or PA-milk during ripening. However, there is no significant variations found in the concentration levels of hydrophobic and hydrophilic peptides in cheeses. According to Buffa et al. (2004), total organic acid content of 1-day old cheese steadily increased from day 1 to 60 which was not affected by milk treatment (HHP or pasteurization). However, both RA and PR milk cheese showed higher levels of organic acids at the end of ripening than those made of PA milk. The cheeses made from PR milk are firmer, less fracturable and more elastic than those made from PA with a cheese matrix microstructure closer to that of RA milk cheeses (Buffa et al., 2001c). Taking into account these results, pressure treatment of cheese milk does not seem to adversely affect the overall cheese quality.

7.3.3.3 MICROORGANISMS REDUCTION OR INACTIVATION IN CHEESE

Application of HHP in cheese may decrease the total number of microbes, including microbiota spoilage or inactivation of pathogenic microbes, increase the shelf life and safety of the product. The reduction of the number of microorganisms by HHP depends on pressure conditions, the type of cheese and the stage of ripening (Trujillo et al., 2000).

Studies made on HHP-treated cheeses from sheep milk have shown an influence of pressure in the viable microorganisms in cheese. At pressures above 400 MPa (4 log units), a large decline in microbial counts was induced and was most seen (5 log units) at 500 MPa. Cheeses which are pressurized for 15 days after ripening are lower than those processed for day 1, showed that the ripeness of cheeses affects the level of microbial inactivation (Juan et al., 2004, 2007a). In agreement, HHP treatments at 300 or 400 MPa at 2 or 50 days of ripening reduced significantly the counts of undesirable contaminating microorganisms such as enterococci, coagulase-positive staphylococci, gram-negative bacteria, and coliforms, improving the microbiological quality and safety of La Serena cheese (sheep milk) immediately after treatment and at the end of the ripening period (Arqués et al., 2006).

Capellas et al. (1996) observed reductions of 7-log units of *Escherichia coli* populations working on inoculated fresh goat cheeses milk HP-treated from 400 to 500 MPa for 5 to 15 min at refrigeration and room temperature. Nevertheless, these treatments provided only a 2-log unit reduction of *Staphylococcus carnosus* population, whereas treatments at 50°C for 5 min caused a reduction of 7-log units. Multiple-cycle treatments of 500 MPa with times of between 15 and 30 min also improved the inactivation rate. Combination of 500 MPa and nisin was the most effective treatment to inactivate cheese indigenous microbiota. Inactivation of *Bacillus subtilis* spores inoculated in fresh cheese has also been studied. Germination treatments of 60 MPa at 40°C for 210 min followed by vegetative cell inactivation treatments of 500 MPa at 40°C for 15 min caused a lethality of 4.9-log units of *B. subtilis*, whereas the same combination of treatments applied at 25°C only caused a 2.7-log unit reduction (Capellas et al., 2000).

HHP processing can be used to boost the shelf life and sensory acceptance of fresh cheese (Capellas et al., 2001). Trujillo et al. (2002b) revealed that treatment with 500 MPa pressure for 5 min at 25°C delayed cheese spoilage for 2 months when applied during 5–30 min. It was found that HP treatment in combination with nisine extende cheese shelf-life.

7.3.3.4 CHEESE RIPENING ACCELERATION OR ARREST

Ripening of cheese is a long and costly process, and consequently various approaches have been used to accelerate the ripening process of goat and sheep milk cheeses including the use of elevated ripening temperatures (Ferrazza et al., 2004), the addition of enzymes (Hernández et al., 2005, 2009), the use of cheese slurries and the selection, attenuation, or genetic modification of starter, and more recently the use of HHP (Law, 2001). Different strategies have been used in the application of HHP to accelerate cheese ripening such as the application of low pressures (50–100 MPa) during long times (2–3 days), or high pressures (400–600 MPa) for short times (5–15 min), the use of an initial high pressure during short times followed by a low-pressure treatment for long times, and the use of HHP combined with other technologies.

Yokohama et al. (1992) described the prospective use of HHP technology to accelerate the maturing of Cheddar cheese in a patent. Cheddar cheese, made using a 10-fold higher level of a proteolytic starter, was exposed to pressure from 0.1 to 300 MPa at 25°C for 3 days. The authors claimed that the flavor of the treated cheese was equivalent to that of six-month-old Cheddar. The treated cheese had higher levels of free amino acids than control cheeses, but this may have been due to the use of a more proteolytic starter in cheese manufacture.

A higher degree of proteolysis was found in goat milk cheeses by the application of 50 MPa when the treatment was applied on 4 days old cheeses (Saldo et al., 1999). This treatment enhanced the primary proteolysis of cheeses, while the subsequent peptidolysis remained unaffected (Saldo et al., 2001). Saldo et al. (2002) subjected goat milk cheeses to pressures of 400 MPa for 5 min and lowered the breakdown of α_s-caseins, explained by the reduction of residual rennet and activity of some proteases. However, the HHP-treatment increased the free amino acids release, reaching twice the value found in untreated cheeses after 28 days, possibly a result of higher peptidase activity. These authors observed that pressurization of cheese increased pH levels, which could enhance the enzymatic activity. On the contrary, HHP treatment of 400 MPa for 5 min decelerated the lipolysis of goat milk cheese, treated cheeses having lower amounts of free fatty acids, and it attributed to diminution of starter counts and inactivation of lipolytic enzymes from the secondary microbiota, which reduced the amount of fatty acids released (Saldo et al., 2003). Levels of proteolysis of goat milk cheeses increased when the previous treatment (400 MPa for 5 min) was followed by 50 MPa for 72 h (Saldo et al., 2000). These results suggest that a combination of shock and high-pressure treatment, which causes an enzyme release, followed by a long and moderate-pressure treatment, which enhances enzymatic activity, would be the best conditions to accelerate the ripening of this goat milk cheese variety (Saldo et al., 2000).

In Hispánico cheese manufactured with a mixture of cow and sheep milks, the HHP treatment of 400 MPa for 5 min accelerated the casein degradation and increased the free amino acid content; however, the treatment did not influence the taste quality of cheeses, but decreased the odor quality, limiting the formation of volatile compounds (Ávila et al., 2006). The possibility to accelerate the ripening of sheep milk cheeses by HHP treatments from 200 to 500 MPa for 10 min applied on 1- and 15-day-old cheeses has been studied by Juan et al. (2004, 2007a). These authors found that primary proteolysis was enhanced by pressures of 300 and 400 MPa applied on the first day of ripening and with 200 to 500 MPa applied at 15 days of manufacturing, whereas secondary proteolysis was only enhanced by the treatment at 300 MPa. Cheeses HHP-treated on 1 day of ripening presented the highest free amino acid levels, and those treated at the same pressure on day 15 of manufacturing had the highest levels of water-soluble nitrogen (Juan et al., 2004). However, Garde et al. (2007), submitting raw sheep milk cheeses at 300 and 400 MPa for 10 min at 10°C on day 2 of ripening, observed a delay on the primary proteolysis but an increase in the free amino acid levels. When the treatment was applied at 50 days of ripening, the proteolysis of cheese was not affected.

HHP treatments increase the lysis of starter, which is important in cheese ripening, as it releases intracellular enzymes into the cheese matrix, which play an important role in the breakdown of large and intermediate size peptides into small peptides and amino acids. Furthermore, HHP causes the destabilization of micelles, which might render caseins more susceptible to the action of proteolytic enzymes. It seems that the application of HHP during the early stages of ripening tends to have a greater effect on ripening process than when HHP is applied at a later stage of ripening.

Lipolysis of sheep milk cheeses is also improved by pressurization at 300 MPa at 1 day of ripening, showing twice the levels of free fatty acids after pressurization, which could be caused by the early lysis of cells and better interaction of microbial lipase with the fat. However, at 15 days of ripening the levels of free fatty acids were similar to untreated cheeses (Juan et al., 2007b). Pressurization modified the formation of volatile compounds, and panelist described HHP-cheeses as lower odor and aroma qualities (Arqués et al., 2007; Juan et al., 2007c).

HHP treatments at ≥500 MPa drastically reduced the proteolysis and lipolysis of sheep milk cheeses, probably due to the reduction of starter bacteria and enzyme inactivation by the pressure, suggesting that it may be a useful method for slowing or arresting cheese ripening at the optimum stage of ripening (Juan et al., 2004, 2007a, b). A HHP at 400 or 600 at 21, 42, or 63 days after manufacture also decelerates lipolysis and formation of volatile compounds in ovine milk blue-veined cheese. The lowest concentrations were found in cheeses treated at 600 MPa on day 21 and were associated with the lower counts of microbial groups (Calzada et al., 2013). The potential application of HHP processing for cheese ripening is evident from the results described; however, it depends on the stage of ripening and the variety of cheese.

In spite of accelerating the ripening of cheeses, HHP treatment can alter the rheological characteristics of cheeses. Torres-Mora et al. (1996) suggested the possibility of the generation of desirable new cheese textures and reduced variability of MCs within blocks of reduced fat Cheddar cheese by use of HHP, which produced a more continuous microstructure in Cheddar cheese curd post-pressurization.

HHP treatment of 50 MPa for 72 h produced a softer texture on goat milk cheeses, with a more uniform structure (Saldo et al., 2001). When these cheeses were pressurized at 400 MPa or 400 MPa followed by 50 MPa, their elasticity and mouth coating increased, and crumbliness decreased, but the acceptability of cheeses was not affected. Working on sheep milk cheeses, Juan et al. (2007d) observed that pressurization of cheeses (from 200 to 500 MPa) on the first day of ripening decreased the fracturability and increased the deformability of cheeses. These phenomena could be explained by the increase in the water retention capacity and the more homogeneous microstructure produced by HHP, reducing the possible areas of fracture. The highest treatment (500 MPa) changed significantly the texture of cheese, producing the less crumbly and the most elastic and softest cheese. When the HHP treatment was done at 15 days of ripening, slight differences were found in the texture of cheeses, but the HHP-treated cheeses were qualified as the favorites by the panelists, showing that it is possible to improve cheese texture by HHP. Nevertheless, when sheep raw milk cheeses were HHP-treated at 300 or 400 MPa for 10 min on day 2 of ripening, the fracturability, hardness, and elasticity values were increased (Garde et al., 2007), and cheeses treated at 400 MPa received the lowest scores for quality of taste from panelists. In this case, pressurized cheeses presented lower breakdown of α_{s1}-CN, which could produce firmer cheeses. According to Garde et al. (2007), when pressure treatment was applied at 50 days of ripening, cheese texture was not affected.

7.3.3.5 ACID COAGULATION PROPERTIES AND YOGHURT MANUFACTURE

Several studies on the effects of HHP on acid coagulation properties and yogurt manufacture have been reported in cow milk but only a few studies in other milks. HHP favors acid coagulation, and therefore HHP-treated milk inoculated with LAB acidifies faster than its homologous without treatment, possibly because of HHP-induced micelle disruption, which would enhance the supply of nitrogen for bacteria (Huppertz et al., 2004). Furthermore, improvements in rigidity, force at the breaking point and resistance to syneresis of acid-induced gels made from HHP-treated milk, have been observed (Johnston et al., 1993).

Ferragut et al. (2000) elaborated sheep milk yogurt (16% total solids) from HHP-treated milk using different combinations of temperature and pressure (10, 25, and 55°C; 200, 350 and 500 MPa for 15 min) and from heat treated milk (95°C, 5 min). Firmness was higher in HHP yogurts than in those made from heat-treated milk and increased as pressure increased. Treatments of 350 MPa at 25°C and 500 MPa at 55°C showed no differences in whey syneresis of centrifuged yogurts compared with those heat-treated. This increase in firmness could be due to disintegration of casein micelles into smaller subunits with improved aggregating characteristics, but also to whey protein denaturation. Yoghurt evolution in storage at 4°C for 20 days showed a good stability in terms of firmness in all treatments but water retention was only maintained in HHP yogurts.

Other authors were studied the effect of using HHP-treated milk to manufacture low-fat yogurts (De Ancos et al., 2000; Penna et al., 2006; Udabage et al., 2010), or the use of HHP to prevent rise of acidity during cold storage (Tanaka and Hatanaka, 1992), or to inactivate yogurt microbiota (Reps et al., 1999) and probiotic bacteria (Jakowska et al., 2004), but all were developed in cow milk. However, these results show that prolongation of the shelf-life of yogurt by HHP processing can be obtained by complete inactivation of yeasts, molds, and LAB, while maintaining probiotic bacteria active in probiotic yogurt.

7.4 ULTRA-HIGH-PRESSURE HOMOGENIZATION

Conventional homogenization using moderate pressures (18–20 MPa) has been extensively used in the dairy industry to stabilize emulsions and then prevent creaming during storage due to the reduction of fat globule size. As previously mentioned, HHP processing is one of the food preservation treatments being developed and applied as a process for the production of safe and nutritious foods One class of high-pressure technology based on homogenization is microfluidization, which is based on the principle of collisions between high-speed liquid jets; the fluid is divided into 2 channels at the valve inlet, which then collide in the reaction chamber. Another type of HPP that has been developed is high pressure homogenization (HPH). The principle of the operation is similar to that of conventional homogenizers used in the dairy industry except that it works at higher pressures (up to 400 MPa) by using new materials and valve designs. This technology is also called ultra-high-pressure homogenization (UHPH) depending on the pressure achieved. However, the pressure limit between HPH and UHPH has never been established. In this chapter, we will refer to UHPH for pressures above 100 MPa. Ultra-high-pressure (UHPH) homogenization technology was first used in the pharmaceutical, chemical, and biochemical industries (Floury et al., 2000). Several applications of UHPH in the food industry have been reported. Possible uses of this technology for the dairy industry are being studied as alternative to heat treatment, with possible applications in milk (Pereda et al., 2007), cheese (Zamora et al., 2007) and yogurt (Serra et al., 2007) production. However, since this is a relatively new technology, most studies have been conducted on cow's milk.

7.4.1 EFFECTS ON MILK MICROBIOLOGY

High-pressure homogenization and UHPH have been commonly used for cell disruption of dense microbial cultures (Saboya et al., 2003), and to cause a reduction of the microbial population, improving the microbial safety of food products (Thiebaud et al., 2003; Hayes et al., 2005). In general, by increasing the pressure, inlet temperature, and number of passes through the machine, more microbial inactivation is achieved. The effects of UHPH on bacterial cells are not well known yet, but microorganisms are probably disrupted by the sudden pressure drop by torsion and shear stresses, and most probably by cavitation shock waves resulting from imploding gas bubbles (Thiebaud et al., 2003).

Different authors like Vachon et al. (2002); and Pereda et al. (2007) have suggested the possible potential of UHPH as a combined pasteurization and homogenization step to obtain commercial milk with a shelf life similar to conventional market milk; however, literature related to the effects of UHPH on milk shelf life is scarce.

Microbiological changes of UHPH-treated milks, at different temperatures and pressures have been studied immediately after treatment. However, most have been performed in cow's milk. Results obtained by Pereda et al. (2007) showed that UHPH treatments of 200 and 300 MPa and inlet temperature (Ti) of 30–40°C drastically reduced or eliminated psychrotrophic, lactococci, lactobacilli, coliforms, and enterococci count in cow milk, showing that these treatments were as efficient (99.99%) in reducing the total bacterial population as high pasteurization treatment (90°C, 15 s), reaching reductions of approximately 3.5 log cfu/mL. Thiebaud et al. (2003) obtained lower microbial reductions (1 to 1.15 log cfu/mL) working at 200 MPa and Ti = 24°C, and Hayes et al. (2005) obtained 1.83 log cycles of inactivation for total counts in whole bovine milk working at 200 MPa and Ti = 45°C. However, Smiddy et al. (2007) found a reduction of ~5 log units for UHPH-treated milk at 250 MPa but working at Ti = 55–70°C.

Ruiz-Espinosa et al. (2013) study multiple-pass UHPH to reduced indigenous spoilage microflora and a *Staphylococcus aureus* inoculated in whole milk. These authors found an optimized UHPH process (five passes from 202 MPa; four passes from 232 MPa) obtaining microbial reduction above 5 log cfu/mL, as a requisite parameter for establishing an alternative method of pasteurization.

The bacteriological quality of goat milk treated by UHPH (200 MPa at Ti = 30°C), by pasteurization (72°C for 15 s) or by homogenization-pasteurization (18 + 2 MPa and 72°C for 15 s) has been evaluated by Quevedo et al. (2010). Overall, UHPH treatment was as efficient as pasteurization in reducing all the bacterial population studied in milk, reaching important reductions in the different microbial groups (~3–4.6 log cfu/ml).

Amador-Espejo et al. (2014a) studied the effect of ultra-high-pressure homogenization on raw whole milk to obtain processing conditions for the sterilization of milk. Treatments at 300 MPa with Ti of 55, 65, 75, and 85°C, and at 200 MPa with Ti = 85°C presented high level of microbial reduction than UHT treatment (138°C for 4 s). Furthermore, UHPH treatments performed at 300 MPa with Ti = 75 and 85°C produced sterile milk after sample incubation (30 and 45°C), showing a real alternative to heat treatment.

To date, only two works in which microbial shelf life of milk during cold storage was assessed, are available. In the study of Smiddy et al. (2007) a microbial shelf life between 4 and 7 d was obtained for UHPH-treated milk at 200 or 250 MPa at the first valve and 5 MPa at the second valve. However, Pereda et al. (2007) showed that UHPH treatments (200 and 300 MPa, Ti = 30–40°C) produced milk with a microbial shelf life between 14 and 18 d, similar or even better to that achieved for high pasteurized milk (90°C, 15 s).

Numerous authors have investigated the inactivation of the microorganisms responsible for the main foodborne diseases from contaminated milk (*Salmonella* spp., *Listeria monocytogenes*, *Staphylococcus aureus*, *Escherichia coli*, etc.), by the use of HPH and UHPH treatments (Vachon et al., 2002; Briñez et al., 2006a–c).

UHPH treatments at 300 MPa with inlet temperatures (Ti) of 55, 65, 75 and 85°C were applied to commercial UHT treated whole milk inoculated with different *Bacillus* spores in order to evaluate the inactivation level achieved. UHPH at 300 MPa with Ti=75 was capable of a spore inactivation of ~5 log cfu/mL, and the treatment at 300 MPa with Ti=85°C produced the complete inactivation at the spore level inoculated (~1×10^6 cfu/mL) leading to the possibility of obtaining commercially sterile milk (Amador-Espejo et al., 2014b).

7.4.2 *EFFECTS ON MAJOR CONSTITUENTS AND CHARACTERISTICS OF MILK*

The UHPH treatment, besides achieving a reduction in microbial counts, generates changes in the main structures (fat globule and casein micelle) and physicochemical properties of milk such as color, viscosity, and creaming. UHPH has been shown to decrease milk fat globule size in bovine and caprine whole milk with increasing pressures up to 200 MPa; however, when pressures in the range of 250–300 MPa are applied an increase in the particle size is observed, probably due to the formation of fat globule clusters (Thiebaud et al., 2003; Pereda et al., 2007; Serra et al., 2007; Trujillo et al., 2007; Zamora et al., 2007). HPH promotes casein micelle dissociation and partial whey protein denaturation. Hayes and Kelly (2003b) observed a 5% decrease in casein micelle size in samples UHPH-treated at 200 MPa. Similar results were obtained by Sandra et al. (2005) subjecting skim milk to 186 MPa, but contrary to HHP, UHPH does not cause complete disruption of casein micelle, although modifications in the casein micelle surface are produced. Zamora et al. (2007) observed denaturation of whey milk proteins by UHPH, being α-La more resistant than β-Lg (~12 and ~35% denaturation, respectively). The effect of UHPH on native milk enzymes (lactoperoxidase, plasmin, and alkaline phosphatase) has been studied by Hayes and Kelly (2003b); and Pereda et al. (2007), who showed that greater inactivation is achieved by increasing pressure.

Rodríguez-Alcalá et al. (2009) studied the effect of HPH (0–350 MPa) of raw cow, goat, and sheep milks on the fatty acid total content and profile. Results indicated that no significant differences are found in the fatty acid composition, especially in the polyunsaturated fatty acids and conjugated linoleic acid isomers profile of raw milk treated by UHPH process up to 350 MPa.

Color, texture, and mouthfeel are important characteristics that determine consumer perception of freshness of milk. An increase in lightness (L*) in UHPH and conventionally-homogenized milk was observed by Hayes and Kelly (2003b); and by Pereda et al. (2007) compared to raw milk, due to the higher number of particles that reflect light more efficiently in homogenized samples. Viscosity is highly dependent on the particle size distribution in food emulsions. Pereda et al. (2007) working on whole bovine milk observed a decrease in viscosity in samples UHPH-treated at 200 MPa, while samples treated at 300 MPa had similar viscosity to RA or PA milk samples. Nevertheless, differences in instrumental color and viscosity measurements between

UHPH treated milks and RA or PA milks were not visually or sensorially obvious. Finally, due to the reduction in particle size achieved, no creaming is observed in UHPH-treated milks during refrigerated storage.

Another important thing to highlight is the reduction in the destruction of hidrosoluble vitamins (nicotinamide, thiamine, riboflavin and vitamin C) by UHPH compared to milks heat-treated. A sample milk treated at 300 MPa with Ti= 45°C presented the closest vitamin content to that of raw fresh milk. In addition, the increase of the Ti in the UHPH treatments produced a significant increment on the vitamin destruction; however, it was less than the destruction produced by the UHT treatment. Additionally, UHPH not affected the content of liposoluble vitamins (Amador-Espejo et al., 2015).

7.4.3 EFFECTS ON TECHNOLOGICAL PROPERTIES OF MILK AND DAIRY PRODUCTS

7.4.3.1 CHEESE MAKING PROPERTIES

As mentioned earlier, most of the studies have been done on cow's milk. In this sense, Hayes, and Kelly (2003a) reported that two-stage high-pressure homogenization was generally more effective in decreasing the average fat globule size and enhancing rennet coagulation properties than single-stage HPH at the same total pressures, especially at 200 MPa. Higher elastic modulus in rennet gels was observed in homogenized samples.

The effect of UHPH at a pressure of 179 MPa on the renneting characteristics of skim milk has been studied by Sandra and Dalgleish (2007) showing that the gelation time of the UHPH-treated milk is shorter than that of the heat-treated milk, presumably because the homogenization removes some of the κ-casein from the surface of the casein micelles.

The effects of single- or 2-stage UHPH (100 to 330 MPa) at an inlet temperature of 30°C on the cheese-making properties of bovine milk have been investigated and compared with those from RA, heat-pasteurized (72°C for 15 s), and conventional homogenized-pasteurized (15 + 3 MPa, 72°C for 15 s) treatments (Zamora et al., 2007). According to these authors single-stage UHPH at 200 and 300 MPa enhances rennet coagulation properties; however, these properties are negatively affected by the use of the UHPH secondary stage. Increasing the pressure led to higher yields and MC of curds. The improvement in the cheese-making properties of milk by UHPH could be explained by changes to the protein-fat structures due to the combined effect of heat and homogenization.

More recently, Quevedo et al. (2011) studied the effect of UHPH (100–300 MPa) on the rennet coagulation properties of goat milk at laboratory scale by using Optigraph system (Ysebaert Inc., Frepillon, France) and by Gelograph®-NT (Gel-Instrumente, Thalwil, Switzerland) directly into the cheese vat, and compared the results with those from raw, pasteurized (72°C for 15 s) and homogenized-pasteurized (18+2 MPa, 72°C for 15 s). UHPH-treated milk at 300 MPa did not coagulate, and the best conditions for goat milk coagulation were 200 MPa at 30°C. UHPH treatment at 200 MPa increased the CF, the RCF and the wet cheese yield, and shortened the rennet coagulation, compared to the other technological treatments assayed. These results could be explained by the particle size reduction and formation of a homogeneous regular and dense molecular structure, which facilitates the aggregation and compaction of the obtained curd.

7.4.3.2 CHEESE FROM HHP-TREATED MILK AND RIPENING CHARACTERISTICS

Zamora et al. (2015) studied the effects of UHPH on the composition and biochemistry of starter-free fresh cheeses and their evolution during cold storage comparing with pasteurization (80°C for 15 s) and homogenization-pasteurization (homogenization at 60°C with 15 and 3 MPa at first and second valves, respectively, followed by pasteurization at 80°C for 15 s). These authors observed that cheeses from UHPH-treated milk (300 MPa, Ti = 30°C) showed lower moisture loss during storage than those from conventionally homogenized-pasteurized milk. UHPH triggered higher water-holding capacity in UHPH-cheeses due to the changes produced in the microstructure and the water typology of cheeses (Zamora et al., 2011). Lipolysis and proteolysis levels in cheeses from UHPH-treated milk were lower than those from conventionally treated milk. Nevertheless, these cheeses presented an increase in the lipid oxidation levels. The incorporation of caseins and whey proteins at the MFGM due to UHPH treatment (Zamora et al., 2012) might render fat globules less susceptible to lipases, however may play a role in enhancing lipid oxidation. The HPH treatment of milk at 100 MPa has proved to increase both proteolytic and lipolytic activities in goat cheese during the ripening compared to those produced from sanitized (61°C for 20 min) or pasteurized (72°C for 15 s) milk (Guerzoni et al., 1999). Also, Crescenza, a traditional Italian soft cheese, when obtained using milk treated at 100 MPa, has shown an early and significant lipolysis attributed to either an indirect effect of the shift of the microbial population or a consequence of different exposure of the fat globules to enzymatic activity as well as to the activation of the naturally occurring lipolytic enzymes (Lanciotti et al., 2004a).

Vanini et al. (2008) compared Pecorino cheeses obtained from sheep milk previously subjected to HPH at 100 MPa with those produced from RA and heat-treated sheep milk. The HPH milk treatment induced a significant increase of the cheese yield and caused a reduction of enterococci, lactococci, and yeasts in the curds. Enterococci cell loads remained at lower levels in cheeses obtained from HPH milk over the ripening period. Analyzes of lipolysis, proteolysis, volatile compounds, and sensory traits evidenced that the pressure treatment can also be regarded as a useful tool for differentiating products obtained from the same raw material. In fact, such a milk treatment induced a marked lipolysis, an early proteolysis, a relevant modification of the volatile molecule profiles and sensory properties of Pecorino cheese.

The effect of UHPH (200 MPa, 30°C) of goat milk on the lipolytic and proteolytic profile of cheese was studied and compared to cheeses made from homogenized-pasteurized (18 + 2 MPa, 72 C for 15 s) and pasteurized (72 C, 15 s) milk (Juan et al., 2015, 2016). Cheeses produced from ultra-high-pressure homogenized milk presented similar levels of lipolysis, lipid oxidation and flavor and aroma to those made from pasteurized milk (Juan et al., 2015). Moreover, UHPH-treated milk cheeses, presented higher hydrolysis of α-, β-, para-k-casein, and whey soluble nitrogen and hydrophobic and hydrophilic peptides, than those made with pasteurized milk, possibly due to a more exposure of proteins to the proteolytic enzymes. Although levels of free amino acids were reduced during the first days of ripening because of the decrease in the aminopeptidase activity, not differences in free amino acids levels were observed at 60 days of ripening (Juan et al., 2016) showing a UHPH as a feasible treatment of goat milk for cheese production.

7.4.3.3 ACID COAGULATION PROPERTIES AND YOGHURT MANUFACTURE

Lanciotti et al. (2004b) studied the effect of moderate homogenization pressures (30–75 MPa) on milk for yogurt production, showing that yogurts presented a high microbiological quality and exhibited a greater variety of textures associated with an increase of viscosity with increasing pressure, which is possibly due to the modifications of the fat globule and casein size. On the other hand, these authors reported that HPH milk treatment enhanced the viability of *St. thermophilus* and *Lb. delbrueckii* spp. *bulgaricus* during refrigerated storage of yogurt and favored the growth of the former compared to the latter, therefore reducing the risk of post-acidification.

Serra et al. (2007, 2009a) compared UHPH-treated milk at 200–300 MPa with an inlet temperature of 30 and 40°C with milk heat-treated (90°C for 90 s) and homogenized at 15 MPa, and with milk treated under the same thermal conditions and also fortified with 3% skim milk powder, in their ability to produce set and stirred yogurt. According to the authors, milk homogenized at 200 or 300 MPa is suitable for set yogurt production showing higher gel strengths on coagulation, higher gel firmness in texture analysis, less syneresis and lower titratable acidity compared with conventionally treated milk fortified with 3% skim milk powder. The disruption of the network from UHPH-treated milk into stirred gels resulted in yogurts with higher consistency, less syneresis but coarser structure than the conventional ones. The same research group also investigated proteolysis, lipolysis volatile profile and lipid oxidation in these yogurts during storage at 4°C. UHPH milk treatment, especially at 300 MPa, resulted in yogurts with similar proteolytic, lipolytic, oxidation, and volatile profiles, compared to those obtained following the conventional process, with small variations depending on the UHPH treatment applied. UHPH milk yogurts presented similar levels of hydrophobic peptides, which are of special interest for their potential biological activities, even though no skim milk powder was added to UHPH milk samples, but lower levels of lactic acid were present. Counts of *St. thermopilus* were higher than those of *Lb. delbrueckii* spp. *bulgaricus* in all yogurts during storage, but the total cell counts were higher in yogurts with added skim milk and heat-treated than in those form UHPH-treated milk (Serra et al., 2008, 2009b, c).

Moreover, considering that yogurts from UHPH-treated milk at 300 MPa at either 30°C or 40°C had lower levels of lactic acid, they could present an interesting alternative for improving the viability of probiotic bacteria in yogurt. These findings make UHPH treatment at 300 MPa a potential alternative to conventional treatment to obtain good quality products without the need for adding non-fat solids. However, this technology has not been explored for goat or sheep milk yogurt production.

7.5 CONCLUSION, FUTURE PROSPECT AND RESEARCH OPPORTUNITIES

Traditionally, goat, and sheep milk products have been a fundamental part of the diets of many Mediterranean countries. These milks and milk products are rich in a large variety of essential nutrients such as minerals, vitamins, proteins, and lipids, they present very peculiar sensory characteristics, and their consumption has also been associated with beneficial health effects. In order to ensure the food safety of these products, several strategies can be applied such as physical, chemical, and biological methods. However, from all these, heat treatment is the most used due to its economic costs and its efficiency against foodborne pathogens. Nevertheless, heat treatment, depending on its severity, also induces changes in the sensory (i.e., flavor, color, texture, etc.), and nutritional characteristics of milk and dairy products.

Of the most investigated new preservation technologies, HHP has proved to be suitable for preparing products such as juices, meat, fish, seafood, and sauces, fruit, and vegetables, herbs, and dressings, which have very attractive feature for consumers, for example: improved taste, color, texture, and shelf life. In the case of dairy products there are a limited number of products treated by this technology, such as cheese-based sandwich fillings, probiotic yogurt with a shelf life of 3 months due to the reduction of lactobacillus but with selected probiotic strains that are not affected by pressure, colostrum-based beverages that preserve the functionality of heat sensitive bioactive components present in colostrum such as Igs, lactoferrin, and growth factors, different snacks made by mozzarella and beef jerky or ham pork and cheddar cheese, yogurt sauce, cream cheese, fresh cheese and cheese jerky. However, there are other very interesting research dairy applications such cheese ripening acceleration or arresting the cheese ripening at the optimal point, which have not been transferred to the market, although they possess numerous potential advantages.

The industrial HHP equipment used at present is discontinuous, from 200 to 600 L of capacity for liquid, solid, viscous, and particulated packaged foods. The processing cost estimated by using this equipment is ~11–16 eurocents per kilogram of product, investment, and operation costs inclusive (Mújica-Paz et al., 2011), and currently HHP machines are able to process from 266 kg to 2.8 tn/h (www.hyperbaric.com), so the implementation of this technology in the dairy industry today is possible with a reasonable price.

Ultra-high-pressure homogenization is a very new technology, based on the same principles as conventional homogenization, but which works at much higher pressures (from 100 to 400 MPa) thanks to the latest developments in material science. It has current applications in the cosmetic and pharmaceutical industries for the elaboration of nanodispersions and for the extraction of enzymes from microbial cells. However, considering the mechanisms involved, UHPH is being studied for the treatment of milk in order to inactivate microorganisms, to produce very fine and stable emulsions and to modify the rheological and/or the coagulation properties of milk and dairy products, thus improving their hygienic quality and their functional properties. Several studies have shown that the treatment of UHPH in milk improved cheese making properties, showing the possibility of using this technology of milk for cheese production. In addition, although it has not been studied in goat or sheep milk, it also seems to be a good alternative for producing good quality yogurts. This technology allows a continuous treatment of liquid products but currently there is a low level of development and research, so commercial applications of this novel technology in the food industry have yet to be realized.

KEYWORDS

- **acid coagulation**
- **casein micelle**
- **cheese making**
- **cow milk**
- **dairy products**
- **functional properties**
- **high hydrostatic pressure**
- **homogenization**
- **non-thermal treatments**
- **sensorial characteristics**
- **shelf-life**
- **volatile compounds**

REFERENCES

Albenzio, M., Corbo, M. R., Rehman, S. U., Fox, E. E., De Angelis, M., Corsetti, A., Sevi, A., & Gobbetti, M., (2001). Microbiological and biochemical characteristics of canestrato Pugliese cheese made from raw milk, pasteurized milk or by heating the curd in hot whey. *International Journal of Food Microbiology, 67*, 35–48.

Amador, E. G. G., Hernanez-Herrero, M. M., Juan, B., & Trujillo, A. J., (2014b). Inactivation of *Bacillus* spores inoculated in milk by ultra high pressure homogenization. *Food Microbiology, 44*, 204–210.

Amador-Espejo, G. G., Gallardo-Chacon, J. J., Nykänen, H., Juan, B., & Trujillo, A. J., (2015). Effect of ultra high-pressure Homogenization on hydro- and liposoluble milk vitamins. *Food Research International, 77*, 49–54.

Amador-Espejo, G. G., Suàrez-Berencia, A., Juan, B., Bárcenas, M. E., & Trujillo, A. J., (2014a). Effect of moderate inlet temperatures in ultra-high-pressure homogenization treatments on physicochemical and sensory characteristics of milk. *Journal of Dairy Science, 97*, 659–671.

Arqués, J. L., Garde, S., Fernández-García, E., Gaya, P., & Núñez, M., (2007). Volatile compounds, odor, and aroma of La Serena cheese high-pressure treated at two different stages of ripening. *Journal of Dairy Science, 90*, 3627–3639.

Arqués, J. L., Garde, S., Gaya, P., Medina, M., & Núñez, M., (2006). Short communication: Inactivation of microbial contaminants in raw milk La Serena cheese by high-pressure treatments. *Journal of Dairy Science, 89*, 888–891.

Asteri, I. A., Robertson, N., Kagkli, D. M., Andrewes, P., Nychas, G., Coolbear, T., Holland, R., Crowb, V., & Tsakalidou, E., (2009). Technological and flavor potential of cultures isolated from traditional Greek cheeses – a pool of novel species and starters. *International Dairy Journal, 19*, 595–604.

Ávila, M., Garde, S., Gaya, P., Medina, M., & Núñez, M., (2006). Effect of high-pressure treatment and a bacteriocion-producing lactic culture on the proteolysis, texture and taste of Hispánico cheese. *Journal of Dairy Science, 89*, 2882–2893.

Balcones, E., Olano, A., & Calvo, M. M., (1996). Factors affecting the rennet clotting properties of ewe's milk. *Journal of Agricultural and Food Chemistry, 44*(8), 1993–1996.

Beuvier, E., & Buchin, S., (2004). Raw Milk Cheeses. In: Fox, P., McSweeney, P., Cogan, T., & Guinee, T., (eds.), *Cheese: Chemistry, Physics and Microbiology* (pp. 319–345). Academic Press, London.

Beuvier, E., Berthaud, K., Cegarra, S., Dasen, A., Pochet, S., Buchin, S., & Duboz, G., (1997). Ripening and quality of Swiss-type cheese made from raw, pasteurized or microfiltered milk. *International Dairy Journal, 7*, 311–323.

Briñez, W. J., Roig-Sagués, A. X., Hernández-Herrero, M. M., & Guamis-López, B., (2006a). Inactivation of *Listeria innocua* in milk and orange juice by ultrahigh-pressure homogenization. *Journal of Food Protection, 69*, 86–92.

Briñez, W. J., Roig-Sagués, A. X., Hernández-Herrero, M. M., & Guamis-López, B., (2006b). Inactivation of two strains of *Escherichia coli* inoculated into whole and skim milk by ultrahigh-pressure homogenization. *Lait, 86*, 241–249.

Briñez, W. J., Roig-Sagués, A. X., Hernández-Herrero, M. M., & Guamis-López, B., (2006c). Inactivation of *Staphylococcus* spp. strains in whole milk and orange juice using ultra high pressure homogenization at inlet temperatures of 6 and 20°C. *Food Control, 18*, 1282–1288.

Buffa, M., Guamis, B., Pavia, M., & Trujillo, A. J., (2001a). Lipolysis in cheese made from raw, pasteurized or high-pressure-treated goats' milk. *International Dairy Journal, 11*, 175–179.

Buffa, M., Guamis, B., Royo, C., & Trujillo, A. J., (2001b). Microbiological changes throughout ripening of goat cheese made from raw, pasteurized and high-pressure treated milk. *Food Microbiology, 18*, 45–51.

Buffa, M., Guamis, B., Saldo, J., & Trujillo, A. J., (2004). Changes in organic acids during ripening of cheeses made from raw, pasteurized or high-pressure-treated goats' milk. *Food Science and Technology, 37*, 247–253.

Buffa, M., Trujillo, A. J., & Guamis, B., (2001d). Rennet coagulation properties of raw, pasteurized and high pressure-treated goat milk. *Milchwissenschaft, 56*, 243–246.

Buffa, M., Trujillo, A. J., Pavia, M., & Guamis, B., (2001c). Changes in textural, microstructural and color characteristics during ripening of cheeses made from raw, pasteurized or high-pressure-treated goats' milk. *International Dairy Journal, 11*, 927–934.

Calvo, M. M., & Balcones, E., (1998). Influence of heat treatment on rennet clotting properties of mixtures of cow's, ewe's and goat's milk and on cheese yield. *Journal of Agriculture and Food Chemistry, 46*, 2957–2962.

Calvo, M. M., & Balcones, E., (2000). Some Factors influencing the syneresis of bovine, ovine, and caprine milks. *Journal of Dairy Science, 83*, 1733–1739.

Calvo, M. M., (2002). Influence of fat, heat treatments and species on milk rennet clotting properties and glycomacropeptide formation. *European Food Research Technology, 214*, 182–185.

Calzada, J., Del, O. A., Picon, A., Gaya, P., & Nuñez, M., (2013). High-pressure processing decelerates lipolysis and formation of volatile compounds in ovine milk blue-veined cheese. *Journal of Dairy Science, 96*, 7500–7510.

Capellas, M., Mor-Mur, M., Gervilla, R., Yuste, J., & Guamis, B., (2000). Effect of high pressure combined with milk heat or nisin on inoculated bacteria and mesophiles of goat's milk fresh cheese. *Food Microbiology, 17*, 633–641.

Capellas, M., Mor-Mur, M., Sendra, E., & Guamis, B., (2001). Effect of high-pressure processing on physico-chemical characteristics of fresh goats' milk cheese (Mató). *International Dairy Journal, 11*, 165–173.

Capellas, M., Mor-Mur, M., Sendra, E., Pla, R., & Guamis, B., (1996). Populations of aerobic mesophils and inoculated *E. coli* during storage of fresh goat's milk cheese treated with high pressure. *Journal of Food Protection, 59*, 582–587.

Chávarri, E., Bustamante, M. A., Santisteban, A., Virto, M., Albisu, M., Barron, L. J. R., and de Renobales, M., (2000). Effect of milk pasteurization on lipolysis during ripening of ovine cheese manufactured at different times of the year. *Lait, 80*, 433–444.

Cheftel, J. C., (1991). Applications des hautes pressions en technologie alimentaire. *News of Food and Agri-Food Industries, 108*, 141–153.

De Ancos, B., Cano, M. P., & Gómez, R., (2000). Characteristics of stirred low-fat yogurt as affected by high pressure. *International Dairy Journal, 10*, 105–111.

De La Fuente, M. A., Olano, A., Casal, V., & Juárez, M., (1999). Effects of high pressure and heat treatment on the mineral balance of goats' milk. *Journal of Dairy Research, 66*, 65–72.

El-Zahar, K., Chobert, J. M., Sitohy, M., Dalgalarrondo, M., & Heartlé, T., (2003). Proteolytic degradation of ewe milk proteins during fermentation of yoghurts and storage. *Molecular Nutrition and Food Research, 47*, 199–206.

Faccia, M., Gambacorta, G., Caponio, F., Pati, S., & Di Luccia, A., (2007). Influence of type of milk and ripening time on proteolysis and lipolysis in a cheese made from overheated milk. *International Journal of Food Science and Technology, 42*, 427–433.

Farnsworth, J. P., Li, J., & Guo, M. R., (2003). Improved structure and consistency of probiotic goat's milk yogurt. *Australian Journal of Dairy Technology, 58*, 187.

Felipe, X., Capellas, M., & Law, A. J. R., (1997). Comparison of the effects of high-pressure treatments and heat pasteurization on the whey proteins in goat's milk. *Journal of Agriculture and Food Chemistry, 45*, 627–631.

Fernández-García, E., Carbonell, M., & Núñez, M., (2002). Volatile fraction and sensory characteristics of Manchego cheese. 1. Comparison of raw and pasteurized milk cheese. *Journal of Dairy Research, 69*, 579–593.

Ferragut, V., Martínez, V. M., Trujillo, A. J., & Guamis, B., (2000). Properties of yogurts made from ewe's milk treated by high hydrostatic pressure. *Milchwissenschaft, 55*, 267–269.

Ferrazza, R. E., Fresno, J. M., Ribeiro, J. I., Tornadijo, M. E., & Mansur, F. M., (2004). Changes in the microbial flora of Zamorano cheese (P.D.O.) by accelerated ripening process. *Food Research International, 37*, 149–155.

Floury, J., Desrumaux, A., & Lardières, J., (2000). Effect of high-pressure homogenization on droplet size distributions and rheological properties of model oil-in-water emulsions. *Innovative Food Science and Emerging Technologies, 1*, 127–134.

Fryer, P. J., & Versteeg, C., (2008). Processing technology innovation in the food industry. *Innovation: Management, Policy and Practice, 10*, 74–90.

Gallot-Lavallée, T., (1998). Effectiveness of high pressure treatment for destruction of *Listeria monocytogenes* in raw milk goat cheese. *Sciences des Aliments, 18*, 647–655.

García, P., Rodríguez, L., Rodríguez, A., & Martínez, B., (2010). Food biopreservation: Promising strategies using bacteriocins, bacteriophages and endolysins. *Trends in Food Science and Technology, 21*, 373–382.

Garde, S., Arqués, J. L., Gaya, P., Medina, M., & Nuñez, M., (2007). Effect of high-pressure treatments on proteolysis and texture of *ewes*' raw *milk* La Serena *cheese*. *International Dairy Journal, 17*, 1424–1433.

Gaucheron, F., Famelart, M. H., Mariette, F., Raulot, K., Michel, F., & Le Graet, Y., (1997). Combined effects of temperature and high pressure treatments on physicochemical characteristics of skim milk. *Food Chemistry, 59*, 439–447.

Gaya, E., Medina, M., Rodríguez-Marín, M. A., & Núñez, M., (1990). Accelerated ripening of ewes' milk Manchego cheese: The effect of elevated ripening temperatures. *Journal of Dairy Science, 73*, 26–32.

Gaya, P., Sánchez, C., Núñez, M., & Fernández-García, E., (2005). Proteolysis during ripening of Manchego cheese made from raw or pasteurized ewes' milk. Seasonal variation. *Journal of Dairy Research, 72*, 287–295.

Gervilla, R., Capellas, M., Ferragut, V., & Guamis, B., (1997a). Effect of high hydrostatic pressure on *Listeria innocua* 910 CECT inoculated into ewe's milk. *Journal of Food Protection, 60*, 33–37.

Gervilla, R., Felipe, X., Ferragut, V., & Guamis, B., (1997b). Effect of high hydrostatic pressure on *Escherichia coli* and *Pseudomonas fluorescens* strains in ovine milk. *Journal of Dairy Science, 80*, 2297–2303.

Gervilla, R., Ferragut, V., & Guamis, B., (2001). High hydrostatic pressure effects on color and milk-fat globule of ewe's milk. *Journal of Food Science, 66*, 880–885.

Gervilla, R., Mor-Mur, M., Ferragut, V., & Guamis, B., (1999a). Kinetics of destruction of *Escherichia coli* and *Pseudomonas fluorescens* inoculated in ewe's milk by high hydrostatic pressure. *Food Microbiology, 16*, 173–184.

Gervilla, R., Sendra, E., Ferragut, V., & Guamis, B., (1999b). Sensitivity of *Staphylococcus aureus* and *Lactobacillus helveticus* in ovine milk subjected to high hydrostatic pressure. *Journal of Dairy Science, 82*, 1099–1107.

Gómez, M. J., Rodríguez, E., Gaya, I., Núñez, M., & Medina, M. M., (1999). Characteristics of Manchego cheese manufactured from raw and pasteurized ovine milk and with defined-strain of commercial mixed-strain starter cultures. *Journal Dairy Science, 82*, 2300–2307.

Gómez-Ruiz, J. A., Ballesteros, C., González-Viñas, M. A., Cabezas, L., & Martínez-Castro, I., (2002). Relationships between volatile compounds and odor in Manchego cheese: Comparison between artisanal and industrial cheeses at different ripening times. *Lait, 82*, 613–628.

Grappin, R., & Beuvier, E., (1997). Possible implications of milk pasteurization on the manufacture and sensory quality of ripened cheese. *International Dairy Journal, 7*, 751–761.

Guerzoni, M. E., Vannini, L., Chaves-López, C., Lanciotti, R., Suzzi, G., & Gianotti, A., (1999). Effect of high-pressure homogenization on microbial and chemico-physical characteristics of goat cheeses. *Journal of Dairy Science, 82*, 851–862.

Guo, M. R., (2003). Goat milk. In: Caballero, B. B., Trugo, L., & Finglas, P., (eds.), *Encyclopedia of Food Sciences and Nutrition* (pp. 2944–2949). London: Academic Press.

Hayes, M. G., & Kelly, A. L., (2003a). High pressure homogenization of milk (b) effects on indigenous enzymatic activity. *Journal of Dairy Research, 70*, 307–313.

Hayes, M. G., & Kelly, A. L., (2003b). High pressure homogenization of raw whole milk; (a) effects on fat globule size and other properties. *Journal of Dairy Research, 70*, 297–305.

Hayes, M. G., Fox, P. F., & Kelly, A. L., (2005). Potential applications of high pressure homogenisation in processing of liquid milk. *Journal of Dairy Research, 72*, 25–33.

Henry, G., Mollé, D., Morgan, F., Fauquant, J., & Bouhallab, S., (2002). Heat induced covalent complex between casein micelles and β-lactoglobulin from goats milk: Identification of an involved disulfide bond. *Journal of Agriculture and Food Chemistry, 50*, 185–191.

Hernández, I., Barrón, L. J. R., Virto, M., Pérez-Elortondo, F. J., Flanagan, C., Rozas, U., Nájera, A. I., et al., (2009). Lipolysis, proteolysis and sensory properties of ewe's raw milk cheese (Idiazábal) made with lipase addition. *Food Chemistry, 116*, 158–166.

Hernández, I., de Renobales, M., Virto, M., Pérez-Elortondo, F. J., Barrón, L. J. R., Flanagan, C., & Albisu, M., (2005). Assessment of industrial lipases for flavor development in commercial Idiazábal (ewe's raw milk) cheese. *Enzyme and Microbial Technology, 36*, 870–879.

Huppertz, T., Fox, P. F., & Kelly, A. L., (2006a). High-pressure-induced changes in ovine milk. 1. Effects on the mineral balance and pH. *Milchwissenschaft, 61*, 285–288.

Huppertz, T., Fox, P. F., & Kelly, A. L., (2006b). High-pressure-induced changes in ovine milk. 2. Effects on casein micelles and whey proteins. *Milchwissenschaft, 61*, 394–397.

Jakowska, A., Wisniewska, K., & Reps, A., (2005). Application of probiotic bacteria in production of yogurt preserved by high pressure. *High Pressure Research, 25*, 57–62.

Johnston, D. E., Austin, B. A., & Murphy, R. J., (1993). Properties of acid-set gels prepared from high pressure treated skim milk. *Milchwissenschaft, 48*, 206–209.

Juan, B., Barron, L. J. R., Ferragut, V., & Trujillo, A. J., (2007c). Effects of high pressure treatment on volatile profile during ripening of ewe milk cheese. *Journal of Dairy Science, 90*, 124–135.

Juan, B., Ferragut, V., Buffa, M., Guamis, B., & Trujillo, A. J., (2007a). Effects of high pressure on proteolytic enzymes in cheese: Relationship with the proteolysis of ewe milk cheeses. *Journal of Dairy Science, 90*, 2113–2125.

Juan, B., Ferragut, V., Buffa, M., Guamis, B., & Trujillo, A. J., (2007b). Effects of high-pressure treatment on free fatty acids release during ripening of ewes' milk cheese. *Journal of Dairy Research, 74*, 438–445.

Juan, B., Ferragut, V., Guamis, B., Buffa, M., & Trujillo, A. J., (2004). Proteolysis of a high pressure-treated ewe's milk cheese. *Milchwissenschaft, 59*, 616–619.

Juan, B., Quevedo, J. M., Zamora, A., Guamis, B., & Trujillo, A. J., (2015). Lipolysis of cheeses made from goat milk treated by ultra-high-pressure homogenization. *LWT-Food Science and Technology, 60*, 1034–1038.

Juan, B., Trujillo, A. J., Guamis, B., Buffa, M., & Ferragut, V., (2007d). Rheological, textural and sensory characteristics of high-pressure treated semi-hard ewes' milk cheese. *International Dairy Journal, 17*, 248–254.

Juan, B., Zamora, A., Quevedo, J. M., & Trujillo, A. J., (2016). Proteolysis of cheese made from goat milk treated by ultra high pressure homogenization. *LWT – Food Science and Technology, 69*, 17–23.

Lanciotti, R., Chaves, L. C., Patrignani, F., Paparella, A., Guerzoni, M. E., Serio, A., & Suzzi, G., (2004a). Effects of milk treatment with dynamic high pressure on microbial population as well as on the lipolytic and proteolytic profiles of Crescenza cheese. *International Journal of Dairy Technology, 57*, 19–25.

Lanciotti, R., Vannini, L., Pittia, P., & Guerzoni, M. E., (2004b). Suitability of high-dynamic-pressure-treated milk for the production of yoghurt. *Food Microbiology, 21*, 753–760.

Law, A. J. R., (1995). Heat denaturation of bovine, caprine and ovine whey proteins. *Milchwissenschaft, 50*, 384–388.

Law, A. J. R., Leaver, J., Felipe, X., Ferragut, V., Pla, R., & Guamis, B., (1998). Comparison of the effects of high pressure and thermal treatments on the casein micelle in goat's milk. *Journal of Agriculture and Food Chemistry, 46*, 2523–2530.

Law, B. A., (2001). Controlled and accelerated cheese ripening: The research base for new technologies. *International Dairy Journal, 11*, 383–398.

Li, J., & Guo, M., (2006). Effects of polymerized whey proteins on consistency and water-holding properties of goat's milk yogurt. *Journal of Food Science, 71*, C34–C38.

López-Fandiño, R., & Olano, A., (1998). Cheese-making properties of ovine and caprine milks submitted to high pressures. *Lait, 78*, 341–350.

López-Fandiño, R., (2006). High pressure-induced changes in milk proteins and possible application in dairy technology. *International Dairy Journal, 16*, 1119–1131.

López-Fandiño, R., De La Fuente, M. A., Ramos, M., & Olano, A., (1998). Distribution of minerals and proteins between the soluble and colloidal phases of pressurized milks from different species. *Journal of Dairy Research, 65*, 69–78.

Lucey, J. A., & Singh, H., (1998). Formation and physical properties of acid milk gels: A review. *Food Research International, 30*, 529–542.

Lucey, J. A., Teo, C. T., Munro, P. A., & Singh, H., (1997). Rheological properties at small (dynamic) and large (yield) deformations of acid gels made from heated milk. *Journal of Dairy Research, 64*, 591–600.

Madrau, M. A., Mangia, N. P., Murgia, M. A., Sanna, M. G., Garau, G., Leccis, L., Caredda, M., & Deiana, P., (2006). Employment of autochthonous microflora in Pecorino Sardo cheese manufacturing and evolution of physicochemical parameters during ripening. *International Dairy Journal, 16*, 876–885.

Masle, I., & Morgan, F., (2001). Compositional factors involved in the variable acidification capacity of goat milk by lactic starters. *Lait, 81*, 561–569.

Mendia, C., Ibáñez, E. C., Torre, E., & Barcina, Y., (1999). Effect of pasteurization on the sensory characteristics of a ewe's-milk cheese. *Journal of Sensory Studies, 14*, 415–424.

Mendia, C., Ibáñez, E. J., Torre, I., & Barcina, Y., (2000). Effect of pasteurization and use of a native starter cultureon proteolysis in a ewes' milk cheese. *Food Control, 11*, 195–200.

Moatsou, G., Kandarakis, I., Moschopoulou, E., Anifantakis, E., & Alichanidis, E., (2001). Effect of technological parameters on the characteristics of kasseri cheese made from raw or pasteurized ewes' milk. *International Journal of Dairy Technology, 54*, 69–77.

Moatsou, G., Katsaros, G., Bakopanos, C., Kandarakis, I., Taoukis, P., & Politis, I., (2008). Effect of high-pressure treatment at various temperatures on activity of indigenous proteolytic enzymes and denaturation of whey proteins in ovine milk. *International Dairy Journal, 18*, 1119–1125.

Montilla, A., Balcones, E., Olano, A., & Calvo, M. M., (1995). Influence of heat treatments on whey protein denaturation and rennet clotting properties of cow's and goat's milk. *Journal of Agriculture and Food Chemistry, 43*, 1908–1977.

Morgan, E., Bodin, J. P., & Gaborit, E., (2001). Lien entre le niveau de lipolyse du lait de chevre et la qualite sensorielle des fromages au lait cru ou pasteurise. *Lait, 81*, 743–756.

Mújica-Paz, H., Valdez-Fragoso, A., Tonello, S. C., Welti-Chanes, J., & Torres, J. A., (2011). High-pressure processing technologies for the pasteurization and sterilization of foods. *Food and Bioprocess Technology, 4*, 969–985.

Mussa, D. M., & Ramaswamy, H., (1997). Ultra-high pressure pasteurization of milk: Kinetics of microbial destruction and changes in psycho-chemical characteristics. *Food Science and Technology, 30*, 551–557.

Needs, E. C., Capellas, M., Bland, A. P., Manoj, P., MacDougal, D., & Paul, G., (2000). Comparison of heat and pressure treatments of skimmed milk, fortified with whey concentrate, for set yogurt preparation: Effects on milk proteins and gel structure. *Journal of Dairy Research, 67*, 329–348.

Nieto-Arribas, P., Poveda, J. M., Seseña, S., Palop, L., & Cabezas, L., (2009). Technological characterization of *Lactobacillus* isolates from traditional Manchego cheese for potential use as adjunct starter cultures. *Food Control, 20*, 1092–1098.

Ordóñez, A. I., Ibáñez, E. C., Torre, P., & Barcina, Y., (1999). Effect of ewe's milk pasteurization on the free amino acids in Idiazabal cheese. *International Dairy Journal, 9*, 135–141.

Ortigosa, M., Torre, P., & Izco, J. M., (2001). Effect of pasteurization of ewe's milk and use of a native starter culture on the volatile components and sensory characteristics of Roncal cheese. *Journal of Dairy Science, 84*, 1320–1330.

Pandya, A. J., & Ghodke, K. M., (2007). Goat and sheep milk products other than cheeses and yoghurt. *Small Ruminant Research, 68*, 193–206.

Patterson, M. F., Linton, M., & Doona, C. J., (2007). Introduction to high pressure processing of foods. In: Donna, C. J., & Feeherry, F. E., (eds.), *High Pressure Processing of Foods* (pp. 1–14). Iowa, USA: Blackwell Publishing Professional.

Penna, A. L. B., Subbarao-Gurram, & Barbosa-Canovas, G. V., (2006). Effect of high hydrostatic pressure processing on rheological and textural properties of probiotic low-fat yoghurt fermented by different starter cultures. *Journal of Food Process Engineering, 29*, 447–461.

Pereda, J., Ferragut, V., Quevedo, J. M., Guamis, B., & Trujillo, A. J., (2007). Effects of ultra-high pressure homogenization on microbial and physicochemical shelf life of milk. *Journal of Dairy Science, 90*, 1081–1093.

Posecion, N. C., Crowe, N. L., Robinson, A. R., & Asiedu, S. K., (2005). The development of a goat's milk yogurt. *Journal of the Science of Food and Agriculture, 85*, 1909–1913.

Quevedo, J. M., Buffa, M., Guamis, B., & Trujillo, A. J., (2010). Effects of ultra-high-pressure homogenization on microbial populations of goat's milk and cheese. In: Fito, P., & Toldrá, F., (eds.), *International Conference on Food Innovation-Food Innova 2010* (p. 75). València, Spain: Editorial Universitat Politècnica de València.

Quevedo, J. M., Buffa, M., Zamora, A., Guamis, B., & Trujillo, A. J., (2011). Influence of ultra-high-pressure homogenization on rennet coagulation properties of goat milk. *IDF International Symposium on Sheep, Goat and Other Non-Cow MILK*. Athens, Greek: International Dairy Federation.

Raynal, K., & Remeuf, F., (1998). The effect of heating on physicochemical and renneting properties of milk: A comparison between caprine, ovine and bovine milk. *International Dairy Journal, 8*, 695–706.

Raynal-Ljutovac, K., Lagriffoul, G., Paccard, P., Guillet, I., & Chilliard, Y., (2008). Composition of goat and sheep milk products: An update. *Small Ruminant Research, 79*, 57–72.

Raynal-Ljutovac, K., Park, Y. W., Gaucheron, F., & Bouhallab, S., (2007). Heat stability and enzymatic modifications of goat and sheep milk. *Small Ruminant Research, 68*, 207–220.

Remeuf, F., & Raynal, K., (2001). Effect of calcium addition, acidification, ultra filtration and storage on renneting properties of heated goat's, ewe's and cow's milks. *Lait, 81*, 381–399.

Reps, A., Warminska-Radyko, I., & Dajnowiec, F., (1999). Effect of high pressure on yoghurt. In: Ludwig, H., (ed.), *Advances in High Pressure Bioscience and Biotechnology* (pp. 453–456). Heidelberg, Germany: Springer.

Ribeiro, A. C., & Ribeiro, S. D. A., (2010). Specialty products made from goat milk. *Small Ruminant Research, 89*, 225–233.

Risch, S. J., (2009). Food packaging history and innovations. *Journal of Agriculture and Food Chemistry, 57*, 8089–8092.

Rodríguez-Alcalá, L. M., Harte, F., & Fontecha, J., (2009). Fatty acid profile and CLA isomers content of cow, ewe and goat milks processed by high pressure homogenization. *Innovative Food Science and Emerging Technologies, 10*, 32–36.

Ruiz-Espinosa, H., Amador-Espejo, G. G., Barcenas-Pozos, M. E., Angulo-Guerrero, J. O., Garcia, H. S., & Welti-Chanes, J., (2013). Multiple-pass high-pressure homogenization of milk for the development of pasteurization-like processing conditions. *Letters in Applied Microbiology, 56*, 142–148.

Saboya, L. V., Maillard, M. B., & Lortal, S., (2003). Efficient mechanical disruption of *Lactobacillus helveticus*, *Lactococcus lactis* and *Propionibacterium freudenreichii* by a new high-pressure homogenizer and recovery of intracellular aminotransferase activity. *Journal of Industrial Microbiology and Biotechnology, 30*, 1–5.

Saldo, J., Fernández, A., Sendra, E., Butz, P., Tauscher, B., & Guamis, B., (2003). High pressure treatment decelerated the lipólisis in a caprine cheese. *Food Research International, 36*, 1061–1068.

Saldo, J., McSweeney, P. L. H., Sendra, E., Kelly, A. L., & Guamis, B., (2002). Proteolysis in caprine milk cheese treated by high pressure to accelerate cheese ripening. *International Dairy Journal, 12*, 35–44.

Saldo, J., Sendra, E., & Guamis, B., (1999). Evaluación de un método para acelerar la maduración de quesos de cabra mediante alta presión hidrostática. *Alimentaria, 36*, 75–78.

Saldo, J., Sendra, E., & Guamis, B., (2000). High hydrostatic pressure for acceleration ripening of goat's milk cheese: Proteolysis and texture. *Journal of Food Science, 65*, 636–640.

Saldo, J., Sendra, E., & Guamis, B., (2001). Hard cheese structure after a high hydrostatic pressure treatment at 50 MPa for 72 h applied to cheese after brining. *Lait, 81*, 625–635.

Sandra, S., & Dalgleish, D. G., (2005). Effects of ultra-high-pressure homogenization and heating on structural properties of casein micelles in reconstituted skim milk powder. *International Dairy Journal, 15*, 1095–1104.

Sandra, S., & Dalgleish, D., (2007). The effect of ultra-high-pressure homogenization (UHPH) on rennet coagulation properties of unheated and heated fresh skimmed milk. *International Dairy Journal, 17*, 1043–1052.

Serra, M., Trujillo, A. J., Guamis, B., & Ferragut, V., (2009a). Evaluation of physical properties during storage of set and stirred yogurts made from ultra-high-pressure homogenization-treated milk. *Food Hydrocolloids, 23*, 82–91.

Serra, M., Trujillo, A. J., Guamis, B., & Ferragut, V., (2009b). Proteolysis of yogurts made from ultra-high-pressure homogenized milk during cold storage. *Journal of Dairy Science, 92*, 72–78.

Serra, M., Trujillo, A. J., Guamis, B., & Ferragut, V., (2009c). Flavor profiles and survival of starter cultures of yogurt produced from high-pressure homogenized milk. *International Dairy Journal, 19*, 100–106.

Serra, M., Trujillo, A. J., Pereda, J., Guamis, B., & Ferragut, V., (2008). Quantification of lipolysis and lipid oxidation during cold storage of yogurts produced from milk treated by ultra-high pressure homogenization. *Journal of Food Engineering, 89*, 99–104.

Serra, M., Trujillo, A. J., Quevedo, J. M., Guamis, B., & Ferragut, V., (2007). Acid coagulation properties and suitability for *yogurt* production of cows' milk treated by high-pressure homogenization. *International Dairy Journal, 17*, 782–790.

Singh, H., & Waungana, A., (2001). Influence of heat treatment of milk on cheesemaking properties. *International Dairy Journal, 11*, 543–551.

Smelt, J. P. P. M., (1998). Recent advances in the microbiology of high pressure processing. *Trends in Food Science y Technology, 9*, 152–158.

Smiddy, M. A., Martin, J., Huppertz, T., & Kelly, A. L., (2007). Microbial shelf life of high-pressure-homogenized milk. *International Dairy Journal, 17*, 29–32.

Sousa, M. J., & Malcata, X., (1997). Ripening of ovine milk cheeses: Effects of plant rennet, pasteurization, and addition of starter on lipolysis. *Food Chemistry, 59*, 427–432.

Tanaka, T., & Hatanaka, K., (1992). Application of hydrostatic pressure to yoghurt to prevent its after-acidification. *Journal of Japanese Society of Food Science and Technology, 39*, 173–177.

Thiebaud, M., Dumay, E., Picart, L., Guiraud, J. P., & Cheftel, J. C., (2003). High-pressure homogenization of raw bovine milk. Effects on fat globule size distribution and microbial inactivation. *International Dairy Journal, 13*, 427–439.

Torres, M. M. A., Soeldner, A., Ting, E. Y., Hawes, A. C. O., Aleman, G. D., Bakshi, G. S., McManus, W. R., et al., (1996). *Early Microstructure Changes in Cheddar Cheese and the Effects of High-Pressure Curd Processing* (p. 9). ITF Annual meeting. New Orlenas, Lousiana, USA. Book of abstracts.

Trujillo, A. J., Buffa, M., Casals, I., Fernández, P., & Guamis, B., (2002a). Proteolysis in goat cheese made from raw, pasteurized or pressure-treated milk. *Innovative Food Science and Emerging Technologies, 3*, 309–319.

Trujillo, A. J., Capellas, M., Buffa, M., Royo, C., Gervilla, R., Felipe, X., Sendra, E., et al., (2000). Application of high-pressure treatment for cheese production. *Food Research International, 33*, 311–316.

Trujillo, A. J., Ferragut, V., Gervilla, R., Capellas, M., & Guamis, B., (1997). High hydrostatic pressure effects on milk and milk products. In: Pandalai, S. G., (ed.), *Recent Research Developments in Agricultural and Food Chemistry* (Vol. 1, pp. 137–159). Trivandrum, India: Research Signpost.

Trujillo, A. J., Royo, C., Ferragut, V., & Guamis, B., (1999a). Influence of pressurization on goat milk and cheese composition and yield. *Milchwissenschaft, 54*, 197–199.

Trujillo, A. J., Royo, C., Guamis, B., & Ferragut, V., (1999b). Ripening profiles of goat cheese produced from milk treated with high pressure. *Journal of Food Science, 64*, 833–837.

Trujillo, A. J., Zamora, A., Pereda, J., Quevedo, J. M., & Guamis, B., (2007). Effect of ultra-high-pressure homogenization on microbial and rennet coagulation properties of goat's

milk. In: 5th *International Symposium on the Challenge to Sheep and Goats Milk Sectors Proceedings* (p. 168). Alghero, Italy: International Dairy Federation.
Trujillo, A. T., Capellas, M., Saldo, J., Gervilla, R., & Guamis, B., (2002b). Applications of high hydrostatic pressure on milk and dairy products: A review. *Innovative Food Science and Emerging Technologies*, *3*, 295–307.
Udabage, P., Augustin, M. A., Versteeg, C., Puvanenthiran, A., Yoo, J. A., Allen, N., McKinnon, I., et al., (2010). Properties of low-fat stirred yoghurts made from high-pressure-processed skim milk. *Innovative Food Science and Emerging Technologies, 11*, 32–38.
Vachon, J. F., Kheadr, E. E., Giasson, J., Paquin, P., & Fliss, I., (2002). Inactivation of food borne pathogens in milk using dynamic high pressure. *Journal of Food Protection, 65*, 345–352.
Van, H. A. C. M., De Koster, P. G., & Boerrigter, I. J., (1987). The renneting properties of heated milks. *Netherlands Milk and Dairy Journal, 41*, 3–18.
Vannini, L., Patrignani, F., Iucci, L., Ndagijimana, M., Vallicelli, M., Lanciotti, R., & Guerzoni, M. E., (2008). Effect of a pre-treatment of milk with high pressure homogenization on yield as well as on microbiological, lipolytic and proteolytic patterns of "Pecorino" cheese. *International Journal of Food Microbiology, 128*, 329–335.
Xanthopoulos, V., Polychroniadou, A., Litopoulou-Tzanetaki, E., & Tzanetakis, N., (2000). Characteristics of Anevato cheese made from raw or heat-treated milk inoculated with a lactic starter. *Food Science and Technology, 33*, 483–488.
Yokohama, H., Sawamura, N., & Motobayashi, N., (1992). *Method for Accelerating Cheese Ripening.* European Patent Application N° 0469857 A1.
Zamberlin, Š., PogaČiĆ, T., Mahnet, S., Golem, Ž., Havranek, J., & SamarŽija, D., (2010). The effect of heat treatment of ovine milk on the compositional and sensory properties of set yoghurt. *International Journal of Dairy Technology, 63*, 587–592.
Zamora, A., Ferragut, V., Guamis, B., & Trujillo, A. J., (2012). Changes in the surface protein of the fat globules during ultra-high-pressure homogenization and conventional treatments of milk. *Food Hydrocolloids, 29*, 135–143.
Zamora, A., Ferragut, V., Jaramillo, P. D., Guamis, B., & Trujillo, A. J., (2007). Effects of ultra-high-pressure homogenization on the cheese-making properties of milk. *Journal of Dairy Science, 90*, 13–23.
Zamora, A., Ferragut, V., Juan, B., Guamis, B., & Trujillo, A. J., (2011). Effect of ultra-high-pressure homogenization of milk on the texture and water-typology of a starter-free fresh cheese. *Innovative Food Science and Emerging Technologies, 12*, 484–490.
Zamora, A., Juan, B., & Trujillo, A. J., (2015). Compositional and biochemical changes during cold storage of starter-free fresh cheeses made from ultra-high-pressure homogenized milk. *Food Chemistry, 176*, 433–440.

CHAPTER 8

Sustainable Use of Food-Grade Microorganisms in Traditional Fermented Food Production

KOLAWOLE BANWO,* OMOTADE R. OGUNREMI, and ADEKEMI T. ADESULU-DAHUNSI

*Corresponding author. E-mail: kolabanwo@yahoo.com

ABSTRACT

Fermented foods have gained interest among consumers because of their possible nutritional and health-promoting characteristics. Microorganisms that carry out the fermentation process of such foods elicit the functionality of the foods. Safety and functional characteristics of traditional fermented foods is paramount to their general acceptability. The quality attributes of indigenous microorganisms are constantly explored to enhance their use as starter cultures for traditional fermentation process. This will assist in the enhancement of the nutritional, bioprotective, and sensory attributes. Food safety issues are essentially reliant on the recognition and adequately imploring the necessities of public health protection. Development of food laws should resolutely be centered on the hazard present in a food, which poses risks to human health and life. The non-compliance of food legislation in developing nations is because of the laxity in applying the principles of hazard analysis and critical control points (HACCP) policies in the cottage food processing industries. Food legislation is a tool that employs enforcement of these quality parameters. Governmental and non-governmental policies aid in the enactment of these legislations for food safety and general well-being of the populace who consume the food. This chapter has contributed to the functional and quality attributes of the employment of indigenous microbes in a fermentation process. In conclusion, the safety and challenges accompanied in the making of fermented foods and how to surmount them to meet today's consumer acceptability through the enactment of food legislation policies is important to the sustainable use of indigenous microorganism.

8.1 INTRODUCTION

The conversion of raw ingredients into food and other forms of food products is referred to as food processing. This involves the transformation of harvested crops to create attractive, marketable food products with long shelf life. The improvement in the quality and safety of foods is dependent on its processing techniques. Foods are considered safe if it is free from hazards related to physical, chemical, and microbiological at a concentration that does not pose a public health risk (Balarebe et al., 2017). Safe food when consumed does not pose a deleterious effect on the consumers. Food processing plays an essential role in food security because it deals with the surplus and shortages in the food chain, thereby increasing food availability. Indigenous microorganisms refer to a group of beneficial microbes that are intrinsic in foods produced in a certain area. Food contamination is the presence of materials or microorganisms that are alien to such foods. Contamination can also be through human beings who handle food in an unhygienic manner and environment. Microorganisms generally cause unwanted spoilage of foods while contamination by disease-causing microorganisms triggers food safety problems.

The three groups of microorganisms commonly found in foods are as follows molds, yeasts, and bacteria. The microorganisms commonly associated with the spoilage of foods are molds with limited usage in the food industry except in the ripening of cheese. The most valuable microorganisms in the food industry are yeasts this is important because of the fermentation of simple and complex sugars into ethanol and carbon dioxide. The Baker's yeasts are employed as sources of protein in the composition of animal feed and this is usually carried out industrially. The importance of bacteria in food/industrial microbiology cannot be over-emphasized; these groups are delineated into the following centered on the product of fermentation, e.g., lactic acid bacteria (LAB), acetic acid bacteria, propionic acid bacteria. Considering the integral part of food for fermentation, microorganisms employed are either proteolytic, saccharolytic, and lipolytic. Sustainable production of food using the indigenous food-grade microorganisms which guarantees benefit beyond nutrition is termed functional. Fermentation involving the activities of microbes such as bacteria, molds, and yeasts elicits this functionality to the food products (Kentish and Feng, 2014; Mao and Miao, 2015; Bastiannello et al., 2016). Fermented food contains some microorganisms that are yet to be fully identified. However, LAB and yeasts are the most predominant microorganisms of beneficial importance isolated from the matrices of foods and beverages (Zuidam and Heinrich, 2010).

This chapter focuses on the potential, health, or nutritional benefits that result from the growth or action of indigenous microorganisms in the area of food fermentations, protective cultures, and bacteriocins. All the desirable properties add up to the sustainability of functional characteristics of food-grade indigenous microorganisms.

8.2 SAFETY POTENTIALS OF FOOD-GRADE MICROORGANISMS IN TRADITIONAL FERMENTATION

Food safety is a technical terminology employed to define the conduct, regulation, and keeping of food in ways that will avert a foodborne illness. It refers to the circumstances that surround the preservation of foods quality in the avoidance of foodborne illnesses (Codex Alimentarius Commission, 2009). It is a very significant element of food security. Fermentation improves the quality of foods' wholesomeness and in several regions of the world, particularly in most parts of Africa; this technology serves as a preservation technique to guarantee food safety (Yasmine, 2000). This technique aids in regulation and proliferation of several pathogens in foods. Thus, it has an important influence on the nutrition of humans, in regions where pecuniary conditions pose a foremost barrier to certifying principles of food safety (FAO, 2010).

Indigenous fermented foods undergo processing through the activities of microorganisms before consumption. Odunfa (1985) reported that the classification of African fermented foods into groups is related to the raw materials employed in their processing. Fermentation is a relatively economical and inexpensive method of food bio-preservation with the involvement of live microbial strains (food-grade microorganisms) which leads to improved food safety and stability. It is a readily available food preservation technology mainly carried out spontaneously, using a previous batch of fermentation or the use of starter cultures in a controlled fermentation process, which are sometimes employed to accelerate the spontaneous fermentation of food products. The involvement of microorganisms in the fermentation or handling of food enables the translation of perishable foods commodities to products that are fit for consumption with a good shelf-life. In Africa, fermentation as a food preservation and preparation method is traditional and widely practiced in many homes on small to large scale level, as many Africans rely on agriculture for their livelihood. These fermented foods formed major diets of the populace and thus played important roles in the nutrition of any developing countries (Adesulu and Awojobi, 2013).

Natural fermentation also considered as spontaneous is the primary production technique of fermented foods, which is because of the actions of the presence of innate microbes in the raw material and at different conditions such as the substrate, duration of fermentation, temperature, pH, etc. This allows for the involvement of diverse microorganisms. The characteristics of the food product solely depend on the bacteriological quality and the food raw materials. Spontaneous fermentation is used to produce varieties of wholesome foods with specific taste, flavor, aroma, and texture. This technique causes biological modifications in the solid or liquid substrate. It is a low-cost food processing technology, which increases food availability and improves food quality by retaining the nutrient contents present in the raw foods, thereby helping to tackle the menace of spoilage of food and ultimately leading to the prevention of foodborne diseases predominant in several developing countries of Africa (Oyewole, 1997; Achi, 2005; Oyewole and Isah, 2012). The associated benefits of food-grade microorganisms during fermentation of food raw materials to produce wholesome and nutritious food is of enormous importance to the host health. This does not eliminate the actions of spoilage and pathogenic microorganisms during production especially fermentation under very poor hygienic conditions.

Fermented foods have been cherished by many societies in the world for their safety potentials and health benefits on the host when consumed. However, these foods face various microbial safety challenges. Therefore, to ensure that food safety techniques are firmly observed, the production of food at the farm to the table level is considered as everyone's responsibility (Ray and Sivakumar, 2009; Park et al., 2016). Numerous potential health or nutritional benefits from food-grade microbes are as a result of the growth and action of food-grade organisms in the intestinal tract following ingestion of foods containing them. In many developing countries in Africa, the prevalence of malnutrition and nutritional deficiencies are commonly encountered which is as a result of lack of quality diet, lack of proper and reliable storage facilities, lesser nutrient bioavailability, and the epidemic of infectious diseases. Food safety is of great importance due to microbial contamination in food, therefore in order to fight against these problems, simple, cost-effective, and beneficial fermentation technology is of paramount significance. The fermentation process improves the nutritional quality, digestibility, and increased bioavailability of nutrients in food, and reduces the anti-nutritive content, prolong the shelf stability and safety of the product. Awareness on the positive roles of these foods on human health and general well-being has resulted in many studies on fermented foods and beverages.

The ingestion of food contaminated by microbial pathogens leads to the condition known as foodborne illness. Considerable efforts must be employed by food producers/handlers during harvesting, transportation, processing conditions, and storage of food in order to ensure food safety. The responsiveness of the handlers and consumers on food safety will assist in strengthening and aiding the implementation of food standards, enhanced hygienic customs associated with food production and processing that may lead to the complete elimination of foodborne illnesses. The World Health Organization (WHO) have defined keys to safer food, which include keeping the food surface clean, separation of raw food from processed foods, maintaining safe temperature for the storage of foods and use of hygienic and safe water. Critical control point is another safety measure that should be considered during food processing. Hazard analysis and critical control point (HACCP) is a process control system that detects points where hazards might occur during the food production process and puts into place precautionary measures to avert these hazards from happening. Microbiological risks have been identified and controlled with these HACCP measures. During food fermentation, storage, handling, cooking, and cleaning procedures HACCP practices should be strictly adhered to in order to minimize the risks associated. During the natural fermentation process of cereal grains and cassava tubers, microbial succession involving majorly food-grade microorganisms takes place with the activity of LAB predominantly dominating the fermenting mix, and later yeast species, also *Bacillus* species have been reported to dominate oil seeds and legumes fermentation (Oguntoyinbo et al., 2010; Banwo et al., 2012; Adesulu-Dahunsi et al., 2017a, b). The understanding of the food-grade microorganisms' metabolism has given insights on their use as starter cultures with functional characteristics, which are categorized by at least one inherent property such as the organoleptic, technological, nutritional, and health benefits, all these have contributed to the food safety parameters in many nations (Leroy and De Vuyst, 2004; Adesulu-Dahunsi et al., 2018). From time immemorial, the act of fermentation was done without the interpretation of the mechanisms of action of the microbial interactions, where the raw foods were naturally fermented by autochthonous food-grade organisms (Mogensen et al., 2002; Hansen, 2004; FAO, 2010). Food-grade microorganisms play essential roles during food fermentation and have resulted in metabolites production, such as organic acids (lactic, acetic, formic, propionic, succinic, and citric acids), antimicrobial substances and bacteriocins, production of food additives, such as enzymes, fragrances, and bioactive peptides (BPs), production of polymers, nutraceuticals, preservation of food through inhibition of pathogens, enhancement of the nutritional value, sensory attribute of the food, and also enzymes and alcohols production by the yeasts which greatly affect the taste and quality of the final food product (Josephsen and Jespersen, 2004).

The potential use of food-grade microorganisms and the tremendous roles played during fermentation process have contributed immensely and provided solutions to the growing world food demand and curbing food insecurity issues in Africa and the world at large, according to the United Nations report, it was estimated that over 821 million people in the world (about one in every nine people) suffered from hunger (FAO, 2019). The beneficial roles of microbial strains have been extensively explored in the development of health-promoting features, which are probiotics, prebiotics, and synbiotics, the use of these adequately characterized and defined starter cultures for the food fermentation process in a controlled environment and the production of functional foods have been documented (Pedersen et al., 2012; Ogunremi et al., 2015).

In Africa, traditional fermented foods constitute the major diet of many populaces, safe history in the consumption of these indigenous foods with LAB have been reported, because of the 'generally recognized as safe' (GRAS) status. LAB is predominantly involved in the fermentation of indigenous beverages and foods, the activity of LAB constitutes certain beneficial health effects because of their metabolic potentials on the host and extended shelf life in when compared with raw materials. LAB produces mainly lactic acid, and other organic acids in minute quantity; acetic and formic acid, production of these substances causes reduction in pH and thus inhibit the growth of many foods spoiling bacteria and killing of pathogenic microorganisms. Many food-grade organisms have been reported for their probiotic effects in food and feed formulations. These food-grade organisms are reportedly used in the production of foods and beverages with enhanced functional, nutritive content and technological features. Microbes of food origin can subsist in the gastrointestinal tract (GIT), adhere to its epithelial cells and extracellular matrices (Banwo et al., 2013a, b; Jia et al., 2017). The application of precisely selected strains with multifunctional potentials as starter or adjunct culture during food processes/fermentation will result in the improvement of food substrate and production of quality food products.

The characterization of these beneficial microorganisms responsible for the fermentative processes have resulted in the development and the use of defined strains as starter cultures in a large-scale production of foods and beverages, which has greatly impacted on the food industries over the years.

8.3 FUNCTIONAL CHARACTERISTICS OF INDIGENOUS MICROORGANISMS IN FERMENTATION

The physiological properties and cellular components of different groups of food-grade microorganisms are continually explored and exploited to improve specific quality attributes of traditional foods. Species and strain-specific diversities in the metabolic activities and metabolites of some strains of yeasts, LAB and some other bacterial species are used to improve the safety, protection, organoleptic appeal, nutritional value, and health benefits of foods. These characteristics are very important in the selection of the probiotic potentials of these microorganisms. The term probiotics has been defined as "Live microorganisms which when administered in adequate amounts confer a health benefit on the host" (FAO/WHO, 2002). The advancement of probiotic studies during the past decade has heralded an important improvement in the food industry, which has translated to the development of foods known as nutraceuticals, which possess added benefits beyond nutrition (Ouwehand et al., 2002; Saad et al., 2013). The use of probiotic microorganisms should possess certain qualities such as the ability to survive in the gastrointestinal (GI) transit; i.e., they must have the ability to tolerate gastric juices, acids, and exposure to bile. Additionally, they must possess the ability to multiply and persist in the digestive tract of mammals. Above all the safety assessment and maintenance of the effectiveness for the period of the shelf-life of the product is of beneficial importance (FAO/WHO, 2002; Saad et al., 2013). Microorganisms which belong to the heterogeneous group of LAB (*Lactobacillus, Enterococcus, Weissella, and Leuconostoc*) and to the genus *Bifidobacterium* are being used as probiotics (Ouwehand et al., 2002; Saad et al., 2013).

Several authors from sub-Saharan Africa had isolated LAB species of probiotic potentials from different fermented foods from their regions such as *ogi, nunu, kulenaoto, music, obushera*, etc. (Mathara et al., 2008; Owusu-Kwarteng et al., 2012; Banwo et al., 2013a; Akabanda et al., 2014; Oguntoyinbo and Narbad, 2015). These microorganisms possessed several technological characteristics such as exopolysaccharide (EPS) production, amylolytic, and heavy metals chelating potentials (Patel et al., 2017), *in vitro* and *in vivo* adhesion potentials (Mathara et al., 2008; Owusu-Kwarteng et al., 2012; Banwo et al., 2013a, b; Sanni et al., 2013; Akabanda et al., 2014; Ogunremi et al., 2015; Oguntoyinbo and Narbad, 2015; Ojekunle et al., 2017; Adesulu-Dahunsi et al., 2018; Ajao et al., 2018; Nnawuhie et al., 2018; Olojede et al., 2020). *Enterococcus faecium* strains that possessed probiotic potentials and exhibited antagonism against food spoilage bacteria were isolated from raw cow's milk (Banwo et al., 2013a). However, other microbes which include yeasts from the traditional fermentation process have been established as potential probiotics of interests (Ouwehand et al., 2002; Ogunremi et al., 2015). These microorganisms with probiotic potentials isolated from traditional fermented foods have been studied *in vitro,* and *in situ*. Ogunremi et al. (2015) reported yeasts of the probiotic potentials of the species *Pichia kluyveri, Issatchenkia Orientalis, Pichia kudriavzevii,* and *Candida tropicalis* from cereal-based traditional fermented foods in Nigeria. Yeasts from traditional fermented foods in Africa demonstrated technological features of importance such as phytase degradation, *in vitro* serum cholesterol reduction and antioxidant capacities (Adegboye et al., 2014; Alakeji et al., 2015; Ogunremi et al., 2015).

"A protective culture connotes the cautious use of microorganisms as a mechanism to control the bacteriological status in a food product without any considerable change to the technological and sensory properties" (Alexandraki et al., 2013). There are approaches to the biological preservation in the enhancement of safety potentials and extension of shelf-life stability of food products. These are the inoculation of microorganisms, which have protective and inhibitory effects in the food matrix, with resultant *in situ* production of the inhibitory substance with competitive influence against spoilage and pathogenic bacteria and the usage of antimicrobial metabolites in its purified form as biopreservatives (Gaggia et al., 2011; Alexandraki et al., 2013). Microorganisms used as protective cultures, especially the bacteriocin producers, possess several benefits, due to the production of antimicrobial metabolites such as organic acids, carbon dioxide, ethanol, hydrogen peroxide, and diacetyl, which are efficient in action. The competitive ability of protective cultures with pathogens is an alternative strategy to combat the proliferation of undesired organisms in food matrices. Furthermore, these microorganisms may possess some additional functional potentials, which can be beneficial to the consumers. Finally, yet importantly, they can contribute to the taste, quality, and nutritional value of the food product. Therefore, the perception of "protective cultures" is of broad-spectrum capacity and not strictly related to the production of bacteriocins and other metabolites only (Gaggia et al., 2011; Alexandraki et al., 2013). Few of such functional and technological significance of strains associated with indigenous fermented food is compiled in Table 8.1.

TABLE 8.1 Functional and Technological Features of Indigenous Fermented Food and Microorganisms

Functional/ Technological Potentials	Fermented Food	Indigenous Microorganism(s)	References
Probiotic potentials	*Ogi*, *Obushera*, *Nunu*/ *Nono*, *Kulenaoto*, *Mursik*, Raw cow milk	*L. plantarum*, *L. fermentum*, *Enterococcus faecium*, *Pichia kudriavzevii*, *Issatchenkia orientalis*, *Candida* spp.	Mathara et al. (2008); Owusu-Kwarteng et al. (2012); Banwo et al. (2013a, b); Akabanda et al. (2014); Oguntoyinbo and Narbad (2015); Ajao et al. (2018)
Exopolysaccharide production	*Garri*, *Ogi*	*L. plantarum*, *Pediococcus acidilactici*, *Weissella cibaria*, *W. confusa*	Malang et al. (2015); Adesulu-Dahunsi et al. (2017b, 2018)
Heavy metals chelating potentials	*Ogi*, *Fufu*	*L. plantarum*, *Weissella cibaria*	Ojekunle et al. (2017)
Reduction of anti-nutritional factors	*Ogi*	*L. plantarum*, *Pichia kudriavzevii*, *Issatchenkia orientalis*, *Streptococcus cremoris*	Onipede et al. (2014); Ogunremi et al. (2017)
Folate/riboflavin production	*Ogi*, *Ben saalaga*, *Iru*	*L. fermentum*, *L. plantarum*, *Pediococcus acidilactici*, *Bacillus subtilis*	Greppi et al. (2017); Olanbiwoninu et al. (2017); Okoroafor et al. (2018)
In vitro serum cholesterol-lowering ability	*Nono*, *Wara*,	Lactic acid bacteria and yeasts	Adegboye et al. (2014); Alakeji et al. (2014); Ogunremi et al. (2015)
Rheological properties	Sorghum bread, Maize dough	Lactic acid bacteria and Yeasts	Annan et al. (2003); Olojede et al. (2020)

8.3.1 BIOPROTECTION AND SAFETY POTENTIALS OF LACTIC ACID BACTERIA (LAB)

LAB are sources of several antimicrobial metabolites that are gradually gaining prominence as suitable alternatives to synthetic preservative for food applications (Khochamit et al., 2015). Broad-spectrum inhibitory metabolites such as bacteriocins, diacetyl, hydrogen peroxide, organic acids, reutericyclin, and some low molecular weight compounds have been characterized from various species of LAB (Fernandez et al., 2013; Khochamit et al., 2015). The use of purified biomolecules or application of such producing strains as starter culture has potentials to confer safety and protection on fermented food products (Cotter et al., 2005).

Bacteriocins are defined as antimicrobial peptides produced by bacteria, which are ribosomally synthesized (Bello et al., 2012). They are bactericidal compounds with either a broad or narrow spectrum of activities; their modes of action are associated with the blockage of the biosynthesizes of the cell wall or with the formation of pores in the cell membrane resulting in leakage of the cell (Inglin et al., 2015). They can control foodborne pathogens and spoilage microorganisms, such as *Clostridium botulinum*, *Staphylococcus aureus* and *Listeria monocytogenes* (Bello et al., 2012).

Crude extract of nisin A from *Lactococcus lactis* 40 FEL3 isolated from Italian fermented foods had considerable killing effect on *Listeria monocytogenes,* resulting in a reduction by approximately 3 log CFU/ml within the first 6 h. Further to this observation, application of the producing strain as a starter during production of Cottage cheese decreased *L. monocytogenes* below the initial inoculum levels (Bello et al., 2012). Bactericidal bacteriocins from *Lactobacillus sakei* inhibited *Enterococcus* spp., *Listeria* spp., *Escherichia coli*, *Klebsiella* spp., *Pseudomonas* spp., *Staphylococcus* spp. and *Streptococcus* spp. (Todorov et al., 2013). Strains of *Lactobacillus* and *Pediococcus* species isolated from some Nigerian traditional fermented foods including, palm wine, *ogi*, *kunun-Zaki*, and yogurt, showed inhibitory activity due to bacteriocin production against wide spectrum of foodborne pathogens (Banwo et al., 2013a, b; Oladipo et al., 2017). Strains of *Lactococcus lactis* and *Pediococcus acidilactici* isolated from reportedly produced nisin and pediocin, respectively, which were active against *L. ivanovii* (Fernandez et al., 2013).

In species/strain specific cases, LAB have been reported to accumulate organic acids, which include lactic, acetic, and propionic acids at different proportions from the metabolism of sugars in the food milieu. The dissociated and undissociated forms of these organic acids interfere with membrane potentials and active transport in spoilage and pathogenic microorganisms (Caplice and Fitzgerald, 1999). In addition, several low molecular weight antifungal compounds have been characterized from supernatants of LAB. Although respective compounds are secreted at concentrations that are significantly below their established minimum inhibitory concentration, they are suggested to exert antifungal effects due to the synergistic activity of multiple compounds released by the LAB (Romanens et al., 2019; Sadeghi et al., 2019). Antifungal compounds, including azelaic acid, phenyllactic acid, (S)-(−)-2-hydroxyisocaproic acid, DL-ρ-hydroxyphenyl lactic acid, vanillic acid and cytidine were detected in cell-free supernatant of antifungal strains of *L. citreum, L. brevis, L. reuteri* (Le Lay et al., 2016a). Around 10 bioactive compounds believed to contribute to the eventual fungal inhibitory effects were detected in the cell-free supernatant of the *Lb. reuteri* strain (Sadeghi et al., 2019). *Lactobacillus plantarum* strain that demonstrated activity against important food spoilage species of *Aspergillus, Penicillium, and Cladosporium* was noted to produce phenylacetic acid (Russo et al., 2017). Some antifungal LAB strains have presented potentials to replace synthetic antifungal additives in foods systems such as bakery products (Le lay et al., 2016b), cheese (Cheong et al., 2014) and cereal beverage (Russo et al., 2017). Reuterin is another important antimicrobial compound produced by *Lactobacillus reuteri,* heterofermentative LAB. It inhibits a broad range of Gram-positive and Gram-negative bacteria, including food spoilage and pathogenic microorganisms (Caplice and Fitzgerald, 1999). Recently, it was demonstrated to inhibit wide spectrum of yeasts and molds (Vimont et al., 2019). It is proposed that reuterin induce oxidative stress in microbial cells, resulting in reduction of glutathione and amendment of functional enzymes. Reuterin and other antimicrobial metabolites derived from LAB, are quickly gaining prominence as food potential food preservatives.

Safety can be further promoted by functional starter cultures in food systems by the removal of toxic contaminants, especially mycotoxins and heavy metals (Zoghi et al., 2014). These toxins are removed either by binding to microbial cell envelopes or degradation to a less/non-toxic product. *L. reuteri* significantly reduced the quantities of the aflatoxins B_1, B_2, G_1 and G_2 in a spiked solution (Sadeghi et al., 2019). This important property was revealed by a significant reduction of the levels of mycotoxins in pan whole-wheat sourdough bread. The incorporation of cadmium and lead-tolerant strains of *W. cibaria* and *Lb. plantarum* into skimmed milk prevented histopathological defects in rats (Ojekunle et al., 2017).

8.3.2 NUTRITIONAL VALUE OF SOME INDIGENOUS FERMENTED FOODS

Selected LAB strains metabolize complex and simple compounds in food, thereby improving the bioavailability of macro- and micronutrients and fortifying some essential and limiting nutritional compounds in foods. Phytase, the enzyme that mediates dephosphorylation of phytate (myo-inositol hexakisphosphate-IP_6), an inherent anti-nutritional factor that chelates dietary minerals in plant foods is widely distributed among species of LAB. Overproducing strains can be exploited to enhance the bioavailability of iron, zinc, and calcium. Phytase production was described for species of *Lactobacillus, Pediococcus, and Streptococcus* (Ragavendra and Halami, 2009; Onipede et al., 2014). *Lb. plantarum* and *S. cremoris* were applied to remove phytate during the production of *ogi* from sorghum (Onipede et al., 2014). LAB strains have been reported to possess a complete pathway for biosynthesis of some essential vitamins, including folate and riboflavin (Greppi et al., 2017). Folate is required for nucleotide synthesis, DNA replication and in turn cell division and growth. Folate intake reduces the possibility of neural tube defects, several cancers, pre-neoplastic lesions, and megaloblastic anemia (Greppi et al., 2017). *In vitro* microbiological assay revealed 59% of 151 LAB isolates from *ben-saalaga* produced folate at species/strain-specific capacities, ranging from 0.3 ng mL^{-1} to 120.9 ng mL^{-1} (Greppi et al., 2017). Folate-producing species are *Lb. fermentum, L. plantarum*, and *P. acidilactici* (Greppi et al., 2017). Riboflavin is an important precursor for cofactors required for some enzymatic functions in the body. It is limited in a plant-based diet and the deficiency is associated with poor vision and growth. Production of riboflavin for diet enrichment was reported in *Lactobacillus* strains (Thakur et al., 2016). High producers of riboflavin were identified among strains of *Staphylococcus* sp. isolated in the course of fermentation of the seeds of locust bean to produce *iru* (a fermented locust bean condiment) (Olanbiwoninu et al., 2017). LAB are potential cell factories for the synthesis of some essential amino acids. Intra- and extracellular accumulation of lysine and methionine up to 247.00 and 15.00 mg/L, respectively was reported in *Lactobacillus* spp. isolated from *ogi* (Odunfa et al., 2001).

8.3.3 ORGANOLEPTIC VALUE OF INDIGENOUS FERMENTED FOODS

Yeasts and LAB are gaining attention as sources of natural and safe compounds to impart appealing organoleptic attributes on foods. Different species and strains transform sugars, proteins, fats, organic acids, and amino acids via varied enzymatic metabolic pathways to a pool of precursor compounds that are further converted to volatile and non-volatile compounds with characteristic tastes and odor (Shoba and Agrawal, 2007; Park and Kim, 2019). These compounds impart a variety of flavors such as fruity, flowery, roasty, sour, and spicy nuances (Annan et al., 2003). In addition, there are numerous reports on the capacity of widespread species of LAB to improve the rheological properties of different food systems by secreting technologically relevant EPS. Inclusion of an EPS-producing LAB in starter culture confers benefits such as improved texture, enhanced mouth-feel and reduced syneresis in food products (Leroy and DeVuyst, 2016).

Aromatic compounds that are already described from LAB isolated from various indigenous food sources include organic acids, diacetyl, and esters (Shoba and Agrawal, 2007; Rodriguez et al., 2019). Production of varied amounts of diacetyl was reported to be distributed among species of *Enterococcus, Lactobacillus, Lactococcus, and Leuconostoc* isolated from fruit and vegetable sources (Rodriguez et al., 2019). The use of *Lb. fermentum* as a single species starter culture to ferment Ghanaian maize dough imparted a unique profile of flavor compounds with a high level of acetic acid compared to the spontaneously fermented dough (Annan et al., 2003). Lactic acid bacterial fermentation of rice beverage using *Lb. fermentum* produced acetoin and 4-ethenyl-2-methoxyphenyl characterized by yogurt-like and smoky odor, respectively (Park and Kim, 2019). *Lb. plantarum* fermentation of similar beverage generated hexanoic acid and acetic acid (Park and Kim, 2019).

Certain species and strains of LAB can produce wide diversity of EPS which may be homopolysaccharides or heteropolysaccharide. Homopolysaccharides described from LAB strains are broadly categorized as glucans, fructans, and galactans (Torino et al., 2015). Glucans are mainly released by species of *Weissella, Lactobacillus, Leuconostoc, and Pediococcus* (Patel and Prajapati, 2013; Patel et al., 2012; Malang et al., 2015; Dertli et al., 2016; Adesulu-Dahunsi et al., 2018). Reports on fructan production by LAB are limited. However, *W. confusa* strains isolated from cassava fermentation were shown to produce levan (a type of fructan) (Malang et al., 2015). Galactan production was reported for *W. confusa* isolate from idli batter, an Indian acidic fermented food (Kavitake et al., 2016). The different behaviors displayed by LAB EPS when released into food systems are determined by their composition, molecular structure, and molecular mass (Lynch et al., 2018b). They are garnering increased interest from beverage and bakery industries as potential natural and sustainable replacement for synthetic hydrocolloids such as hydroxyl propyl methylcellulose (HPMC) (Lynch et al., 2018a). EPS produced by LAB act as vicious, stabilizing, gelling or emulsifying agents in the food industry (Adesulu-Dahunsi et al., 2018). The application of *Weissella cibaria,* which possesses the ability to produce dextran when used as starter-culture during the fermentation of quinoa-based beverage enhanced the water binding capacity, viscosity, and overall taste perception of the product (Zannini et al., 2018). EPS-producing starter culture of *Streptococcus thermophilus* increased consistency and viscosity, lowered flow and hindered syneresis in yogurt (Hassan et al., 2003).

8.4 QUALITY AND CHALLENGES IN THE SUSTAINABLE USE OF MICROORGANISMS DURING TRADITIONAL FERMENTATION

In Africa, and the world at large, the beneficial roles microbes play during food fermentation and their applications in medicines, food, and pharmaceutical industries have significantly affected the diets and culture of many communities. Fermentation possesses the potentials of controlling the proliferation of several pathogens in foods, which aid in the improvement of food safety (FAO, 2010). Thus, it makes an important contribution to the nutrition of the general populace in the developing countries, where the problems related to the economy poses a major obstacle to ensuring food safety. To certify the quality and safety of food during production and fermentation processes, farmers, and household/artisanal producers should be able to guarantee wholesomeness of the crops harvested and processed, as the misuse of agro-chemicals such as pesticides and chemical preservatives may have harmful effect on human health in the end. Many fermented foods are processed by the activities of the microflora naturally present in the raw food materials. Fermented foods are designated as edible, wholesome, and are generally valued for attributes such as improved cooking time and processing properties (Holzapfel, 2002). Numerous benefits associated with fermented foods makes these foods important, as they form an essential part of people's diets, lifestyle, security, and they are readily available. The positive attributes derived from these foods guarantees the well-being of many people in the rural and urban settlements. The microbes that carry out the fermentation processes make the following available in the food: enriched nutrients, enhancement of the organoleptic qualities, bio-preservative properties, safety, degradation of toxic substances and residues (e.g., linamarin in cassava, mycotoxins, etc.), degradation of antinutritive factors, probiotics, and beneficial potentials (Holzapfel, 2002). The involvement of food-grade microbes during food fermentation gives assurance of food security and safety to the people even in the period of famine, harsh environmental and climatic conditions, or disasters. These foods assist small-scale food processors in their income generation, which gives them a means of livelihood (Adesulu and Awojobi, 2013).

However, there are some challenges in the spontaneous fermentation of food which are as follows: involvement of a diversity of native microorganisms with unknown characteristics, improper storage system, climatic change affecting agricultural production, deficiencies in organization and creation of network among food handlers and providers who rely on microbial systems for food processing. The sustainable use of microorganisms in food fermentation processing is based on the intrinsic and extrinsic factors of microorganisms innate in the food and the conditions in which the foods are been stored. Indigenous fermented products assist to promote the safety of foods due to the mechanisms of action of the microorganisms. However, if the hygienic content during processing is poor, it can create a problem of public health concern to the consumers. Upgrading of the process of fermentation may result in the production of fermented foods, with suitable quality and safety and these can be achieved through: The use of appropriate starter culture, certification of raw materials, optimization of the process of fermentation, implementation of the HACCP system and manufacturing practices of good standard, development of better equipment and specification for commercialization. A good hygienic practice, which entails the understanding of the fermenting microbes, is also an important factor to be considered during a fermentation process. Enlightenments of the public about good agricultural practices should be employed to decrease microbial and chemical hazards in food from the farm to the table (Josephsen and Jespersen, 2004; FAO, 2010).

8.5 FOOD SAFETY LEGISLATION

Food legislation is a terminology applied to the "food law, which standardizes the production, trade, and handling of food and involves the regulation of food control, food safety and relevant aspects of food trade" (Vapnek, 2002). The establishment of legal frameworks ensures the pillar of a proper food control system. In most countries, the enactment of food laws and regulations are done by the government, enforced on food handlers in order to certify that food is safe and reaches the consumers in an acceptable condition

(Hansen, 2004). The Ministry of Health, which is under the federal government agency, ensures the enactment of guiding principles and protocols on food safety, monitoring, and assessment at the national level. It is also accountable for the evaluation of the nutritive content of food, food handlers' hygiene and their environment, foodborne disease control and all regulations regarding food (Oloo et al., 2018). The basic process involving four pivotal arms of food legislation is depicted in Figure 8.1.

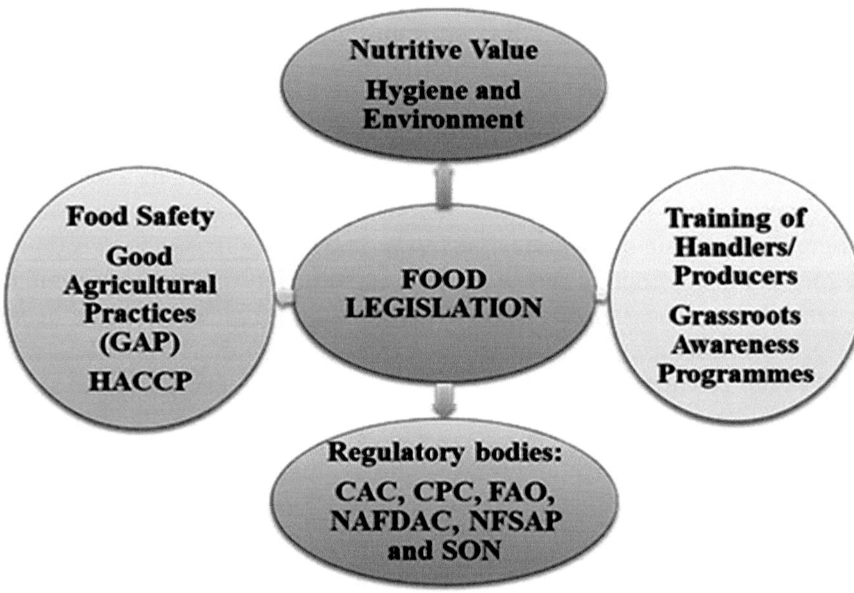

FIGURE 8.1 Four pivotal arms of food legislation.

The food legislation in Nigeria faces the challenges of the issue of food safety and discrepancies in the enforcement of Good Agricultural Practices, which include the misuse of agrochemicals, practice of pesticides for fishing in the fresh and marine waters, abuse of pesticides to stored grains and the abuse of additives such as nitrates/nitrites. In addition are the ignorance of the populace, ineptitude, and shortage of technical expertise, which leads to the inadequate implementation of the existing legislation. The conditions of droughts and porosity of some borders in some regions force infiltration of contaminated food into the value chain (Oloo et al., 2018).

There is an assertion that Nigeria possesses several food legislations. These legislations focus on protection of consumer rights, coordination, and improvement of significant policies on food. These policies ensure strict adherence to HACCP guidelines and its implications as the bedrock of all food production and processing operations. The governmental policymakers are the National Agency for Food and Drug Administration and Control (NAFDAC), Standards Organization of Nigeria (SON), and Consumer Protection Commission (CPC). In addition, these policymakers and regulatory bodies emphasize and acknowledge the duties of university research communities and local governments' councils in the creation of responsiveness and teaching of workforces for programs focused on food safety. The important factors to achieving the effective impacts in Nigeria on food legislation are through the amplification of thorough public consciousness and customer training about the risks in the neglect of food safety procedures. There should be synergy among the agencies responsible for food safety (Omojokun, 2013; Oloo et al., 2018).

In Ghana, the food production system is managed by the informal settings such as the small and medium-scale enterprises (SMEs). These enterprises do not apply strict adherence to the HACCP systems. Large food processors' adherence to the rules of HACCP is in order to avert a threatening situation related to food safety issues or usually at the request of consumers in foreign markets, which are the export countries of the food products. The cottage food industries are not included in the standards, which leaves a gap in food safety programs because they contribute to the quantity and quality of food consumed in the country (Sefa-Dedeh, 2009; Oloo et al., 2018). The negligence observed in the food supply chains especially from the cottage industries could be the one of the main sources of mortality and morbidity experienced due to foodborne illnesses. In comparison to other limitations such as power outages, non-availability of good motorable roads, and insecurity of food due to short supplies, food safety issues are not a main concern to developing countries. This is the reason why adequate capital is not allocated to that sector of the economy (Sefa-Dedeh, 2009; Oloo et al., 2018). In Ghana, the ministries involved in food safety legislation are the Ministries of Environment and Science, Trade, and Industry and Health while the ministry of Food and Agriculture takes the prominent roles (Sefa-Dedeh, 2009; Oloo et al., 2018). These systems are devoid of the enforcement of the food safety rules and regulations. Nevertheless, Ghana has made an indelible landmark stride in enlisting the strategy which constitutes the National Food Safety Action Plan (NFSAP) to reform the safety concerns of food and health institutions to improve its productivity and authority. This has generated a synergy among the Ministries, Departments, and Agencies, allowing for efficient processes, and avoidance of replication of duties, eradicating conflicts and reassuring improved implementation of legislation of food safety (Sefa-Dedeh, 2009; Oloo et al., 2018).

In comparison to the food safety legislation in West Africa, East Africa had taken the issue in a different dimension. This is with Kenya taking the lead in the African countries with 22 food safety-related legislation under the auspices of different Departments and Agencies. The enactment of Kenya into Codex Alimentarius Commission in 1969, had since played some crucial roles in various Codex Committees (CC), and is currently presiding over the CCAFRICA region (Kilonzo and Gathura, 2016; Oloo et al., 2018). With Kenya been a central member of the East African Communities, there is an incorporation of the management of food safety with several organizations across the different segments delegated to ensuring quality and advancement of trade for definite products. The main concern is to expedite trade and eliminate obstacles permitting for prompter dissemination of goods and services across boundaries, which invariably leads to the reduction of losses in food. In the face of the efforts focused on securing food safety legislative concerns in Kenya, illnesses due to foodborne infections, outbreaks, and other ills are constantly been reported (Kilonzo and Gathura, 2016; Oloo et al., 2018).

Finally, the training of handlers, producers, and regulators will be effective toward attaining the realization in the enactment of food safety legislation. The grassroots should be encouraged to form organizations with the support of government to make training and creation of awareness more lucid (Omojokun, 2013; Oloo et al., 2018).

8.6 CONCLUSION

Indigenous microorganisms are important in conserving, preserving, and ensuring safety potentials of traditional fermented foods. The use of these improved strains as starter cultures in food fermentation is projected to have a great economic impact on the general populace in developing countries. Genomics and biotechnological applications are providing information about the functional and safety potentials of microorganisms in addressing public health challenges, which include foodborne diseases and malnutrition in developing economies. The realization of the sustainability of microorganism in indigenous fermentation process needs the enforcement of Federal agencies, research institutions and Universities in order to maximize safety and ensure benefit to the public. In order to enforce and guarantee food safety regulations, the government should be proactive and make sanctions available to the public on hygiene, safety, and nutritional value of fermented foods before accessibility to the public. There should be training and education of the public to create awareness of the essential importance of fermented food and sustainability of the food systems. The existing international and domestic organizations such as Food and Agricultural Organization (FAO), Codex Alimentarius Committees, NAFDAC, SON, and ministries should possess the expertise, leadership, and coordinate training programs to establish food safety as well as integrating biotechnologies to the dynamic networks. There should be enforcement of establishments of appropriate food safety regulation to intensify consumer awareness about the benefits and risks of these regulations.

KEYWORDS

- **antinutritional properties**
- **fermented foods**
- **food laws**
- **food legislation**
- **food safety**
- **functional foods**
- **indigenous microorganisms**
- **nutritional values**
- **organoleptic properties**
- **regulatory bodies**

REFERENCES

Achi, O. K., (2005). The potential for upgrading traditional fermented foods through biotechnology. *African Journal of Biotechnology, 4*, 375–380.

Adegboye, B. D., Banwo, K., Ogunremi, O. R., & Sanni, A. I., (2014). Probiotic potentials of yeasts isolated from Nono (African fermented milk) and *Wara* (African soft cheese). *Advances in Food Sciences, 36*(3), 115–124.

Adesulu, A. T., & Awojobi, K. O., (2014). Enhancing sustainable development through indigenous fermented food products in Nigeria. *African Journal of Microbiology Research, 8,* 1338–1343.

Adesulu-Dahunsi, A. T., Sanni, A. I., & Jeyaram, K., (2017a). Rapid differentiation among *Lactobacillus, Pediococcus,* and *Weissella* species from some Nigerian indigenous fermented foods. *LWT- Food Science and Technology, 77*, 39–44.

Adesulu-Dahunsi, A. T., Sanni, A. I., & Jeyaram, K., (2018). Production, characterization, and *In vitro* antioxidant activities of exopolysaccharide from *Weissella cibaria* GA44. *LWT-Food Science and Technology*, 432–442.

Adesulu-Dahunsi, A. T., Sanni, A. I., Jeyaram, K., & Banwo, K., (2017b). Genetic diversity of *Lactobacillus plantarum* strains from some indigenous fermented foods in Nigeria. *LWT-Food Science and Technology, 82*, 199–206.

Ajao, O. S., Banwo, K., Ogunremi, O. R., & Sanni, A. I., (2018). Antimicrobial properties and probiotic potentials of lactic acid bacteria isolated from raw beef in Ibadan, Nigeria. *Journal of Microbiology, Biotechnology and Food Science, 8*(2), 770–773.

Alakeji, T. P., Banwo, K., Ogunremi, O. R., & Sanni, A. I., (2015). Functional properties of yeasts isolated from some Nigerian traditional fermented foods. *Journal of Microbiology, Biotechnology and Food Sciences, 4*(5), 437–441.

Alexandraki, V., Tsakalidou, E., Papadimitriou, K., & Holzapfel, W., (2013). Status and trends of the conservation and sustainable use of microorganisms in food processes. In: *Commission on Genetic Resources for Food and Agriculture* (p. 172). Food and Agriculture Organization of the United Nations.

Annan, N. T., Poll, L., Sefa-Dedeh, S., Plahar, W. A., & Jakobsen, M., (2003). Influence of starter culture combinations of *Lactobacillus fermentum, Saccharomyces cerevisiae* and *Candida krusei* on aroma in Ghanaian maize dough fermentation. *European Journal of Food Research and Technology, 216*, 377–384.

Balarebe, M. M., Mohammed, S. S. D., & Idris, S., (2017). The role of biotechnology in food production and processing. *Industrial Engineering, 1*(1), 24–35.

Banwo, K., Sanni, A. I., & Tan, H. R., (2013b). Functional properties of *Pediococcus* species isolated from traditional fermented cereal gruel and milk in Nigeria. *Food Biotechnology, 27*(1), 14–38.

Banwo, K., Sanni, A. I., Tan, H., & Tian, Y. Q., (2012). Phenotypic and genotypic characterization of lactic acid bacteria isolated from some Nigerian traditional fermented foods. *Food Biotechnology, 26*, 124–142.

Banwo, K., Sanni, A., & Tan, H., (2013a). Technological properties and probiotic potential of *Enterococcus faecium* strains isolated from cow milk. *Journal of Applied Microbiology, 114*, 229–241.

Bastianello, E., Montemurro, F., Fasolato, L., Balzan, S., Marchesini, G., Contiero, B., & Novelli, E., (2016). Volatile compounds and microbial development in sonicated cloudy apple juices: Preliminary results. *CyTA-Journal of Food, 14*(1), 65–73.

Bello, B. D., Cocolin, L., Zeppa, G., Field, D., Cotter, P. D., & Hill, C., (2012). Technological characterization of bacteriocin producing *Lactococcus lactis* strains employed to control *Listeria monocytogenes* in Cottage cheese. *International Journal of Food Microbiology, 153*, 58–65.

Caplice, E., & Fitzgerald, G. F., (1999). Food fermentations: Role of microorganisms in food production and preservation. *International Journal of Food Microbiology, 50*, 131–149.

Cheong, E. Y. L., Sandhu, A., Jayabalan, J., Kieu, L. T. T., Nhiep, N. T., My Ho, H. T., Zwielehner, J., et al., (2014). Isolation of lactic acid bacteria with antifungal activity against the common cheese spoilage mould *Penicillium commune* and their potential as biopreservatives in cheese. *Food Control, 46*, 91–97.

Codex Alimentarius Commission (CAC), (2009). *Basic Texts Food Hygiene* (4th edn., pp. 142–156). Secretariat of the Joint FAO/WHO Food Standards Programme, Rome. FAO.

Cotter, P. D., Hill, C., & Ross, R. P., (2005). Bacteriocins: developing innate immunity for food. *Nature Review Microbiology, 3*, 777–788.

Dertli, E., Mercan, E., Arıcı, M., Yılmaz, M. T., & Sağdıç, O., (2016). Characterization of lactic acid bacteria from Turkish sourdough and determination of their exopolysaccharide (EPS) production characteristics. *LWT-Food Science and Technology, 71*, 116–124.

Domingos-Lopes, M. F. P., Nagy, A., Stanton, C., Ross, P. R., Gelencsér, E., and Silva, C.C.G., (2017). Immunomodulatory activity of exopolysaccharide producing *Leuconostoc citreum* strain isolated from Pico cheese. *Journal of Functional Foods, 33*, 235–243.

European Commission, (2010). *Functional Foods* (p. 55). Publications Office of the European Union, Luxembourg.

FAO, (2010). *Agricultural Biotechnologies in Developing Countries: Options and Opportunities in Crops, Forestry, Livestock, Fisheries, and Agro-Industry to Face the Challenges of Food Insecurity and Climate Change (ABDC-10): Current Status and Options for Biotechnologies in Food Processing and in Food Safety in Developing Countries*. FAO International Technical Conference, Guadalajara, Mexico.

FAO, (2019). In: Bélanger, J., & Pilling, D., (eds.), *The State of the World's Biodiversity for Food and Agriculture* (p. 572). FAO Commission on Genetic Resources for Food and Agriculture Assessments. Rome.

Fernandez, B., Le Lay, C., Jean, J., & Fliss, I., (2013). Growth, acid production and bacteriocin production by probiotic candidates under simulated colonic conditions. *Journal of Applied Microbiology, 114*, 877–885.

Gaggia, F., Di Gioia, D., Baffoni, L., & Biavati, B., (2011). The role of protective and probiotic cultures in food and feed and their impact on food safety. *Trends in Food Science and Technology, 22*(1), 58–66.

Greppi, A., Saubade, F., Botta, C., Humblot, C., Guyot, J. P., & Cocolin, L., (2017). Potential probiotic *Pichia kudriavzevii* strains and their ability to enhance folate content of traditional cereal-based African fermented food. *Food Microbiology, 62*, 169–177.

Hansen, E. B., (2004). Microorganisms. In: Hui, Y. H., Meunier-Goddik, L., Hansen, Å. S., Josephsen, J., Nip, W. K., Stanfield, P. S., & Toldrá, F., (eds.), *Handbook of Food and Beverage Fermentation Technology* (pp. 9–21). Marcel Dekker, Inc., 270 Madison Avenue, New York, 2. ISBN: 0-8247-4780-1.

Hassan, A. N., Ipsen, R. T., Janzen, T., & Qvist, K. B., (2003). Microstructure and rheology of yogurt made with cultures differing only in their ability to produce exopolysaccharides. *Journal of Dairy Sciences, 86*, 1632–1638.

Hellström, A. M., Almgren, A., Carlsson, N. G., Svanberg, U., & Andlid, T. A., (2012). Degradation of phytate by *Pichia kudriavzevii* TY13 and *Hanseniaspora guilliermondii* TY14 in *Tanzanian togwa. International Journal of Food Microbiology, 153*, 73–77.

Hjortmo, S., Patring, J., Jastrebova, J., & Andlid, T., (2005). Inherent biodiversity of folate content and composition in yeasts. *Trends in Food Science Technology, 16*, 311–316.

Inglin, R. C., Stevens, M. J. A., Meile, L., Lacroix, C., & Meile, L., (2015). High-throughput screening assays for antibacterial and antifungal activities of *Lactobacillus* species. *Journal of Microbiological Methods, 114*, 26–29.

Jia, F. F., Zhang, L. J., Pang, X. H., et al., (2017). Complete genome sequence of bacteriocin-producing *Lactobacillus plantarum* LDS1. 0391, a probiotic strain with gastrointestinal tract resistance and adhesion to the intestinal epithelial cells. *Genomics, 109*, 432–437.

Josephsen, J., & Jespersen, L., (2004). Handbook of food and beverage fermentation technology. In: Hui, Y. H., Meunier-Goddik, L., Hansen, Å. S., Josephsen, J., Nip, W. K., Stanfield, P. S., & Toldrá, F., (eds.), *Starter Cultures and Fermented Products* (pp. 23–49). Marcel Dekker, Inc., 270 Madison Avenue, New York, 3. ISBN: 0-8247-4780-1.

Kavitake, D., Devi, P. B., Singh, S. P., & Shetty, P. H., (2016). Characterization of a novel galactan produced by *Weissella confusa* KR780676 from an acidic fermented food. *International Journal of Biological Macromolecules, 86*, 681–689.

Kentish, S., & Feng, H., (2014). Applications of power ultrasound in food processing. *Annual Review of Food Science and Technology, 5*, 263–284.

Khochamit, N., Siripornadulsil, S., Sukon, P., & Siripornadulsil, W., (2015). Antibacterial activity and genotypic-phenotypic characteristics of bacteriocin-producing *Bacillus subtilis* KKU213: Potential as a probiotic strain. *Microbiological Research, 170*, 36–50.

Kilonzo, R. M., & Gathura, M., (2016). *The Kenya Food Control System*. From: http://www.fao.org/fileadmin/user_upload/agns/news_events/Pre_CCAFRICA_KenyaEN.pdf (accessed on 27 September 2021).

Le Lay, C., Coton, E., Le Blay, G., Chobert, J. M., Haertlé, T., Choiset, Y., Van, L. N. N., et al., (2016a). Identification and quantification of antifungal compounds produced by lactic acid bacteria and *Propionibacteria*. *International Journal of Food Microbiology, 239*, 79–85.

Le Lay, C., Mounier, J., Vasseur, V., Weill, A., Le Blay, G., Barbier, G., & Coton, E., (2016b). *In vitro* and *in situ* screening of lactic acid bacteria and *Propionibacteria* antifungal activities against bakery product spoilage molds. *Food Control, 60*, 247–255.

Leroy, F., & De Vuyst, L., (2004). Lactic acid bacteria as functional starter cultures for the food fermentation industry. *Trends in Food Science and Technology, 15*, 67–78.

Leroy, F., & De Vuyst, L., (2016). Advances in production and simplified methods for recovery and quantification of exopolysaccharides for applications in food and health. *Journal of Dairy Sciences, 99*, 3229–3238.

Lynch, K. M., Aidan, C. A., & Arendt, E. K., (2018a). Exopolysaccharide producing lactic acid bacteria: Their techno-functional role and potential application in gluten-free bread products. *Food Research International, 110*, 52–61.

Lynch, K. M., Zannini, E., Aidan, C. A., & Arendt, E. K., (2018b). Lactic acid bacteria exopolysaccharides in foods and beverages: Isolation, properties, characterization, and health benefits. *Annual Review of Food Science and Technology, 9*, 155–176.

Malang, S. K., Maina, N. H., Schwab, C., Tenkanen, M., & Lacroix, C., (2015). Characterization of exopolysaccharide and ropy capsular polysaccharide formation by *Weissella*. *Food Microbiology, 46*, 418–427.

Mao, L., & Miao, S., (2015). Structuring food emulsions to improve nutrient delivery during digestion. *Food Engineering Reviews, 7*(4), 439–451.

Mathara, M. J., Schillinger, U., Kutima, P. M., Mbugua, K. S., Guigas, C., Franz, C. M. A. P., & Holzapfel, W. H., (2008). Functional Properties of *Lactobacillus plantarum* strains isolated from Maasai traditional fermented milk products in Kenya. *Current Microbiology, 56*, 315–321.

Mogensen, G., Salminen, S., O'Brien, J., Ouwehand, A., Holzapfel, W., Shortt, C., Fonden, R., et al., (2002). Inventory of microorganisms with a documented history of use in food. *Bullet of IDF, 377*, 10–19.

Nnawuihe, E. I., Banwo, K., & Sanni, A. I., (2018). Characterization and probiotic potentials of lactic acid bacteria isolated from ingesta of selected ruminants. *Acta Alimentaria. An International Journal of Food Science, 47*(1), 61–69.

Odunfa, S. A., (1985). African fermented foods. In: Wood, B. J. B., (ed.), *Microbiology of Fermented Foods* (Vol. 2, pp. 155–199). Elsevier Science, London and New York.

Odunfa, S. A., Adeniran, S. A., Teniola, O. D., & Nordstrom, J., (2001). Evaluation of lysine and methionine production in some lactobacilli and yeasts from Ogi. *International Journal of Food Microbiology, 63*, 159–163.

Ogunremi, O. R., Sanni, A. I., & Agrawal, R., (2015). Probiotic potentials of yeasts isolated from some cereal-based Nigerian traditional fermented food products. *Journal of Applied Microbiology, 119*, 797–808.

Oguntoyinbo, F. A., Huch, M., Cho, G., Schillinger, U., Holzapfel, W. H., Sanni, A. I., *et al., (*2010*).* Diversity of *Bacillus* species isolated from Okpehe, a traditional fermented soup condiment from Nigeria. *Journal of Food Protection, 73*, 870–878.

Ojekunle, O., Banwo, K., & Sanni, A., (2017). *In vitro* and *In vivo* evaluation of *Weissella cibaria* and *Lactobacillus plantarum* for their protective effect against cadmium and lead toxicities. *Letters in Applied Microbiology, 64*, 379–385.

Okoroafor, I., Banwo, K., Olanbiwoninu, A. A., & Odunfa, S. A., (2019). Folate enrichment of Ogi (a Fermented Cereal Gruel) using folate producing starter cultures. *Advances in Microbiology, 9*, 177–193.

Oladipo, I. C., Anwooko, D. O., Daramola, D. I., Babalola, F. B., Abidoye, G. T., Olawale, O. D., Iyanda, K. M., et al., (2017). Bioprotective potential of bacteriocins from some *Lactobacillus* species isolated from foods. *Journal of Microbiology Biotechnology and Food Sciences, 6*(3), 900–904.

Olanbiwoninu, A. A., Irokosu, O., & Odunfa, S. A., (2017). Riboflavin enriched iru: A fermented vegetable protein. *African Journal of Microbiology Research, 11*(13), 546–552.

Olojede, O. A., Sanni, A. I., & Banwo, K., (2020). Effect of legume addition on the physicochemical and organoleptic attributes of sorghum-based sourdough bread. *LWT – Food Science and Technology, 118*, 108769.

Oloo, B., Daisy, L., & Oniang'o, R., (2018). Food safety legislation in some developing countries. In: *Food Safety: Some Global Trends* (pp. 19–35). intechopen. Chapter 3. http://dx.doi.org/10.5772/intechopen.75587.

Omojokun, J., (2013). Regulation and enforcement of legislation on food safety in Nigeria. Chapter from the book: *Mycotoxin and Food Safety in Developing Countries*. Available from: http://www.intechopen.com/books/mycotoxin-and-food-safetyindeveloping-countries (accessed on 27 September 2021).

Onipede, G. O., Banwo, K., Ogunremi, O. R., & Sanni, A. I., (2014). Influence of starter culture lactic acid bacteria on the phytic acid content of sorghum-ogi (An indigenous cereal gruel). *Annals of Food Science and Technology, 15*(1), 121–134.

Ouwehand, A. C., Salminen, S., & Isolauri, E., (2002). Probiotics: An overview of beneficial effects. *Antonie Van Leeuwenhoek, 82*, 279–289.

Owusu-Kwarteng, J., Akabanda, F., Nielsen, D. S., Tano-Debrah, K., Glover, R. L., & Jespersen, L., (2012). Identification of lactic acid bacteria isolated during traditional fura processing in Ghana. *Food Microbiology, 32*, 72–78.

Oyewole, O. A., & Isah, P., (2012). Locally fermented foods in Nigeria and their significance to national economy: A review. *Journal of Recent Advances in Agriculture, 1*, 92–102.

Oyewole, O. B., (1997). Lactic fermented foods in Africa and their benefits. *Food Control, 8*, 289–297.

Park, M. K., & Kim, Y. S., (2019). Distinctive formation of volatile compounds in fermented rice inoculated by different molds, yeasts, and lactic acid bacteria. *Molecules, 24*(11), 2123.

Park, S., Ji, Y., Park, H., et al., (2016). Evaluation of functional properties of lactobacilli isolated from Korean white kimchi. *Food Control, 69*, 5–12.

Patel, A., & Prajapati, J. B., (2013). Food and health applications of exopolysaccharides produced by lactic acid bacteria. *Advances in Dairy Research, 1*, 107–114. doi: 10.4172/2329-888X.1000107.

Patel, A., Aparna, S. V., Shah, N., & Verma, D. K., (2017). Lactic acid bacteria as metal quenchers to improve food safety and quality. *Agro Life Journal, 6*(2), 146–154.

Patel, A., Lindström, C., Patel, A., Prajapati, J. B., & Holst, O., (2012). Probiotic properties of exopolysaccharide producing lactic acid bacteria isolated from vegetables and traditional Indian fermented foods. *International Journal of Fermented Foods, 1*(1), 87–101.

Pedersen, L. L., Owusu-Kwarteng, J., Thorsen, L., & Jespersen, L., (2012). Biodiversity and probiotic potential of yeasts isolated from *Fura*, a West African spontaneously fermented cereal. *International Journal of Food Microbiology, 159*, 144–151.

Ray, R. C., & Sivakumar, P. S., (2009). Traditional and novel fermented foods and beverages from tropical root and tuber crops: Review. *International Journal of Food Science and Technology, 44*, 1073–1087.

Rodríguez, L. G., Mohamed, F., Bleckwedel, J., Medina, R., De Vuyst, L., Hebert, E. M., & Mozzi, F., (2019). Diversity and functional properties of lactic acid bacteria isolated from wild fruits and flowers present in Northern Argentina. *Frontiers in Microbiology, 10*, 1091. doi: 10.3389/fmicb.2019.01091.

Romanens, E., Leischtfeld, S. F., Volland, A., Stevens, M., Krahenmann, U., Isele, D., Fischer, B., et al., (2019). Screening of lactic acid bacteria and yeast strains to select adapted antifungal co-cultures for cocoa bean fermentation. *International Journal of Food Microbiology, 290*, 262–272.

Russo, P., Arena, M. P., Fiocco, D., Capozzi, V., Drider, D., & Spano, G., (2017). *Lactobacillus plantarum* with broad antifungal activity: A promising approach to increase safety and shelf-life of cereal-based products. *International Journal of Food Microbiology, 247*, 48–54.

Saad, N., Delattre, C., Urdaci, M., Schmitter, J. M., & Bressollier, P., (2013). An overview of the last advances in probiotic and prebiotic field. *LWT Food Science Technology, 50*, 1–16.

Sadeghi, A., Ebrahimi, M., Mortazavi, S. A., & Abedfar, A., (2019). Application of the selected antifungal LAB isolate as a protective starter culture in pan whole-wheat sourdough bread. *Food Control, 95*, 298–307.

Sanni, A. I., Franz, C. M. A. P., Schillinger, U., Huch, M., Guigas, C., & Holzapfel, W. H., (2013). Characterization and technological properties of lactic acid bacteria in the production of 'sorghurt', a cereal-based product. *Food Biotechnology, 27*, 178–198.

Sefa-Dedeh, S., (2009). *Ghana: Overview of Food Safety Situation in Africa-Country Position Presented at the Go-Global Conference.* Accra.

Shoba, R. P., & Agrawal, R., (2007). Volatile compounds of therapeutic importance produced by *Leuconostoc paramesenteroides*, a native laboratory isolate. *Turkish Journal of Biology, 31*, 35–40.

Thakur, K., Lule, V. K., Rajni, C. S., Kumar, N., Mandal, S., Anand, S., Kumari, V., & Tomar, S. K., (2016). Riboflavin producing probiotic Lactobacilli as a biotechnological strategy to obtain riboflavin-enriched fermented foods. *Journal of Pure and Applied Microbiology, 10*(1), 161–166.

Todorov, S. D., Vaz-Velho, M., Gombossy De, M. F. B. D., & Holzapfel, W. H., (2013). Partial characterization of bacteriocins produced by three strains of *Lactobacillus sakei*, isolated

from salpicao, a fermented meat product from North-West of Portugal. *Food Control, 30*, 111–121.
Torino, M. I., Font De, V. G., & Mozzi, F., (2015). Biopolymers from lactic acid bacteria: Novel applications in foods and beverages. *Frontiers in Microbiology, 6*, 834. http://dx.doi.org/10.3389/fmicb.2015.00834.
Vimont, A., Fernandez, B., Ahmed, G., Fortin, H. P., & Fliss, I., (2019). Quantitative antifungal activity of reuterin against food isolates of yeasts and moulds and its potential application in yogurt. *International Journal of Food Microbiology, 289*, 182–188.
Williams, M., Pehu, E., & Ragasa, C., (2006). Functional foods: Opportunities and challenges for developing countries. *Agricultural and Rural Development Notes,* 19.
Yasmine, M., (2000). Impact of small scale fermentation technology on food safety in developing countries. *International Journal of Food Microbiology, 75*(3), 213–229.
Zannini, E., Mauch, A., Galle, S., Ganzle, M., Coffey, A., Arendt, E. K., Taylor, J. P., & Waters, D. M., (2013). Barley malt wort fermentation by exopolysaccharide forming *Weissella cibaria* MG1 for the production of a novel beverage. *Journal of Applied Microbiology, 115*, 1379–1387.
Zoghi, A., Khosravi-Darani, E. K., & Sohrabvandi, S., (2014). Surface binding of toxins and heavy metals by probiotics. *Mini-Reviews in Medicinal Chemistry, 14*, 84–98.
Zuidam, N. J., & Heinrich, E., (2010). Encapsulation of aroma. In: Zuidam, N. J., & Heinrich, E., (eds.), *Encapsulation Technologies for Active Food Ingredients and Food Processing*. New York, Springer.

CHAPTER 9

Advanced Molecular Tools and Techniques for Assessment of Microbial Diversity in Fermented Food Products

DAMANPREET KAUR, SUSHMA GURUMAYUM, PRASAD RASANE,* SAWINDER KAUR, JYOTI SINGH, NAVNEET KAUR, KAJAL DHAWAN, and ASHWANI KUMAR

**Corresponding author. E-mails: rasaneprasad@gmail.com; prasadrasane4u@gmail.com*

ABSTRACT

All over the world, fermented food products form an integral part of the human diet. For centuries, the preparation of fermented foods involved spontaneous fermentation by indigenous microbial activity present in the environment and in the food ingredients. Knowledge of these microbial communities is essential for shaping the final characteristic of the product and also provides an insight into the fermentation product. Earlier, traditional culture-dependent methods were employed for enumeration of microbial diversity. These methods required the cultivation of microorganisms on a synthetic medium. The advancements in food microbiology have led to the application of culture-independent techniques for elucidating the microbial ecology associated with the fermented food products. Culture-independent approaches such as PCR-DGGE and sequencing have helped in studying microbial activities during fermentation. Among culture-independent techniques, sequencing techniques have been widely used to exploit the microbial population. Various NGS techniques like Ion torrent, Roche pyrosequencing and Illumina have resulted in in-depth analysis of microbial species. Besides this, metatranscriptome, and metagenomics also helps in studying microbial consortia in several fermented products. The knowledge gained about the microbial diversity and the changes that are taking place during fermentation process via the use of sequencing techniques is vital in monitoring the process and improving the food quality. Besides these approaches, an emerging field in the 'omics' sciences, *viz*., metabolomics has been applied to the fermentation process for the study of metabolites generated. This chapter deals with techniques associated with the study of microbial diversity present in fermented food products.

9.1 INTRODUCTION

Fermented foods form an integral part of diet since time immemorial. Preparation and consumption of fermented foods is practiced in many places all across the globe. Fermented foods have several desirable characteristics like prolonged shelf life, improved digestibility, enhanced aroma, and flavor, increased nutrient content and even remove toxins and antinutritional compounds. Therefore, it is important to learn about the microbial community which brings about such desirable changes and impart the above qualities in food. Knowledge about the microorganisms involved and the microbial interaction within the fermenting ecosystem will help to deliver desired quality product (van Hijum et al., 2013). For centuries, fermented food preparation relied on spontaneous fermentation executed by the indigenous microflora present in the surrounding environment and raw ingredients used (de Angelis et al., 2015). The development of microbiota during the fermentation process has been examined to a great extent and it has been observed that these populations contribute in a major way to the fermentation process. Knowledge about the microbial diversity involved in fermentation is an integral factor responsible for shaping the final characteristic of the product (Polka et al., 2015). The study of microbial diversity gives an insight into the fermentation process. It helps us to understand the relationship among microbial communities involved, dynamic changes occurring, and special properties of the final fermented product.

In the last few decades, advances in the identification and detection of microorganisms have been made through the application of culture-independent approaches (Greppi et al., 2015). Such understanding of the microbial diversity through traditional culturing method and culture-independent technique help to optimize the fermentation process.

Culture-dependent techniques involves the growth of microorganisms on microbial culture media. This technique proves useful in determining the concentration of live microorganisms present in a substrate. However, these techniques sometimes may result in the underestimation of microflora that fail to grow on regular culture media. In certain cases, it also fails to separate the dominant microflora (Wei et al., 2013). The limitations associated with the application of culture-dependent techniques have been partially circumvented with the application of molecular-based culture-independent techniques among which PCR-DGGE is widely used. It has been extensively used for analyzing microbial diversity in unculturable complex microbial populations. The ability of PCR-DGGE to detect both viable and culturable as well as viable and non-culturable organisms have led to its wide application when compared with the results of culture-dependent techniques (Mukisa et al., 2012). However, this technique also suffers from limitations arising from incomplete DNA extraction, PCR biases, co-migration, poor resolution of low abundance microbes and sometimes fail to analyze the microbial community within multiple samples (Park et al., 2012; Yan et al., 2013b). This has led to a shift in interest of researchers to next-generation sequencing techniques, such as pyrosequencing in elucidating the diversity and dynamics of microbial population of fermented foods. Such sequencing techniques have brought a revolution in food microbiology through more precise identification of microflora and has thus, replaced the PCR-DGGE approach (Liu et al., 2015). Yet, another recent approach dealing with assessment of quality, processing technique, safety, and microbiology of both the raw ingredients and final product is Metabolomics or Metabolite profiling. The use of metabolomics in fermented products enable to interpret metabolic changes taking place during fermentation and also predict the nutritional quality and sensory properties of final fermented food (Mozzi et al., 2013).

9.2 CULTURE-DEPENDENT METHODS

The microbial community in fermented foods was initially determined by culture-dependent methods which involve isolation, enumeration, biochemical characterization, and identification of microbial population. This is done by cultivation on various selective synthetic culture media and identification of predominant population using phenotypic and molecular techniques (Wei et al., 2013). In this method, both the routine and modified culture media are used to culture, identify, and maintain the microbial population. Media used for culturing technique contains necessary nutrients required to support the growth of the microorganisms. It has carbon as an energy source which is required for incorporation into biomass. Other compounds, such as nitrate ions, ammonium ions, reduced iron and elemental sulfur can also be used as energy source for cultivation. Microbial growth on culture media also requires nitrogen, which is sometimes supplied as inorganic nitrogen compounds, but most commonly it is supplied as amino acids, peptones, or proteins.

There is a wide choice of culture media used for culturing microorganisms due to their diversity and diverse metabolic pathways. Some of these media are Rogosa agar, M17 Agar, de Man Rogosa Sharpe (MRS) agar, Lactobacilli selective agar, trypticase soy agar and plate count agar with bromocresol purple (PCA-BCP agar) for lactic acid bacteria (LAB) isolation (Kostinek et al., 2005; Kormin et al., 2001; Lee and Lee, 2008), Slanetz and Bartley Medium (SAB), Kanamycin Esculin Azide agar for *Enterococcus* spp. (Sengun et al., 2009), yeast glucose chloramphenicol (YGC) agar for yeasts (Yam et al., 2015), yeast extract potato dextrose agar (YPD) for cultivation of yeasts and malt extract agar (MEA) for mycelial fungi.

The methods of culturing microorganisms were first developed in the 19th century by Robert Koch and his colleagues. During that period, plant, and animal tissues were largely used as a source of nutrient to support the microbial growth. Basically, the microorganisms are cultivated in two types of media-selective media and differential media. Selective media helps with the isolation of a particular microorganism from a mixed population and inhibit the growth of non-targeted species. This media includes only carbon or nitrogen as a sole source of energy and facilitate the growth of only a few types of microorganisms. The incorporation of selective toxic compounds into the culture media suppresses the growth of background microorganisms and permits the cultivation of specific species only. On the other hand, the differential media permits the visual detection of microbes from the changes in pH resulting from the metabolic reactions. Besides, the changes in pH, some media detect microbes from color changes taking place during enzymatic reactions due to the presence of chromogenic dyes.

Culture-based approaches have proved useful for comprehending the physiological potential of the microorganisms isolated from culturing plates (Yan et al., 2013a). For decades, and till now, antimicrobial activities associated with microbe isolates of the fermented products have been determined by the conventional culturing plate method followed by phenotypic and genotypic identification of randomly selected isolates (Nguyen et al., 2013a). These properties influence the nutritive value, texture, and flavor of the product by metabolizing the chemical components of the raw material. Various compounds produced by bacteria like bacteriocins, acids, hydrogen peroxide, and bacteriocin like inhibitory substances (BLIS) exhibit antibacterial activity. The compounds produced by *L. plantarum* exhibit antibacterial activities in fermented vegetables which contributes to the overall improvement in their quality (Hashemi et al., 2017). These antimicrobial properties aid in the selection of appropriate industrial traits to be used as a starter culture in different fermented products (Abriouel et al., 2012). Marty et al. (2012) reported that the technological properties of LAB are used to standardize, control microbial risk and production process of fermented meat product in the modern meat manufacturing practice.

In spite of the above advantages, the cultivation of microbes in nutrient media proves to be biased because it results in the cultivation of microorganisms only when their requirements (metabolic and physiological) are reproduced *in-vitro*. Culturing microorganisms using artificial homogenous methods allow the growth of only a small number of species. This method fails to isolate injured and stressed cells which are present in small numbers. It is unreliable as it does not give an accurate representation of the diversity of microbial community and the dynamic changes taking place during fermentation (Adewumi et al., 2013; Yan et al., 2013a). The conventional culture-dependent approaches are incorrectly assumed for certain microorganisms. Lack of proper knowledge regarding selectivity of appropriate culture medium can also give filtered results and deceptive information (van Hijum et al., 2013). This technique, besides being laborious and time-intensive, often fails to isolate the strains that are unable to grow on particular culture media (Wei et al., 2013).

The limitations associated with this method therefore necessitates researchers to find alternative approaches for the complete and accurate assessment of microbial diversity in the microbial ecological niches of fermented foods. This has resulted in the application of molecular culture-independent approaches which target variable regions of the ribosomal RNA (18S rRNA for fungi, 16S rRNA for bacteria and 26S rRNA gene for yeast and mold).

9.3 CULTURE INDEPENDENT METHODS

The nutritional availability, organoleptic quality, safety, and sensorial properties of fermented food consumed worldwide are attributed to its microbial diversity and the type of starter associated with that product (Park et al., 2012). Recently, culture-independent techniques have been extensively used to study the diversity and dynamics of the microbiota in the fermented foods without the need for preceding culturing. It helps to analyze multiple samples simultaneously without any previous knowledge of the microbial environment (Nguyen et al., 2013a). These techniques partially provide a way for overcoming the limitations associated with the use of conventional culturing techniques. The detection of viable but non-culturable (VBNC) population in the food system has been done by culture-independent methods. Most of these techniques are PCR based and the analysis of the PCR product highlights the differences in amplified DNA sequence (Cocolin et al., 2013). A wide variety of molecular tools are used identifying and quantifying isolated microorganisms from fermented food products. Some of them are species-specific PCR, rep-PCR, qPCR, AFLP, MLSA, RAPD, mCold-PCR, mtDNA-RFLP, ARDRA, DGGE, MLST, and TGGE. Species-specific PCR primers help in identifying characteristic species of the genus, whereas, the quantification of LAB isolated from fermented vegetables is done with the use of real-time quantitative PCR (qPCR). Techniques like Amplified-fragment length polymorphism (AFLP), random amplification of polymorphic DNA (RAPD), Multilocus sequence analysis (MLSA) using housekeeping genes as molecular markers and Multilocus sequence typing (MLST) results in the amplification and discrimination of LAB strains in fermented foods. Uncultivable microorganisms present in the fermented foods are identified using amplified ribosomal DNA restriction analysis (ARDRA) whereas low abundant microorganisms present in fermented foods have been detected using modified CO-amplification at low denaturation temperature PCR (mCOLD-PCR) method (Tamang et al., 2015). Also culture-independent techniques like restriction fragment length polymorphism (RFLP) and single-stranded conformation polymorphisms (SSCP have been extensively used in exploring the microbial population in cheese and milk (Lusk et al., 2012).

These methods are reliable, rapid, reproducible, economical, and can analyze multiple samples simultaneously (Lv et al., 2012; Wei et al., 2013). One such approach which has proved successful in examining the microbial diversity is the analysis of PCR products using denaturing gradient gel electrophoresis.

9.3.1 POLYMERASE CHAIN REACTION-DENATURING GRADIENT GEL ELECTROPHORESIS (PCR-DGGE)

The introduction of DGGE to food microbiology dated back to the late 90s, and since then, numerous studies have been performed to exploit DGGE as a tool to explore the diversity of microflora in food (Cocolin et al., 2013). PCR-DGGE is a molecular method which has been extensively used to study the microflora of various fermented foods, such as, *Hong Qu* glutinous rice wine (Lv et al., 2012), *inyu* (Wei et al., 2013), *meat products* (Nguyen et al., 2013b), *sausages* (Greppi et al., 2015), *iru* (Adewumi et al., 2013), *oscypek* (Alegria et al., 2012) and fermented soybean products (Chettri and Tamang, 2015; Yan et al., 2013a). The DGGE approach in *inyu* (fermented soy sauce) fermentation revealed the presence of LAB in *inyu* brine samples which the conventional plating technique failed to detect (Wei et al., 2013). The use of this approach in food microbial ecology aids not only in expanding the previous knowledge obtained from conventional culturing methods, but also helps to get the most accurate information of microbial species associated with fermented food products (Polka et al., 2015). This assay can also accelerate the formation of starter cultures and gives comprehensive information of important micro-organisms involved in flavor development as well as spoilage of traditional fermented foods. Besides providing rapid and broad information on microbial population dynamics, the DGGE gels offer the possibility to extract and sequence bands making it act as a priceless tool for identification of predominant species (Lv et al., 2013; Nguyen et al., 2013b; Wei et al., 2013). The steps involved in PCR-DGGE approach as outlined in Figure 9.1 are explained in detail.

9.3.1.1 SELECTION OF SAMPLES

Fermented food samples for DGGE analysis are collected from different geographical areas. Sample collection is done under aseptic conditions to prevent microbial contamination. Samples are refrigerated at 4°C before transporting them to the laboratory for extraction process, and analysis of samples is done within 2–3 days (Devi et al., 2015) or within a week (Chettri and Tamang, 2015).

9.3.1.2 DNA EXTRACTION

As DNA carries the hereditary messages (Cocolin et al., 2013), it acts as a template for amplification of gene, particularly 18S rRNA, 16S rRNA, genes, and 26S rRNA (Lv et al., 2012; Xiu et al., 2012). The quality of DNA extracted from the samples greatly influences the success of this technique. Insufficient DNA shearing from the samples may interfere with the amplification process and may cause biased amplification. This leads to the use of various commercial DNA extraction kits for increasing the efficiency of the extraction process and reduce the chances of error (Ariefdjohan et al., 2010). DNA extraction can be done by Soil Genomic DNA Rapid Extraction Kit (Lv et al., 2012), Power Soil DNA Isolation Kit (Tofalo et al., 2013) and Universal Genomic DNA Extraction Kit Ver: 3.0 (Lv et al., 2013). Tanaka et al. (2012) used a combination of freeze-thaw method, bead-beating method and CTAB method to extract DNA from koji and mash samples used in soy sauce fermentation. Lysozyme-heat lysis method was also reported for the DNA extraction from *Ngari*, dry fermented fish product by Devi et al. (2015). A combination of DNA extraction kits, Plant DNeasy kit and Power

Food DNA isolation kit, was used by Gonzalez et al. (2013) for DNA isolation from samples of Rioja wine. Nam et al. (2012c) employed Ultra Clean Microbial DNA Isolation Kit for extraction of DNA strands from *cheonggukjang*, a Korean fermented soybean paste. The extracted DNA strands help in determining the microbial ecology of the fermented food samples.

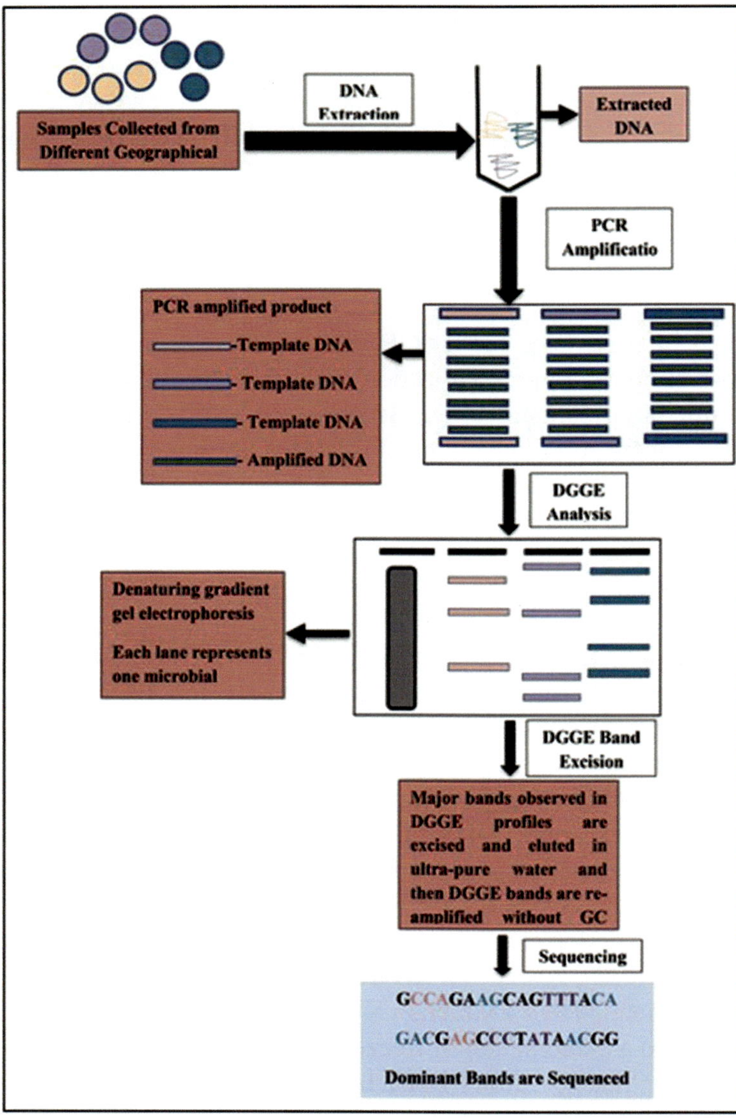

FIGURE 9.1 Steps involved in PCR-DGGE molecular technique.

9.3.1.3 *PCR AMPLIFICATION*

The bacterial diversity of samples is analyzed by PCR amplification of 16S rRNA encoding genes (Lv et al., 2012). The amplification assay is performed by using appropriate primers as shown in Table 9.1. Use of direct PCR amplification result in low product yield. Instead, a nested PCR approach is useful in detecting those microorganisms not detected by PCR reaction by using PCR product as a template (Lv et al., 2013). Nested PCR amplification targets 16S rDNA in order to create DNA fragment suitable for DGGE analysis (Wei et al., 2013).

9.3.1.4 *DGGE ANALYSIS*

Denaturing gradient gel electrophoresis involves analysis of amplified PCR product for the identification of microbial ecology. There is difference between two DGGE analysis methods *viz.*, direct PCR-DGGE and nested PCR-DGGE. Direct PCR-DGGE analysis showed 36 bands in DGGE profiles revealing 11 bacterial species, as compared to nested PCR-DGGE assay yielding 46 bands with 18 bacterial species in *Daqu* starter fermentation process in Maotai liquor. This revealed that nested PCR-DGGE provided comprehensive in-depth analysis of microbial community resulting from amplification of DNA extracted from those population present in low abundance as compared to direct PCR-DGGE (Xiu et al., 2012). Similarly, Lv et al. (2013) employed nested PCR-DGGE for better enumeration and resolution of microbial diversity during brewing of Wuyi Hong Qu glutinous rice wine. Denaturing gradient gel of polyacrylamide containing a gradient of chemical urea and formamide result in the separation of same length amplicons but with different nucleotide composition, based on their differential denaturing profiles. During denaturation, the double-stranded DNA molecule becomes sequence-dependent, partially denatured, resulting in the reduced migration of the DNA sequence in the gel (Cocolin et al., 2013; Duarte et al., 2012). Employing this method for microbial profiling after amplification not only result in the differentiation of complex DNA molecules, but also help in studying the dynamics of microbiota over time. Each band in the gel electrophoresis represents components of the microbial community. After gel staining, these bands are excised and re-amplified which is then followed by sequencing to obtain corresponding microbial community (Cocolin et al., 2013).

TABLE 9.1 Primers Used in PCR-DGGE for Amplification of DNA Extracts

Fermented Foods	Primer	Sequence (5′-3′)	Target	References
Kimchi	V1-27F	(B-AC-GAGTTTGATCMTGGCTCAG)	Amplification of 16S rRNA genes from genomic DNA.	Jung et al. (2012)
	V3-541R	(A-X-AC-WTTACCGCGGCTGCTGG)		
Hong Qu	27F	(AGAGTTTGATCCTGGCTCAG)	For bacterial analysis, amplification of 16S rRNA encoding gene under conventional PCR condition.	Lv et al. (2012)
Glutinous rice wine	1492R	(GGCTACCTTGTTACGACTT)		
	NS1	(GTAGTCATATGCTTGTCTC)	For fungal diversity analysis, PCR amplification of 18S rRNA gene.	
	FR1	(AICCATTCAATCGGTAIT)		
Obushera	ObsF	(CTCCTACGGGAGGCAGCA)	Targeting V3 region of bacterial 16S rRNA gene.	Mukisa et al. (2012)
	Obst	(ATTTCACCGCTACACATGGA)		
	NL1GC	(GCCATATCAATAAGCGGA-GGAAAAGCGCCCGCCGCG-CGCGGCGGGCGGGGCGGGG)	Amplifying D1 region of 26S rRNA of eukaryotes.	
	LS2	(TTCCCAAACAACTCGACTC) GC clamp is underlined		
Any	F-968	(AACGCGAAGAACCTTAC)	Generation of bacterial 16S rDNA fragment by PCR amplification of genomic DNA	Wei et al. (2013)
	Bac-1492R-II	(ACGGYTACCTTGTTACGCGA-CTTCGCCCGGGGCGCGCCC-CGGGCG)		
	F-968-GC	(GGGCGGGGGCACGGGGGG-AACGCGAAGAACCTTAC)	Target bacterial 16S rDNA to create DNA fragment for DGGE analysis	
	R-1401	(CGGTGTGTACAAGACCC)		
For Yeast				
Shanxi aged vinegar	Delta12	(TCAACAATGGAATCCCAAC)	PCR amplification of extracted genomic DNA.	Wu et al. (2012)
	delta21	(CATCTTAACACCGTATATGA)		

TABLE 9.1 *(Continued)*

Fermented Foods	Primer	Sequence (5´-3´)	Target	References
	NL1	(GCATATCAATAAGC-GGAGGAA)	Amplification of D1/D2 region of 26S rRNA.	
	NL4	(GGTCCGTGTTTCAAGACGG)		
For Lactic Acid Bacteria				
–	616v	(TACGGGAGGCAGCAG)	Amplification of 16S rRNA.	Wu et al. (2012)
	1492r	(GGCTACCTTGTTACGACT)		
Fermented table olives	FR1-GC	(AICCATTCAATCGGTAIT)	Targeted fungal 18S rRNA gene.	Tofalo et al. (2013)
	FF 390	(CGATAACGAACGAGACCT)		
Soy sauce	357F	(CCTACGGGAGGCAGCAG)	DGGE analysis of bacterial communities at 16S rRNA V3 region.	Tanaka et al. (2012)
	NL1	(GCATATCAATAAGCGGA-GGAAAAG)	DGGE analysis of fungal communities at 26S rRNA gene.	
	LS2	(ATTCCCAAACAACTCGCTC)		
Ngari	Forward	(AGAGTTTGATCCTGGCTCAG)	PCR amplification of 16S rRNA gene.	Devi et al. (2015)
	Reverse	(AAGGAGGTGATCCAGCCGCA)		
Italian salami	343F	(TACGGRAGGCAGCAG)	PCR amplification of bacterial V3-V4 region of 16S rRNA gene.	Polka et al. (2015)
	802R	(TACNVGGGTWTCTAATCC)		
Nem Chua	pheS-21-F	(CAYCCNGCHCGYGAYATGC)	Amplification and sequencing of phenlyalanyl-tRNA and RNA polymerase α subunit gene	Nguyen et al. (2013b)
	pheS-23-R	(GGRTGRACCATVCCNGCHCC)		
	rpoA-23-R	(ACHGTRTTRATDCCDGCRCG)		
	357F	(ATTACCGCGGCTGCTGG)	PCR amplification of 194 bp fragment of V3 region of 16S rRNA gene.	
	518R	(ATTACCGCGGCTGCTGG)		

TABLE 9.1 *(Continued)*

Fermented Foods	Primer	Sequence (5′-3′)	Target	References
Tungrymbai Bekang	Forward	(AGAGTTTGATCCTGGCTCAG)	PCR amplified separation of 16S rDNA gene.	Chettri and Tamang, (2015)
	Reverse	(AAGGAGGTGATCCAGCCGCA)		
Swiss meat	2643F	(CAATTCATGGACCAAGC)	Generation of partial fragments of rpoB gene.	Marty et al. (2012)
	3241R	(GCDACDTGDTCCATACCTGT)		
Cheonggukjang	GC338f	(CGCCCGCCGCGCCCCGCGC-CCGGCCCGCCGCCCCCG-CCCCCTACGGGAGGCAGCAG)	PCR amplification of V3 region of bacterial 16S rRNA gene for bacterial diversity analysis.	Nam et al. (2012c)
	518r	(ACCGCGGCTCTGGCAC)		
Manzanilla Alorena green table olives	GTG5	GTG GTG GTG GTG GTG	REP-PCR analysis.	Abriouel et al. (2012)
Oscypek	27F	AGAGTTTGATYMTGGCTCAG	PCR amplification of 16S rRNA gene of DNA template.	Alegria et al. (2012)
	1492R	GGTTACCTTGTTACGACTT		
	F357	TACGGGAGGCAGCAG	PCR amplification of V3 variable region of prokaryotic 16S rRNA gene.	
	LB2-GC with LAC3	GATTYCACCGCTACACATG AGCAGTAGGGAATCTTCGG	PCR group-specific amplification of 16S rRNA gene of Lactococcus-enterococcus-streptococcus group and *lactobacillus-leuconostoc-pediococcus* group respectively.	
	LB2-GC with LB1	AGCAGTAGGGAATCTTCC		
Maotai liquor	27f	AGAGTTTGATCCTGGCTCAG	Amplification of bacterial 16S rRNA gene.	Xiu et al. (2012)
	1492R	GGTTACCTTGTTACGACT		

TABLE 9.1 *(Continued)*

Fermented Foods	Primer	Sequence (5′-3′)	Target	References
	338f	CGCCCGCCGCGCGCGGCGGG-CGGGGCGGGGGCACGGGGGG-ACTCCTACGGGCGGCAGCAG	Nested PCR analysis.	
	518r	ATTACCGCGGCTGCTGG		
	NS1	GTAGTCATATGCTTGTCT	Amplification of fungal 18S rRNA gene.	
	FR1	AICCATTCAATCGGTAIT		
	NS1 and GCFung	CGCCCGCCGCGCCCGCGCC-CGGCCCGCCGCCCCCGCCC-ATTCCCCGTTACCGTTG	Nested PCR analysis.	
Doenjang	27F	AGAGTTTGATCCTGGCTCAG	Amplification of 16S rRNA gene.	Jeong et al. (2014)
	492R	GGTTACCTTGTTACGACTT		
Fermented vegetables	375F	ATTACCGCGGCTGCTGG	Amplification of 194 bp fragment of V3 region of 16S rRNA gene.	Nguyen et al. (2013a)
	518R	ATTACCGCGGCTGCTGG		
Mawe and Tchoukoutou	$(GTG)_5$	GTGGTGGTGGTGGTG	Amplification of yeast species.	Greppi et al. (2013)
Saeu-jeot	bac340F	CCTACGGGAGGCAGCAG	Measure bacterial 16S rRNA gene.	Jung et al. (2013)
	bac758R	CTACCAGGGTATCTAATCC		
	arc109F	ACKGCTCAGTAACACGT	Measure archaeal 16S rRNA gene.	
	arc344R	TCGCGCCTGCTGCTCCCCGT		
Naturally fermented yak milk	27F	AGAGTTTGATCCTGGCTCAG	Amplification of V1-V3 variable region of 16S rRNA gene.	Liu et al. (2015)
	533R	TTACCGCGGCTGCTGGCAC		
	ITS1F	CTTGGTCATTTAGAGGAAGTAA	Amplification of fungal sequence.	
	ITS2R	GCTGCGTTCTTCATCGATGC		

9.3.1.5 DGGE-BAND EXCISION

The major bands in DGGE profiles are excised from the gel after staining and are re-amplified for the identification of dominant microbial community present in the samples (Lv et al., 2012; Tofalo et al., 2013). Extracted DNA from the gel is re-amplified with the same primers (Table 9.1) and is compared with the original sample to confirm that single bands are present in the same position (Tanaka et al., 2012). This excision results in the tentative identification of the microflora through band position analysis exhibiting the highest percentage of similarity (Nguyen et al., 2013b).

9.3.1.6 SEQUENCING

The excised bands are sequenced and assembled into contiguous sequences. These sequences are then analyzed for closest phylogenetic relatives by comparison with the gene sequences in the GeneBank database (Lv et al., 2012).

The application of the DGGE approach in several fermented foods have been done to study the changes occurring during fermentation and to highlight the predominant microbial population. A study of microbiota of fermented table olives revealed the main species to be *Candida ethanolica* and *Candida stellimalicola* besides *S. cerevisiae.* These were not detected by conventional culturing methods (Tofalo et al., 2013). Fermented foods, such as *Hong Qu,* glutinous rice wine (Lv et al., 2012), *soy sauce* (Tanaka et al., 2012; Yan et al., 2013a), *Shanxi,* aged vinegar (Wu et al., 2012), *Ngari,* dry fermented fish product (Devi et al., 2015), *Nemchua,* fermented meat product (Nguyen et al., 2013b), *Tungrymbai*, and *Bekang,* fermented soybean foods (Chettri and Tamang, 2015), *Taleggio Cheese* (Feligini et al., 2012), *Maotai liquor* (Xiu et al., 2012), *Rioja wines* (González-Arenzana et al., 2013), fermented sausages (Greppi et al., 2015), *Iru,* fermented *Parkia biglobosa* seeds (Adewumi et al., 2013), *Obushera,* millet, *and* sorghum beverages (Mukisa et al., 2012), *Mawe*, and *tchoukoutou,* traditional fermented foods (Greppi et al., 2013) have also been analyzed using PCR-DGGE. This method has been largely applied to study the dynamics of LAB in many kinds of food products. Nguyen et al. (2013b) used this approach to identify LAB species from *nemchua*, Vietnamese fermented meat product. DGGE analysis also helped to identify species such as *L. helveticus/crispatus*, *L. garvieae* and *Vagococcus* which were not isolated from culture-dependent methods using MRS agar plates.

Microbial diversity enumeration through PCR-DGGE however often suffers from some limitations. The major one being low resolution power and difficulty in understanding quantitative data (Polka et al., 2015). GC-clamped primers used in the analysis of microbial population in this approach presents sensitivity limitations. It fails to target low abundance population in the natural environment, i.e., the one which contribute to less than 1% of the total community (Lv et al., 2013; Wei et al., 2013). It sometimes leads to the overestimation of actual microbial diversity by producing multiple DGGE bands due to 16S rRNA gene operon heterogeneity. Conversely, it also results in underestimation of some species which are represented on bands that fails to reach to the same position in gradient gel (Nguyen et al., 2013b).

9.3.2 SEQUENCING TECHNIQUES

The limitations associated with the application of PCR-DGGE can be overcome by using high throughput sequencing technology for the complete analysis of microflora in fermented products. For the past several decades, Maxam-Gilbert, and Sanger sequencing techniques, also called first-generation sequencing techniques, have been used to sequence the bacterial genomics. However, with time several novel approaches emerged which were characterized as 'next-generation sequencing' (NGS) and replaced the existing first-generation techniques (Mayo et al., 2014). The recent advances in these techniques are changing the ways the microbial species are analyzed and are revolutionizing the evaluation and understanding of microbial communities. The sequencing methods enable the analysis of multiple nucleotide sequences at one time. These techniques help to clarify the molecular basis of how organisms react to different conditions associated with foods. It also provides DNA/RNA sequence data for understanding the microbial behavior in a food ecosystem (Liu et al., 2015; Nam et al., 2012a; Polka et al., 2015; van Hijum et al., 2013). Figure 9.2 illustrates a brief overview of applying sequencing technology to understand the microbial diversity in fermented products. NGS has enabled the recognition of complex microflora present in food through high-throughput analysis of short amplicons, particularly hypervariable domains of 16S rDNA.

This technique allows the analysis of thousands of samples together using short DNA sequence barcodes and provides in-depth analysis of samples (Bokulich et al., 2012). These sequencing technologies provide high accuracy in results as these are dependent on automation and high throughput technologies which are able to process a large number of sequences reads at one time (Metzker, 2010). Next-generation sequencing involve shearing of the sample DNA into size compatible for sequencing, and the extracted DNA samples are singled out, enabling them to be sequenced in parallel. These NGS technologies are divided into two platforms viz., 2nd generation in which template molecules are amplified before sequencing and 3rd generation in which single DNA molecules are sequenced without an amplification of the template molecule. Second generation NGS includes various sequencing platforms like 454/Roche pyrosequencing, Solexa (Illumina) and SOLiD (ABI/Life Technologies). While 454/Roche pyrosequencing deals with detecting pyrophosphates released during DNA synthesis, SOLiD sequencing is dependent on multiple cycles of ligation, detection, and cleavage. Solexa (Illumina) sequencing platform is based on detecting single bases that are introduced into growing DNA strands. Although these three sequencing platforms have different technological structure, but their working for the production and detection of sequencing libraries is different. Third generation NGS technologies holds promise of providing cost-effective gene sequencing from single DNA molecule without any PCR amplification step and these include sequencing platform available from Helicos Biosciences and Pacific Biosciences (Solieri et al., 2013).

Although 454 pyrosequencing was the first commercially available system of NGS technology, but this system is now no longer in use (Claesson et al., 2010). However, Illumina and Ion Torrent are the latest next-generation technologies currently used. Both Illumina MiSeq and Ion Torrent PGM platforms are now increasingly being used to provide in-depth analysis of microbial diversity using 16S rRNA gene. Illumina sequencing amplifies the already prepared DNA fragments by isothermic bridge PCR and covalently links amplicons to solid surface and thus results in arrayed cluster formation. Sequencing of these clusters then take place through continuous cycles of single-base extension. This results in identification of incorporated base and the cleavage by terminators enable further cycles. In contrast, Ion Torrent sequencing first prepares DNA fragment using emulsion PCR and then amplifies the individual fragment and results in profiling of bacterial community. However, among both these platforms, ion torrent exhibits high sequencing error rates as compared to Illumina (Salipante et al., 2014). Beside these, PacBio sequencing is also a method of real-time sequencing and is commonly referred to as third-generation sequencing. PacBio sequencing offers the benefit of longer read lengths and long sequence runs. The long sequence reads help in identification of precise location and repetitive sequence regions within single reads. However, due to high error rates, high cost per base and low throughout, the application of this method is hindered. During this sequencing method, sequence information is captured during the replication process of the target DNA molecule (Rhoads and Au, 2015). The bacterial community of various fermented food products have been evaluated using different platforms of NGS technology. The microbial ecology of products like daqu, sourdough, pu-erh tea, Shanxi aged vinegar and Tibetian kefir has been exploited using platforms like Illumina MiSeq, Illumina HiSeq 2000 and pyrosequencing (He et al., 2017).

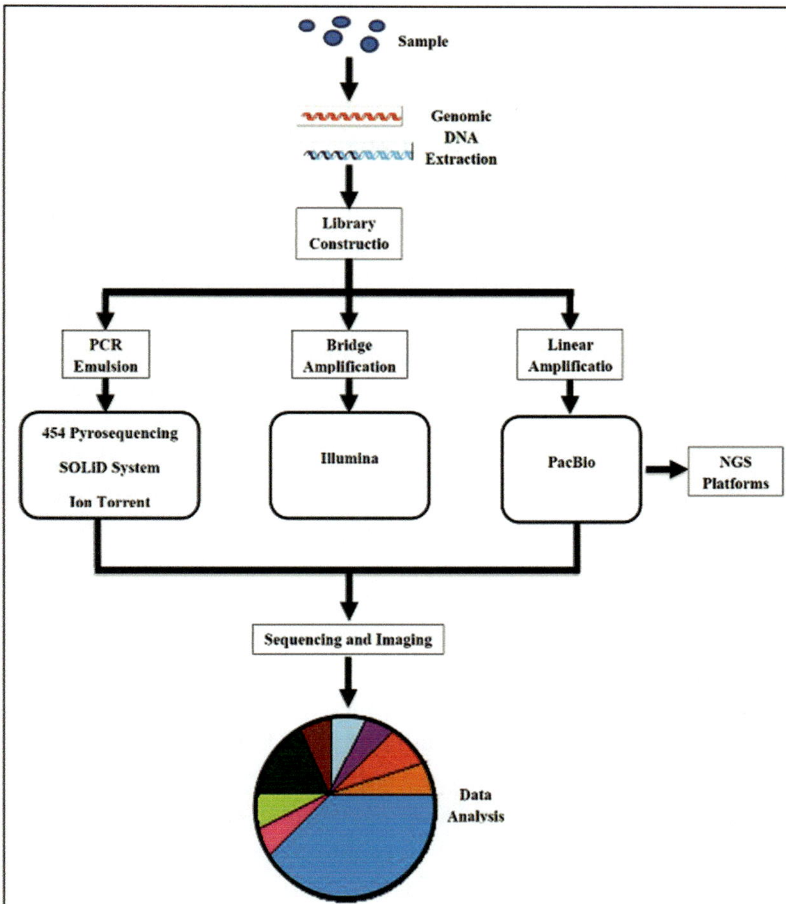

FIGURE 9.2 Working of NGS technology (Mayo et al., 2014).

Applications of NGS technologies includes monitoring the dynamics of microbial species, of which phylobiomics is the most explored one, while metatranscriptomics and metagenomics are still being underexploited. Phylobiomics refer to amplicon-based sequencing which targets the genes of taxonomic importance particularly 16S rRNA. The operational taxonomic unit (OTU) identification of food microbes to species level as provided by phylobiomics is used to obtain in-depth information on microbial succession during food production and spoilage. The identification of sequence reads with the same OTU gives quantitative estimation of the OTU present in the samples analyzed through sequencing. Therefore, the accurate taxonomic analysis of the food microbes in high throughput sequencing depends upon the read lengths of the DNA/RNA molecule. High sequence reads of 16S rRNA having hypervariable regions are used in sequencing for obtaining useful information about microbiota succession during the fermentation process. 16S rRNA gene pyrosequencing application in the production of mozzarella cheese aided in determining the microbial diversity produced in the intermediates as well as provided the abundance of bacterial taxonomic levels detected during the fermentation process. The shotgun sequencing has also been applied to food for understanding the microbial behavior during the fermentation process. Shotgun sequencing has resulted in the detection of viral communities in foods like kimchi, sauerkraut, and fermented shrimp. The study of microbiomes by shotgun library sequencing provides an insight into microbial activity present in food sample. These sequencing techniques are used to study the genomics and phylobiomes of microbial ecology in various fermented beverages and foods (Ercolini et al., 2012; Ercolini, 2013; de Filippis et al., 2017). These techniques have been appreciably used to identify the microbial community in several fermented products such as, *cheonggukjang*, a traditional Korean fermented food (Nam et al., 2012c), *chichi,* maize-based fermented beverage (Elizaquivel et al., 2015), mozzarella cheese (Ercolini et al., 2012), *narezushi*, fermented fish with rice (Kiyohara et al., 2012), *saeu-jeot,* fermented seafood (Jung et al., 2013b) and *kimchi* (Jung et al., 2012; Park et al., 2012). The massive sequencing technique revealed not only the region-specific LAB species and the *Bacillus* species, but also resulted in the identification of unclassified *Bacillus* species involved in the *cheonggukjang* fermentation process (Nam et al., 2012b). Earlier studies on *Doenjang* (Korean fermented soybean paste) by Kim et al. (2010) using both culture-independent and culture-dependent techniques showed the presence of *Bacillus subtilis*, *Bacillus licheniformis* and *Bacillus amyloliquefaciens* as the predominant members. Later, studies using sequencing techniques revealed that the above-mentioned *Bacillus* species contribute to less than 10% in a variety of *Doenjang* samples and found that 51 *Bacillus* species-level phylotypes, including 33 unclassified *Bacillus*, were involved in the entire process of fermentation (Nam et al., 2012a; Jeong et al., 2014). High-throughput sequence (HTS) approach has also been applied to mozzarella cheese for deeper analysis of microbiota which is not given by culture-independent approaches. The sequencing approaches proved useful to identify the source of bacteria in the process of cheese making. The quantitative distribution of OTUs obtained during the experiment suggested that the microflora associated with the fermentation process is carried by natural whey cultures without the development of microorganisms in the raw milk during fermentation (Ercolini et al., 2012). Phylogenetic analysis of kimchi samples based on 16S rRNA sequencing showed the possible changes that were taking place in microbial population throughout the fermentation process. In the early stages of fermentation, samples were found to be dominated by genera *Leconostoc*. However, after 7–10 days of fermentation, there was the relative difference in the abundance of genera Leconostoc and bacterial communities were found to be associated with genera *Leuconostoc, Weisella, and Lactobacillus*. Table 9.2 gives a summary of sequencing information from some selected fermented foods.

Metagenomics and metatranscriptome are the emerging platforms of NGS, which has been used to understand the microbial consortia in several fermented products. These techniques provide sequencing of genomes without PCR amplification unless specifically required and thus avoid the chances of amplification biases (de Filippis et al., 2017). Metagenomics refers to sequencing of environmental DNA without any PCR amplification whereas, metatranscriptomics involve sequencing of total mRNA for exploring the functionality of microbial diversity (Jung et al., 2011, 2013a). DNA-seq metagenomics provides an opportunity to perform in-depth compositional analysis of gene pool of existing microbiota along with its taxonomy. However, only a few are available that explore the microbial community of fermented foods using metagenomics. The use of metagenomics in several fermented foods like kimchi (Ahn et al., 2014; Jung et al., 2011), fermented sausages (Ferrocino et al., 2018), kefir (Marsh et al., 2013), sourdough (Liu et al., 2016) and Cotija cheese (Escobar-Zepeda et al., 2016) have not only helped to decipher the gene content and the functional properties of the microbiota but also elucidated metabolic capabilities of microbial consortia within these fermented products. Metatranscriptomic analysis using complete genome expression of kimchi samples presented a detailed picture of active population, phage infection and gene expression of LAB contributing to overall quality and flavor of kimchi. This technique revealed the presence of carbohydrate metabolism and lactic acid fermentation. Decreased abundance of some species in the later stages of fermentation was found to the effect of bacteriophage infection of *Lactobacillus* species. The genes coding for enzymes involved in various process viz; carbohydrate transport and hydrolysis and metabolite production (mannitol and fructose) were successfully expressed by this technique (Jung et al., 2013a).

TABLE 9.2 Major Findings from Sequencing-based Community Profiling of Indigenous Fermentation

Fermented Food Type	Molecular Analysis Technique	Major Findings	References
Naturally fermented yak milk	High throughput pyrosequencing	Sequence analysis grouped the gene sequence into 11 bacterial and 5 fungal phyla of which Firmicutes and Ascomycota dominate. Dominated bacterial genus was *Lactobacillus* whereas dominated fungal genus was *Saccharomyces*	Liu et al. (2015)
Kimchi (cabbage, radish)	16S rRNA gene amplicon sequencing	Bacterial communities differed between starter-inoculated and non-inoculated kimchi at the early stages of fermentation, but overall, there was no significant differences in the late phases. Kimchi is dominated by bacterial species as *Leuconostoc, Lactobacillus, and Weisella*	Park et al. (2012)
Narezushi (fish)	16S rRNA gene amplicon sequencing	*Lactobacillus* population increased during fermentation while other bacterial population remained unchanged. Time-dependent changes in bacterial community were evaluated. *Lactobacillus plantarum* and *Lactobacillus brevis* dominate the process as identified by culturing.	Kiyohara et al. (2012)
Doenjang (soy bean paste)	16S rRNA gene amplicon sequencing	*Bacillus siamenses* was the populous species during fermentation whereas *Bacillus licheniformis* exhibited salt tolerance during doenjang ripening. Major acid-producing bacteria during fermentation were *Tetragenococcus* and CNS (coagulase-negative staphylococci)	Jeong et al. (2014)
Gochujang (red pepper, rice, soybean mix)	16S rRNA gene amplicon sequencing	*Bacillus subtilis* and *Bacillus licheniformis* were dominant bacterial species. Region-specific bacterial species found in abundance were *Bacillus sonorensis, Bacillus pumilus* and *Weissella salpiscis.*	Nam et al. (2012b)

TABLE 9.2 *(Continued)*

Fermented Food Type	Molecular Analysis Technique	Major Findings	References
Saeu-jeot	16S rRNA gene amplicon sequencing DGGE	Proteobacteria replaced by halophilic Firmicutes as fermentation progressed. Dominant members were *Pseudoalteromonas, Staphylococcus, Salimicrobium, and Alkalibacillus*. *Halanerobium* predominated after 66 days of fermentation	Jung et al. (2013)
Nukadoko (rice bran)	16S rRNA gene amplicon sequencing	Phylogenetic analysis determined *Lactobacillus johnsonii* as dominant strain. Molecular analysis revealed that *Lactobacillus* species including *Lactobacillus namurensis* and *Lactobacillus acetotolerans* predominate. *Lactobacillus plantarum* has also been isolated.	Doi et al. (2013)
Nurul (traditional natural starter used in brewing of Makgeolli)	454 pyrosequencing	Phylogenetic analysis revealed *Aspergillus* and *Mucorales* as dominant fungal genera.	Bal et al. (2016)
Cocoa bean	Metagenomic approach involving 454 pyrosequencing	The metagenomic sequencing analysis identified *Hanseniaspora uvarum, Hanseniaspora opuntia, Saccharomyces cerevisiae, Lactobacillus fermentum*, and *Acetobacter pasteurianus* as the dominant species. Sequence reads associated with the viral communities were dominated by Myoviridae and Siphoviridae reflecting *Lactobacillus* as the dominant host.	Illegghems et al. (2012)
Kombucha	Metagenomic approach involving high-throughput sequencing	The bacterial population was dominated by *Gluconacetobacter* (>85%) while *Acetobacter* represent only a minor population of <2%. Yeast population was found to be dominated by *Zygosaccharomyces* (>95%)	Marsh et al. (2014)

Although, metagenomics and metatranscriptomics provide potentially important information regarding molecular diversity of fermented products, but their cost of operation is sustainably high as compared to amplicon-based sequencing. Also, the need to freeze gene expression at sampling during sequencing of mRNA in metatranscriptome provides technical hindrance. The bioinformatic analysis in metagenomics is complex and intense when compared with amplicon-based sequencing. Thus, this limits the application of these sequencing techniques only to few pioneer studies (de Filippis et al., 2017).

None of the culture-independent method employed has as high efficiency as HTS. The ability of HTS to analyze and sequence multiple samples at the same time provides advantages over other approaches used. This method does not expose the samples to unsafe reagents as done in the case of electrophoresis. It provides insights into taxonomy and activities in case of microbiome studies. However, the need for bioinformatics for analysis of data and translating the sequences into useful information provides a drawback and depending upon the specific conditions, the overall cost of analysis is also high. Taxonomic resolution required in the food also varies due to the complex matrices associated with foods that are deficient in microbial taxa. Also, the short sequence reads in HTS requires 25–30-fold coverage of genome to gather complete genetic information. However, using longer read length along with extension reagents provide an alternative to circumvent this limitation (Ercolini, 2013; Solieri et al., 2013). Despite the few limitations, this is still the best approach for enumeration of microbial diversity.

9.4 METABOLOMICS

Metabolomics, also referred to as metabolic profiling, is an emerging field in the 'omics' sciences which got introduced a little over 10 years ago. It involves analyzing and determining the intracellular metabolites which are low molecular mass compounds. Metabolite profiling is an upcoming field that studies and investigates the dynamic changes taking place in metabolites under various genetic, physiological, and environmental conditions. Metabolomics proves to be an effective way of characterizing and verifying the authenticity of food products as the metabolites identified by this approach contribute to several properties of final products like taste, rheology, and flavor. Metabolomics analyzes has been classified as targeted and untargeted analyzes. Targeted metabolomic identifies and analyzes specific metabolite from many metabolites within-group whereas untargeted metabolomics results in the detection of as many groups of metabolites as possible without identifying any particular compound. The detection and separation of metabolites are very important steps in metabolic profiling. A range of analytical platforms, including capillary electrophoresis mass spectrometry (CE-MS), nuclear magnetic resonance (NMR), liquid chromatography-mass spectrometry (LC-MS) and gas chromatography-mass spectrometry (GC-MS) are used in the metabolomics field for identifying and quantifying the functional properties of samples. Among these, MS-based techniques have been widely used to identify the flavor compounds, organic compounds, and protein digests. Recently, metabolomics has been applied to food science for monitoring safety, quality, processing, and microbiology of both raw materials and final processed product. The application of metabolomics to fermented foods help in the analysis of different food metabolites such as carbohydrates, nucleic acid, peptides, amino acids, and organic acids. These metabolites are responsible for the particular flavor and taste of fermented foods (Kim et al., 2012; Mozzi et al., 2013; Pisano et al., 2016).

Metabolomics analysis is classified as targeted and untargeted analysis. The detection and separation techniques involved in the analysis of metabolite is illustrated in Figure 9.3. Lee et al. (2012) applied mass-spectrometry (MS) based metabolite profiling to characterize the bacterial population of *Meju*. The study demonstrated the changes in metabolites taking place during fermentation and provided the knowledge about the relationship between metabolites and bacterial population. LAB and *Bacillus* predominated *Meju* microflora and microbial analysis of *Meju* fermentation revealed a higher proportion of LAB than *Bacillus* during fermentation. However, the LAB species were present in less proportion than *Bacillus* as lactic acid decreased sharply after 36 days of fermentation. These MS-based metabolite profiling showed that the amino acids (alanine, pyroglutamic acid, leucine, glutamic acid, and tyrosine) responsible for the taste of the product are increased during fermentation. The status of low molecular weight oligosaccharides (sucrose, raffinose, and stachyose) was also studied during metabolite profiling. Metabolomic studies on *cheonggukanj* provided complete information on the metabolic changes that has taken place during the whole process of fermentation (Kim et al., 2012). Microbial profile of Italian mozzarella cheese produced from cow and buffalo milk using metabolomics was studied by Pisano et al. (2016). GC-MS was employed to analyze the metabolite changes occurring during fermentation in *doenjang*. The study concluded that metabolites like leucine, aminoadipic acid, isoleucine, malic acid, lysine, glucosamine, and oxalic acid were the major contributors to distinguish the samples during the entire fermentation process (Namgung et al., 2010). H-NMR spectroscopy detection technique has been used to study the metabolite profile of fermented soymilk. Study Yang et al. (2009) revealed that the inoculated microbial strains during fermentation process resulted in decrease sugars content along with simultaneous increase in lactic acid and succinic acid concentration.

Buffalo mozzarella cheese produced using natural whey cultures as starters were found to be rich in polar metabolites such as valine, serine, and threonine and contain urea and orotic acid in lower concentration. In contrast, cow mozzarella cheese produced using commercial starter cultures particularly *Streptococcus thermophilous*, exhibited high concentration of galactose and phenylalanine. While, cow mozzarella cheese obtained by direct acidification was found to contain high levels of urea and sugar. Cow mozzarella cheese produced using commercial starters was rich in microbial count as compared to that obtained by direct acidification. The presence of different concentration of microbes in both the samples and the identification of discriminant metabolites indicated the association of microbial complexity with the metabolite pool. It is also useful in assessing the quality of soy sauce by providing the comprehensive understanding of the metabolites during fermentation (Zhu and Tramper, 2013). The metabolites in *kimchi*, such as organic acids (lactic and acetic acid), free sugars (glucose and fructose), and other flavoring compounds (mannitol and amino acids) contribute to its taste and flavor. LAB arising during *kimchi* fermentation accords to the metabolites of *kimchi*. The clear relationship between these metabolites and the microbiota produced within fermenting ecosystem, however, was not known. Park et al. (2016) applied the metabolomics approach to *kimchi* for understanding the effect of LAB on kimchi fermentation. Similarly, metabolomics approach was applied to *fuzhuan tea* (fermented tea) and the results were compared to non-fermented green tea for exploring the chemical differences between the two (Keller et al., 2013). This 'omics' approach has also been used to assess the lactobacilli and yeast ability to improve the fermentation process of *Bella di Cerignola* table olives by de Angelis et al. (2015).

Thus, metabolite profiling proves a powerful approach in investigating the chemical changes taking place in fermented foods and is considered as an appropriate method for observing metabolic changes during fermentation. It also ensures the quality and safety of fermented products. Improved strategies by the combination of various analytical methods for identification of wide spectrum of important metabolites can be obtained and the real-time information of fermentation can also be well understood (Keller et al., 2013; Kim et al., 2012; Lee et al., 2012; Mozzi et al., 2013).

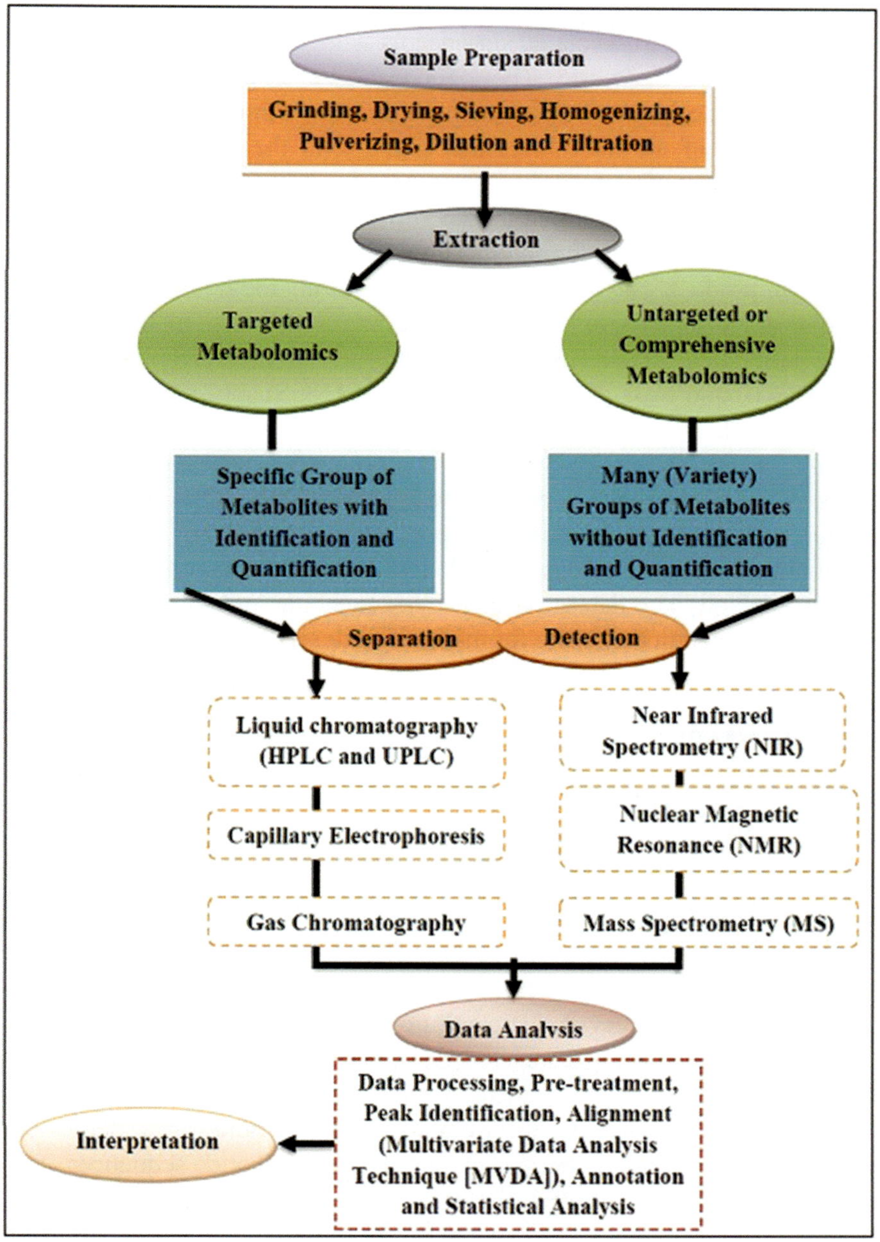

FIGURE 9.3 Separation and detection techniques involved in metabolomics.

9.5 CONCLUSION

Fermented foods are associated very deeply with social and cultural practices of people across the world. These food and beverages have many health benefits. A wide range of substrates are used for these food fermentations. The activities of the microbiota associated with the fermentation process contribute to taste, flavor, aroma, texture, and nutritive value of the product. Therefore, an understanding of the microbial community which brings about desirable changes in the fermenting food is important to ensure quality and consistent product. The traditional routine culturing methods used earlier fail to isolate the stressed or injured cells or the organism present in low abundance in the fermented foods. The use of culture-independent techniques such as PCR-DGGE and Next-generation sequencing (NGS) shows a lot of promise in this regard. The various sequencing platforms like Illumina, Ion-Torrent, SOLiD, PacBio are currently providing a promising result in understanding the behavior of microbial population within food. The application of metabolomics in elucidating the microbial population by studying the changes in metabolites provides further knowledge about the microbial ecology of fermented food. The data generated by sequencing and 'omics' techniques can be combined in the future for in-depth analysis of food samples.

9.6 SUMMARY

Fermented food products have been an important part of human diet. They have several desired characteristics which are owned to microflora produced during the process of fermentation. The study of these microbial species has been carried out extensively using various molecular-based culture-independent and culture-dependent approaches. The conventional culture-dependent technique results in the underestimation of several microbial species and was thus replaced by new holistic culture-independent approaches. Amongst these, PCR-DGGE has been used for a much longer period for enumerating the microbiota present in fermented products. With advancement, several NGS techniques are now used for elucidating microflora. These sequencing techniques have resulted in analysis of multiple nucleotide sequences at one time. Various NGS platforms like 454/ Roche pyrosequencing, ion torrent, Illumina, metagenomics, and metatranscriptomics have proved useful in studying the microbial consortia in fermented products. Recently, metabolomics has also been extensively applied to fermented products for exploring its microbial diversity and various changes taking place during the fermentation process. By use of various analytical platforms like GC-MS, NMR, CE-MS, and LC-MS metabolomics have been able to identify and quantify functional properties of fermented products. Thus, all these techniques have provided promising results in understanding the dynamics of microbiota produced during the fermentation process.

KEYWORDS

- **culture-dependent methods**
- **culture-independent techniques**
- **fermentation**
- **high-throughput sequencing**
- **Illumina**
- **ion torrent**
- **lactic acid bacteria**
- **mass spectrometry**
- **metabolites**
- **metabolomics**
- **metagenomics**
- **metatranscriptome**
- **microbial enumeration**
- **next-generation sequencing**
- **operational taxonomic unit**
- **PacBio**
- **PCR-DGGE**
- **pyrosequencing**
- **sequencing techniques**
- **SOLiD**

REFERENCES

Abriouel, H., Benomar, N., Cobo, A., Caballero, N., Fuentes, M. Á. F., Pérez-Pulido, R., & Gálvez, A., (2012). Characterization of lactic acid bacteria from naturally-fermented manzanilla aloreña green table olives. *Food Microbiology, 32*(2), 308–316.

Adewumi, G. A., Oguntoyinbo, F. A., Keisam, S., & Jeyaram, K., (2013). Combination of culture-independent and culture-dependent molecular methods for the determination of bacterial community of iru, a fermented *Parkia biglobosa* seeds. *Frontiers in Microbiology, 3*, 436.

Ahn, J. Y., Min, J., Lee, S. H., Jang, A., Park, C. K., Kwon, S. D., Park, S. K., Lee, K., & Kim, Y. H., (2014). Metagenomic analysis for identifying kimchi sp. during the industrial-scale batch fermentation. *Toxicology and Environmental Health Sciences, 6*(1), 8–15.

Alegría, Á., Szczesny, P., Mayo, B., Bardowski, J., & Kowalczyk, M., (2012). Biodiversity in traditional polish cheese Oscypek determined by culture-dependent and independent approaches. *Applied and Environmental Microbiology.* AEM-06081.

Ariefdjohan, M. W., Savaiano, D. A., & Nakatsu, C. H., (2010). Comparison of DNA extraction kits for PCR-DGGE analysis of human intestinal microbial communities from fecal specimens. *Nutrition Journal, 9*(1), 23.

Atlas, R. M., (2010). *Handbook of Microbiological Media* (4th edn.). CRC Press.

Bal, J., Yun, S. H., Yeo, S. H., Kim, J. M., & Kim, D. H., (2016). Metagenomic analysis of fungal diversity in Korean traditional wheat-based fermentation starter nuruk. *Food Microbiology, 60*, 73–83.

Bokulich, N. A., Joseph, C. L., Allen, G., Benson, A. K., & Mills, D. A., (2010). Next-generation sequencing reveals significant bacterial diversity of botrytized wine. *PloS One, 7*(5), e36357.

Chettri, R., & Tamang, J. P., (2015). *Bacillus* species isolated from Tungrymbai and Bekang, naturally fermented soybean foods of India. *International Journal of Food Microbiology, 197*, 72–76.

Claesson, M. J., Wang, Q., O'sullivan, O., Greene-Diniz, R., Cole, J. R., Ross, R. P., & O'toole, P. W., (2010). Comparison of two next-generation sequencing technologies for resolving highly complex microbiota composition using tandem variable 16S rRNA gene regions. *Nucleic Acids Research, 38*(22), e200–e200.

Cocolin, L., Alessandria, V., Dolci, P., Gorra, R., & Rantsiou, K., (2013). Culture-independent methods to assess the diversity and dynamics of microbiota during food fermentation. *International Journal of Food Microbiology, 167*(1), 29–43.

De Angelis, M., Campanella, D., Cosmai, L., Summo, C., Rizzello, C. G., & Caponio, F., (2015). Microbiota and metabolome of un-started and started Greek-type fermentation *of Bella di Cerignola* table olives. *Food Microbiology, 52*, 18–30.

De Filippis, F., Parente, E., & Ercolini, D., (2017). Metagenomics insights into food fermentations. *Microbial Biotechnology, 10*(1), 91–102.

Devi, K. R., Deka, M., & Jeyaram, K., (2015). Bacterial dynamics during yearlong spontaneous fermentation for production of ngari, a dry fermented fish product of Northeast India. *International Journal of Food Microbiology, 199*, 62–71.

Doi, K., Phuong, O. T. A., Kawatou, F., Nagayoshi, Y., Fujino, Y., & Ohshima, T., (2013). Identification and characterization of lactic acid bacteria isolated from fermented rice bran product. *Advances in Microbiology, 3*(03), 265.

Duarte, S., Cássio, F., & Pascoal, C. (2012). Denaturing gradient gel electrophoresis (DGGE) in microbial ecology-insights from freshwaters. In: Duarte, S., Cássio, F., & Pascoal, C., (eds.), *Gel Electrophoresis-Principles and Basics*. InTech.

Elizaquível, P., Pérez-Cataluña, A., Yépez, A., Aristimuño, C., Jiménez, E., Cocconcelli, P. S., Vignolo, G., & Aznar, R., (2015). Pyrosequencing vs. culture-dependent approaches to analyze lactic acid bacteria associated to Chicha, a traditional maize-based fermented beverage from Northwestern Argentina. *International Journal of Food Microbiology, 198*, 9–18.

Ercolini, D., (2013). High-throughput sequencing and metagenomics: Moving forward in the culture-independent analysis of food microbial ecology. *Applied Environmental Microbiology, 79*, 3148–3155.

Ercolini, D., De Filippis, F., La Storia, A., & Iacono, M., (2012). "Remake" by high-throughput sequencing of the microbiota involved in the production of water buffalo Mozzarella cheese. *Applied and Environmental Microbiology, 78*(22), 8142–8145.

Escobar-Zepeda, A., Sanchez-Flores, A., & Baruch, M. Q., (2016). Metagenomic analysis of a Mexican ripened cheese reveals a unique complex microbiota. *Food Microbiology, 57*, 116–127.

Feligini, M., Panelli, S., Buffoni, J. N., Bonacina, C., Andrighetto, C., & Lombardi, A., (2012). Identification of microbiota present on the surface of Taleggio cheese using PCR–DGGE and RAPD–PCR. *Journal of Food Science, 77*(11), M609–M615.

Ferrocino, I., Bellio, A., Giordano, M., Macori, G., Romano, A., Rantsiou, K., Decastelli, L., & Cocolin, L., (2018). Shotgun metagenomics and volatile profile of the microbiota of fermented sausages. *Applied and Environmental Microbiology, 84*(3), e02120–17.

González-Arenzana, L., López, R., Santamaría, P., & López-Alfaro, I., (2013). Dynamics of lactic acid bacteria populations in Rioja wines by PCR-DGGE, comparison with culture-dependent methods. *Applied Microbiology and Biotechnology, 97*(15), 6931–6941.

Greppi, A., Ferrocino, I., La Storia, A., Rantsiou, K., Ercolini, D., & Cocolin, L., (2015). Monitoring of the microbiota of fermented sausages by culture-independent rRNA-based approaches. *International Journal of Food Microbiology, 212*, 67–75.

Greppi, A., Rantisou, K., Padonou, W., Hounhouigan, J., Jespersen, L., Jakobsen, M., & Cocolin, L., (2013). Yeast dynamics during spontaneous fermentation of mawè and tchoukoutou, two traditional products from Benin. *International Journal of Food Microbiology, 165*(2), 200–207.

Hashemi, S. M. B., Mousavi, K. A., Kontominas, M. G., Eş, I., Sant'Ana, A. S., Martinez, R. R., & Drider, D., (2017). Fermentation of sarshir (kaymak) by lactic acid bacteria: Antibacterial activity, antioxidant properties, lipid and protein oxidation and fatty acid profile. *Journal of the Science of Food and Agriculture, 97*(13), 4595–4603.

He, G. Q., Liu, T. J., Sadiq, F. A., Gu, J. S., & Zhang, G. H., (2017). Insights into the microbial diversity and community dynamics of Chinese traditional fermented foods from using high-throughput sequencing approaches. *Journal of Zhejiang University-Science B, 18*(4), 289–302.

Illeghems, K., De Vuyst, L., Papalexandratou, Z., & Weckx, S., (2012). Phylogenetic analysis of a spontaneous cocoa bean fermentation metagenome reveals new insights into its bacterial and fungal community diversity. *Plus One, 7*(5), e38040.

Jeong, D. W., Kim, H. R., Jung, G., Han, S., Kim, C. T., & Lee, J. H., (2014). Bacterial community migration in the ripening of doenjang, a traditional Korean fermented soybean food. *Journal of Microbiology and Biotechnology, 24*(5), 648–660.

Jung, J. Y., Lee, S. H., Jin, H. M., Hahn, Y., Madsen, E. L., & Jeon, C. O., (2013a). Metatranscriptomic analysis of lactic acid bacterial gene expression during kimchi fermentation. *International Journal of Food Microbiology, 163*(2, 3), 171–179.

Jung, J. Y., Lee, S. H., Kim, J. M., Park, M. S., Bae, J. W., Hahn, Y., Madsen, E. L., & Jeon, C. O., (2011). Metagenomic analysis of kimchi, the Korean traditional fermented food. *Applied and Environmental Microbiology*, (2011).

Jung, J. Y., Lee, S. H., Lee, H. J., & Jeon, C. O., (2013b). Microbial succession and metabolite changes during fermentation of saeu-jeot: Traditional Korean salted seafood. *Food Microbiology, 34*, 360–368.

Jung, J. Y., Lee, S. H., Lee, H. J., Seo, H. Y., Park, W. S., & Jeon, C. O., (2012). Effects of *Leuconostoc mesenteroides* starter cultures on microbial communities and metabolites during kimchi fermentation. *International Journal of Food Microbiology, 153*(3), 378–387.

Keller, A. C., Weir, T. L., Broeckling, C. D., & Ryan, E. P., (2013). Antibacterial activity and phytochemical profile of fermented *Camellia sinensis* (fuzhuan tea). *Food Research International, 53*(2), 945–949.

Kim, J., Choi, J. N., John, K. M., Kusano, M., Oikawa, A., Saito, K., & Lee, C. H., (2012). GC–TOF-MS-and CE–TOF-MS-based metabolic profiling of cheonggukjang (fast-fermented bean paste) during fermentation and its correlation with metabolic pathways. *Journal of Agricultural and Food Chemistry, 60*(38), 9746–9753.

Kim, T. W., Kim, Y. H., Kim, S. E., Lee, J. H., Park, C. S., & Kim, H. Y., (2010). Identification and distribution of *Bacillus* species in doenjang by whole-cell protein patterns and 16S rRNA gene sequence analysis. *Journal of Microbiology and Biotechnology, 20*(8), 1210–1214.

Kiyohara, M., Koyanagi, T., Matsui, H., Yamamoto, K., Take, H., Katsuyama, Y., Tsuji, A., et al., (2012). Changes in microbiota population during fermentation of narezushi as revealed by pyrosequencing analysis. *Bioscience, Biotechnology, and Biochemistry, 76*(1), 48–52.

Kormin, S., Rusul, G., Radu, S., & Ling, F. H., (2001). Bacteriocin-producing lactic Acid bacteria isolated from traditional fermented food. *Malaysian Journal of Medical Science, 8*, 63.

Kostinek, M., Specht, I., Edward, V. A., Schillinger, U., Hertel, C., Holzapfel, W. H., & Franz, C. M., (2005). Diversity and technological properties of predominant lactic acid bacteria from fermented cassava used for the preparation of gari, a traditional African food. *Systematic and Applied Microbiology, 28*(6), 527–540.

Lee, H. M., & Lee, Y., (2008) A differential medium for lactic acid-producing bacteria in a mixed culture. *Letters in Applied Microbiology, 46*(6), 676–681.

Lee, S. Y., Kim, H. Y., Lee, S., Lee, J. M., Muthaiya, M. J., Kim, B. S., & Lee, C. H., (2012). Mass spectrometry-based metabolite profiling and bacterial diversity characterization of Korean traditional meju during fermentation. *Journal of Microbiology and Biotechnology, 22*(11), 1523–1531.

Liu, T., Li, Y., Chen, J., Sadiq, F. A., Zhang, G., Li, Y., & He, G., (*2016*). Prevalence and diversity of lactic acid bacteria in Chinese traditional sourdough revealed by culture-dependent and pyrosequencing approaches. *LWT-Food Science and Technology*, 91–97.

Liu, W., Xu, X., Sudu, Q., Kwok, L., Guo, Z., Hou, Q., & Zhang, H., (2015). High-throughput sequencing reveals microbial community diversity of Tibetan naturally fermented yak milk. *Annals of Microbiology, 65*(3), 1741–1751.

Lusk, T. S., Ottesen, A. R., White, J. R., Allard, M. W., Brown, E. W., & Kase, J. A., (2012). Characterization of microflora in Latin-style cheeses by next-generation sequencing technology. *BMC Microbiology, 12*(1), 254.

Lv, X. C., Huang, R. L., Chen, F., Zhang, W., Rao, P. F., & Ni, L., (2013). Bacterial community dynamics during the traditional brewing of Wuyi Hong Qu glutinous rice wine as determined by culture-independent methods. *Food Control, 34*(2), 300–306.

Lv, X. C., Weng, X., Zhang, W., Rao, P. F., & Ni, L., (2012). Microbial diversity of traditional fermentation starters for Hong Qu glutinous rice wine as determined by PCR-mediated DGGE. *Food Control, 28*(2), 426–434.

Marsh, A. J., O'Sullivan, O., Hill, C., Ross, R. P., & Cotter, P. D., (2013). Sequencing-based analysis of the bacterial and fungal composition of kefir grains and milks from multiple sources. *PloS One, 8*(7), e69371.

Marsh, A. J., O'Sullivan, O., Hill, C., Ross, R. P., & Cotter, P. D., (2014). Sequence-based analysis of the bacterial and fungal compositions of multiple kombucha (tea fungus) samples. *Food Microbiology, 38*, 171–178.

Marty, E., Buchs, J., Eugster-Meier, E., Lacroix, C., & Meile, L., (2012). Identification of staphylococci and dominant lactic acid bacteria in spontaneously fermented Swiss meat products using PCR–RFLP. *Food Microbiology, 29*(2), 157–166.

Mayo, B., Rachid, C. T. C. C., Alegría, Á., Leite, A. M. O., Peixoto, R. S., & Delgado, S., (2014). Impact of next generation sequencing techniques in food microbiology. *Current Genomics, 15*(4), 293–309.

Metzker, M. L., (2010). Sequencing technologies—the next generation. *Nature Reviews Genetics, 11*(1), 31.

Mozzi, F., Ortiz, M. E., Bleckwedel, J., De Vuyst, L., & Pescuma, M., (2013). Metabolomics as a tool for the comprehensive understanding of fermented and functional foods with lactic acid bacteria. *Food Research International, 54*(1), 1152–1161.

Mukisa, I. M., Porcellato, D., Byaruhanga, Y. B., Muyanja, C. M., Rudi, K., Langsrud, T., & Narvhus, J. A., (2012). The dominant microbial community associated with fermentation of Obushera (sorghum and millet beverages) determined by culture-dependent and culture-independent methods. *International Journal of Food Microbiology, 160*(1), 1–10.

Nam, Y. D., Lee, S. Y., & Lim, S. I., (2012a). Microbial community analysis of Korean soybean pastes by next-generation sequencing. *International Journal of Food Microbiology, 155*(1, 2), 36–42.

Nam, Y. D., Park, S. L., & Lim, S. I., (2012b). Microbial composition of the Korean traditional food "kochujang" analyzed by a massive sequencing technique. *Journal of Food Science, 77*(4), M250–M256.

Nam, Y. D., Yi, S. H., & Lim, S. I., (2012c). Bacterial diversity of cheonggukjang, a traditional Korean fermented food, analyzed by barcoded pyrosequencing. *Food Control, 28*(1), 135–142.

Namgung, H. J., Park, H. J., Cho, I. H., Choi, H. K., Kwon, D. Y., Shim, S. M., & Kim, Y. S., (2010). Metabolite profiling of doenjang, fermented soybean paste, during fermentation. *Journal of the Science of Food and Agriculture, 90*(11), 1926–1935.

Nguyen, D. T. L., Van, H. K., Cnockaert, M., De Brandt, E., Aerts, M., & Vandamme, P., (2013). A description of the lactic acid bacteria microbiota associated with the production of traditional fermented vegetables in Vietnam. *International Journal of Food Microbiology, 163*(1), 19–27.

Nguyen, D. T. L., Van, H. K., Cnockaert, M., De Brandt, E., De Bruyne, K., Le, B. T., & Vandamme, P., (2013). A culture-dependent and independent approach for the identification of lactic acid bacteria associated with the production of nem Chua, a Vietnamese fermented meat product. *Food Research International, 50*(1), 232–240.

Park, E. J., Chun, J., Cha, C. J., Park, W. S., Jeon, C. O., & Bae, J. W., (2012). Bacterial community analysis during fermentation of ten representative kinds of kimchi with barcoded pyrosequencing. *Food Microbiology, 30*(1), 197–204.

Park, S. E., Yoo, S. A., Seo, S. H., Lee, K. I., Na, C. S., & Song, H. S., (2016). GC–MS-based metabolomics approach of Kimchi for the understanding of *Lactobacillus plantarum* fermentation characteristics. *LWT-Food Science and Technology, 68*, 313–321.

Pisano, M. B., Scano, P., Murgia, A., Cosentino, S., & Caboni, P., (2016). Metabolomics and microbiological profile of Italian mozzarella cheese produced with buffalo and cow milk. *Food Chemistry, 192*, 618–624.

Połka, J., Rebecchi, A., Pisacane, V., Morelli, L., & Puglisi, E., (2015). Bacterial diversity in typical Italian salami at different ripening stages as revealed by high-throughput sequencing of 16S rRNA amplicons. *Food Microbiology, 46*, 342–356.

Rhoads, A., & Au, K. F., (2015). PacBio sequencing and its applications. *Genomics, Proteomics & Bioinformatics, 13*(5), 278–289.

Salipante, S. J., Kawashima, T., Rosenthal, C., Hoogestraat, D. R., Cummings, L. A., Sengupta, D. J., & Hoffman, N. G., (2014). Performance comparison of Illumina and ion torrent next-generation sequencing platforms for 16S rRNA-based bacterial community profiling. *Applied and Environmental Microbiology*. AEM-02206.

Sengun, I. Y., Nielsen, D. S., Karapinar, M., & Jakobsen, M., (2009). Identification of lactic acid bacteria isolated from Tarhana, a traditional Turkish fermented food. *International Journal of Food Microbiology, 135*(2), 105–111.

Solieri, L., Dakal, T. C., & Giudici, P., (2013). Next-generation sequencing and its potential impact on food microbial genomics. *Annals of Microbiology, 63*(1), 21–37.

Stanbury, P. F., Whitaker, A., & Hall, S. J., (2013). *Principles of Fermentation Technology*. Elsevier.

Tamang, J. P., Thapa, N., Tamang, B., Rai, A., & Chettri, R., (2015). Microorganisms in fermented foods and beverages. Chapter 1. In: Tamang, J. P., Thapa, N., Tamang, B., Rai, A., & Chettri, R., (eds.), *Health Benefits of Fermented Foods and Beverages* (pp. 1–110). CRC Press, Taylor & Francis Group, New York (USA).

Tanaka, Y., Watanabe, J., & Mogi, Y., (2012). Monitoring of the microbial communities involved in the soy sauce manufacturing process by PCR-denaturing gradient gel electrophoresis. *Food Microbiology, 31*(1), 100–106.

Tofalo, R., Perpetuini, G., Schirone, M., Suzzi, G., & Corsetti, A., (2013). Yeast biota associated with naturally fermented table olives from different Italian cultivars. *International Journal of Food Microbiology, 161*(3), 203–208.

Van, H. S. A., Vaughan, E. E., & Vogel, R. F., (2013). Application of state-of-art sequencing technologies to indigenous food fermentations. *Current Opinion in Biotechnology, 24*(2), 178–186.

Wei, C. L., Chao, S. H., Tsai, W. B., Lee, P. S., Tsau, N. H., Chen, J. S., & Tsai, Y. C., (2013). Analysis of bacterial diversity during the fermentation of inyu, a high-temperature fermented soy sauce, using nested PCR-denaturing gradient gel electrophoresis and the plate count method. *Food Microbiology, 33*(2), 252–261.

Wu, J. J., Ma, Y. K., Zhang, F. F., & Chen, F. S., (2012). Biodiversity of yeasts, lactic acid bacteria and acetic acid bacteria in the fermentation of "Shanxi aged vinegar", a traditional Chinese vinegar. *Food Microbiology, 30*(1), 289–297.

Xiu, L., Kunliang, G., & Hongxun, Z., (2012). Determination of microbial diversity in Daqu, a fermentation starter culture of Maotai liquor, using nested PCR-denaturing gradient gel electrophoresis. *World Journal of Microbiology and Biotechnology, 28*(6), 2375–2381.

Yam, B. Z., Khomeiri, M., Mahounak, A. S., & Jafari, S. M., (2015). Isolation and identification of yeasts and lactic acid bacteria from local traditional fermented camel milk, Chal. *Journal of Food Processing & Technology, 6*(7), 1.

Yan, Y. Z., Qian, Y. L., Chen, J. Y., & Han, B. Z., (2013). Microbial composition during Chinese soy sauce koji-making based on culture-dependent and independent methods. *Food Microbiology, 34*(1), 189–195.

Yan, Z., Zheng, X. W., Han, B. Z., Han, J. S., Nout, M. J., & Chen, J. Y., (2013). Monitoring the ecology of Bacillus during Daqu incubation, a fermentation starter, using culture-dependent and culture-independent methods. *Journal of Microbiology and Biotechnology, 23*, 614–622.

Yang, S. O., Kim, S. H., Cho, S., Lee, J., Kim, Y. S., Yun, S. S., & Choi, H. K., (2009). Classification of fermented soymilk during fermentation by 1H NMR coupled with principal component analysis and elucidation of free-radical scavenging activities. *Bioscience, Biotechnology, and Biochemistry, 73*(5), 1184–1188.

Zhu, Y., & Tramper, J., (2013). Koji-where East meets West in fermentation. *Biotechnology Advances, 31*(8), 1448–1457.

CHAPTER 10

Industrial Prospects of Bacterial Microcompartment Technologies

SHAGUN RASTOGI and CHIRANJIT CHOWDHURY*

**Corresponding author. E-mails: cchowdhury@amity.edu; cchowdhury82@gmail.com*

ABSTRACT

There are two major challenges that biocatalyst-based industries often face. First, the reaction pathways may involve volatile, unstable, or toxic intermediates. Secondly, the reacting enzymes sometimes encounter extreme environments. Both the cases may hinder successful enzymatic catalysis, which decreases the overall reaction rate and thereby reduce the turnover. Encapsulation could be a useful strategy in order to protect enzymes from harsh environmental impact and limit the diffusion of volatile and/or toxic intermediate(s). In nature, cells often alleviate these issues by encapsulating reacting partners in closed container. The process of encapsulation enhances overall kinetics by sequestering toxic or volatile metabolic intermediates and minimizes cellular injury and/or loss of essential carbon. A distinct example of this approach is found in organelles of eukaryotes; whereas in bacteria, this strategy is mediated through the deployment of microcompartments (MCPs). Bacterial MCPs are proteinaceous complexes composed of selectively-porous shells that encase a specific metabolic process. The use of MCPs have great prospect compared to liposome and other nano-compartments of similar class in biotechnology and synthetic biology applications because of their unmatched versatility in terms of modulating substrate access, the ability to encapsulate diverse biochemical cargos and their ease of genetic modification. These intriguing features of MCPs have attracted considerable interest in biocatalyst-based industries. A recent study demonstrates that MCPs shells are robust and protective against proteolysis, oxidative stress, and temperature. Therefore, MCPs have great promise to serve as molecular chambers in terms of cargo delivery in the food and pharmaceutical industries. In this chapter, we focus on engineering approaches undertaken to aim to repurpose MCPs for several biotechnological perspectives.

10.1 INTRODUCTION

Enzyme-based technologies are being exploited in a broad range of applications, such as in bio-energy, paper, textile, agriculture, and household use. During industrial processes, the enzymes are often exposed to high temperatures, extremes of pH, or oxidizing agents, which drastically effect on enzyme stability and thereby reducing product yield (Littlechild, 2015). Moreover, slower reaction rates, production of toxic or volatile intermediate are too great obstacle for efficient product formation (Conrado et al., 2008; Dueber et al., 2009; Lee et al., 2012; Giessen and Silver, 2016). A strategy of confinement of reacting partners in a closed compartment by sequestering them from cytoplasmic milieu perhaps increases substrate concentration in close proximity to metabolic enzymes without significant buildup of pernicious intermediate, which may otherwise be detrimental to the cell (Dueber et al., 2009; Hurtley, 2009; Lopez-Gallego and Schmidt-Dannert, 2010; Held et al., 2013; Frank et al., 2013; Hagen et al., 2018a). In nature, cells often undertake this strategy in order to optimize metabolic reactions and prevent unwanted side reactions. A hallmark example of this strategy is found in peroxisome, a eukaryotic membrane-bound organelles (Chowdhury et al., 2014). Bacteria, on the other hand, confine problematic metabolic pathways within a protein shell, forming a structure known as bacterial microcompartment (MCP) (Figure 10.1). The protein shell effectively confines toxic or volatile pathway intermediates (Cheng et al., 2008; Kerfeld et al., 2010; Chowdhury et al., 2014; Bobik et al., 2015). These proteinaceous organelles serve as tiny reaction chambers, sequester toxic metabolite, and stabilize unstable proteins or biosynthetic intermediates (Kerfeld et al., 2010; Rae et al., 2013; Chowdhury et al., 2014).

Unlike the most of other organic/inorganic nanostructures, these proteinaceous MCPs attribute many exceptional properties like self-assembly and robustness (Kim et al., 2014). Moreover, being biological origin MCPs seem to be biocompatible to serve as drug delivery vehicles or to use as scaffold for the development of vaccine (Chowdhury et al., 2014). Further, the proteinaceous compartments are amenable to precise control in terms of expression, metabolite transport and encapsulation of desired cargo through protein engineering (Fan et al., 2010; Chowdhury et al., 2015; Li et al., 2018). Engineered MCPs could be exploited for the production of biorenewables chemicals that could otherwise be toxic to host bacterial cell. Hence, the current endeavor is focused on developing synthetic MCPs to serve as intracellular nanobioreactors for the production of useful chemicals or toxic proteins or to generate as molecular scaffolds in several synthetic biology and biotechnology applications (Yung et al., 2017; Hagen et al., 2018a). Nevertheless, production, and assembly MCPs in heterologous systems suffer numerous experimental challenges due to the size and complexity of the polyhedral body. In this chapter, progresses and prospects of MCP-based technology are discussed.

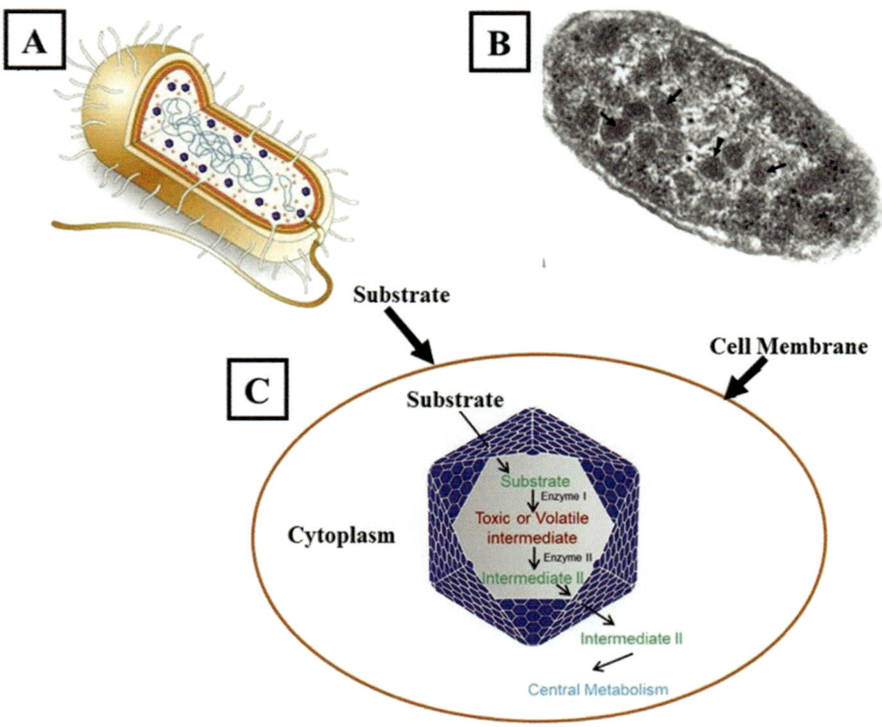

FIGURE 10.1 Schematic representation of microcompartment localization and function: (A) Schematic representation of distribution of microcompartment organelles inside bacterial cell; (B) transmission electron micrograph of thin section of *Salmonella* expressing Pdu microcompartments (Chowdhury et al., 2014); and (c) schematic representation of function of bacterial microcompartments.

10.2 BACTERIAL MICROCOMPARTMENTS (MCPS): A POLYHEDRAL ORGANELLES COMPARTMENTING BACTERIAL METABOLIC PROCESS

Bacterial MCPs are complex protein containers that are widely distributed across bacterial phyla. These protein containers encase serially acting metabolic enzymes and optimize metabolic pathways (Figure 10.1) (Bobik, 2006; Kerfeld et al., 2010; Rae et al., 2013; Chowdhury et al., 2014). The outer shells of the MCP polyhedra are comprised of hexagonal BMC domain proteins that form facets and BMV pentamers that create the vertices (Kerfeld et al., 2010; Rae et al., 2013; Chowdhury et al., 2014; Lehman et al., 2017). MCPs show morphological resemblance to viral capsids; although, their size and shape vary among bacterial species. Typical MCPs are 80–200 nm across the diameter and are visible under electron microscope (Bobik et al., 1999; Chowdhury et al., 2014) (Figure 10.2). Recently, new methods of visualization of polyhedral shell through the crystal structure, atomic force microscopy and fluorescent labeling have provided new insights into subcellular localization, assembly, and structure of MCPs (Savage et al., 2010; Sutter et al., 2016, 2017; Kerfeld et al., 2018). MCPs polyhedral bodies are nearly icosahedral in shape with 20 flat triangular facets and 12 pentagonal vertices (Schmid et al., 2006; Iancu et al., 2007; Yeates et al., 2013). Although, carboxysomal MCPs appears to be regular icosahedron, the other type of MCPs is less regular, often conform structural polymorphisms (Figure 10.2).

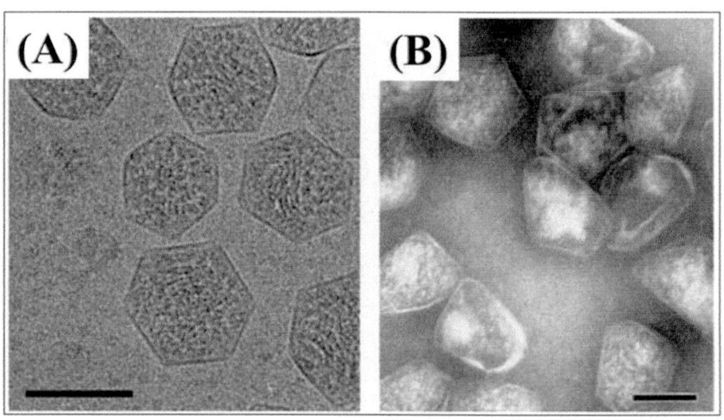

FIGURE 10.2 Purified bacterial microcompartments. (A) Purified carboxysomes from *Halothiobacillus neapolitanus*; and (B) purified propanediol utilization (Pdu) microcompartments from *Salmonella enterica* as observed under transmission electron microscope. The scale bars are 100 nm.

Source: Reprint from Yeates et al. 2011. Open access. https://creativecommons.org/licenses/by/4.0/

The main function of MCP is to seclude certain parts of metabolic reactions cytoplasmic environment by compartmentalizing within a protein shell. MCP thus helps mitigate the toxicity of the intermediate and/or limit the loss of carbon by trapping the intermediate inside its lumen (Havemann et al., 2002; Penrod and Roth, 2006; Chowdhury et al., 2014). In this model (shown in Figure 10.1), a metabolic substrate after crossing through bacterial cell wall enters the lumen of MCP by diffusing across its selectively permeable protein shell (Chowdhury et al., 2015) (Figure 10.1). Inside MCP lumen, the substrate gets converted to a volatile or toxic intermediate. The outer protein shells of MCP serve as diffusion barrier to the intermediate formed and help passage this intermediary metabolic product to the downstream enzyme(s), which further gets converted to nontoxic and/or nonvolatile compound for subsequent downstream reactions (Figure 10.1) (Chowdhury et al., 2014, 2015). Thus, MCPs enhance catalytic efficiencies by creating high local concentrations of enzymes and substrates, confine pathway intermediates that are toxic or rapidly excreted out of the bacterial cell, and enable the use of private cofactor pools (Kerfeld et al., 2005; Rae et al., 2013; Chowdhury et al., 2014; Jakobson et al., 2017; Lehman et al., 2017). A recent study demonstrates that MCP shells are robust in nature and stable at high temperature and pH (Kim et al., 2014). Furthermore, because MCPs are completely made up of protein subunits, these are amenable to genetic modification. The studies of MCPs are gaining attention in a way that will opened up exciting prospects for designing MCPs in synthetic biology research and applications.

10.3 DIVERSE MICROCOMPARTMENT (MCP) ORGANELLES

The three types of MCPs are well studied and those are: carboxysomes, ethanolamine utilization (Eut) and 1,2-propanediol utilization (Pdu) MCPs. Carboxysomes are used for autotrophic carbon fixation. Whereas, the Eut and Pdu MCPs are involved in the coenzyme B_{12}-dependent catabolism of ethanolamine and 1,2-propanediol (1,2-PD), respectively (Kerfeld et al., 2010; Chowdhury et al., 2014). Carboxysomes are mostly found in cyanobacteria that use the Calvin cycle for CO_2 fixation (Bobik, 2006; Rae et al., 2013). The carboxysome enhances the productivity of RuBisCO by raising the local concentration of CO_2 at the vicinity of the enzyme and restrict the outward diffusion of CO_2 (Price et al., 2008) (Figure 10.3). On the other hand, ethanolamine and 1,2-PD metabolism occur in a number of enteric pathogens, including *Klebsiella*, *Salmonella*, *Clostridium*, and some strains of *E. coli* (Figure 10.3).

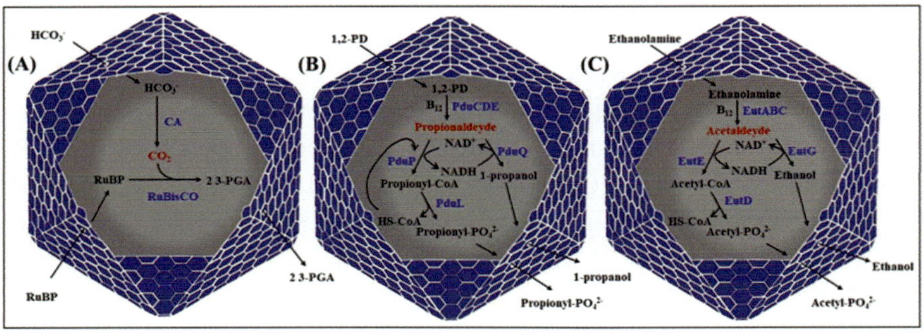

FIGURE 10.3 Representative diagram of (A) CO_2 fixation in the carboxysome; (B) propanediol metabolism in the Pdu microcompartment; and (C) ethanolamine metabolism in the Eut microcompartment.

Several enteric bacteria degrade rhamnose and fucose, the common plant sugars under an anaerobic environment, and produce 1,2-PD; which is then taken up as carbon and energy source by these bacteria (Obradors et al., 1988). During 1,2-PD degradation, the substrate gets converted to propionaldehyde by diol dehydratase, a coenzyme B_{12}-dependent enzyme (Bobik et al., 1997; Chowdhury et al., 2014) (Figure 10.3). Propionaldehyde dehydrogenase converts propionaldehyde to propionyl-CoA. In aerobic respiration, propionyl-CoA feeds into TCA (tricarboxylic acid cycle) cycle via the methyl citrate pathway (Horswill and Escalante-Semerena, 1999; Chowdhury et al., 2014). Therefore, 1,2-PD serves as both carbon and energy source in aerobic conditions (Jeter, 1990). However, an earlier study reveals that *Salmonella* can respire 1,2-PD anaerobically in the presence of tetrathionate. In anaerobic conditions, tetrathionate serves as the terminal electron acceptor (Price-Carter et al., 2001). It has been shown that breakdown of 1,2-PD occurs inside MCP (Bobik et al., 1997; Chowdhury et al., 2014) (Figure 10.3). The function of the Pdu MCP is to isolate propionaldehyde to mitigate cytotoxicity and DNA damage caused by toxic aldehyde intermediate. It was shown that propionaldehyde is built up to a toxic level in the mutants that disrupt shell formation during growth on 1,2-PD (Havemann et al., 2002; Sampson and Bobik, 2008).

Alternatively, ethanolamine is formed by the breakdown of phosphatidylethanolamine, a major phospholipid in both mammalian and bacterial cell membranes. The host diet, the bacterial and epithelial cells in the intestine are thought to provide rich sources of ethanolamine in nature (Garsin, 2010; Held et al., 2013). It is proposed that ethanolamine breakdown occurs inside the MCP shell. Ethanolamine enters MCP shell where it gets converted to acetaldehyde and ammonia with the help of coenzyme-B_{12}-dependent ethanolamine ammonia-lyase. The Eut MCP (EutM) shell is presumed to improve ethanolamine catabolism by protecting cells from aldehyde toxicity and simultaneously minimizing the diffusional loss of volatile acetaldehyde substrate for downstream enzyme, acetaldehyde dehydrogenase (Figure 10.3) (Rondon et al., 1995; Penrod and Roth, 2006). Further studies indicated that the main function of the EutM is to reduce the rate of diffusion of volatile acetaldehyde through the bacterial cell membrane (Penrod and Roth, 2006). The latter model is widely accepted as mutational analyzes show that complete deletion of EutM shell proteins of *Salmonella* results in no growth in ethanolamine (Penrod and Roth, 2006).

In past decades, there is an increase in the availability of the number of bacterial genome sequences. It has now been understood MCPs are not limited certain type of bacterial phyla, rather it spans to about 23% of bacteria distributed across 11 phylum-level taxa (Bobik, 2006; Abdul-Rahman et al., 2013; Jorda et al., 2013; Rae et al., 2013; Chowdhury et al., 2014; Axen et al., 2014; Bobik et al., 2015; Zarzycki et al., 2015) and are involved in 10 or more metabolic processes including glycerol, choline, fucose, rhamnose, and fucoidan (Shively et al., 1973; Talarico et al., 1990; Bobik et al., 1999; Kofoid et al., 1999; Price-Carter et al., 2001; Brinsmade et al., 2005; Craciun and Balskus, 2012; Petit et al., 2013; Erbilgin et al., 2014; Moore and Escalante-Semerena, 2016; Kerfeld et al., 2018).

However, many diverse MCPs are still uncharacterized and have unknown functions. Our understanding of fundamentals of MCP ecology and metabolism would perhaps uncover novel structural and functional paradigms that may be helpful in designing molecular scaffold for several synthetic biology applications.

10.4 NOTABLE FEATURES OF BACTERIAL MICROCOMPARTMENT (MCP)

The most intriguing feature of MCP is the presence of a protein shell at the outer surface covering the metabolic enzymes. This differentiates them from the most common membrane-lipid bound organelles in eukaryotes (Bobik, 2006; Tanaka et al., 2008; Yeates et al., 2008; Kerfeld et al., 2010; Chowdhury et al., 2014). A recent study demonstrates that protein shells are robust and protective against proteolysis, oxidative stress, etc. (Kim and Tullman-Ercek, 2014; Lee et al., 2016). A protection of encapsulated enzymes from proteases was recently monitored *in vivo* by using *C*-terminal SsrA proteolysis tags (Lee et al., 2016). Later on, transmission electron microscopic analyzes revealed that Pdu MCP shells can retain their morphology for several days and can bear temperatures up to 60°C. Pdu MCP shells are also stables at high pH and high ionic strength (up to 400 mM salt) (Kim et al., 2014; Faulkner et al., 2019). Therefore, it would be possible to carry out labile or sensitive reactions inside MCP lumen, which would definitely open up broader applicability of MCP as nanobioreactor in several industrial aspects. However, further studies are needed to put forward this idea.

A closer view of the outer protein shell reveals that the shell is comprised of thousands of self-assembled protein subunits containing either single or doubly fused BMC domains (Yeates et al., 2011, 2013; Chowdhury et al., 2014; Lee et al., 2019). Single (Pfam: PF00936) and fused BMC domain (two fused PF00936 domains) proteins usually adopt hexameric and pseudohexameric structure respectively that tile together into a two-dimensional sheet generating facets of the shell. The vertices of the MCP shell are generally capped by pentameric BMV (bacterial MCPs vertex) domain protein (Pfam: PF03319 domain) (Yeates et al., 2011, 2013; Chowdhury et al., 2014; Lee et al., 2019). Interestingly, MCP shells are not impervious; these are porous that may allow the transport of various metabolites while confining reacting enzymes and intermediates inside their lumen (Crowley et al., 2010; Tanaka et al., 2010; Pang et al., 2011; Cai et al., 2015; Chowdhury et al., 2015). Hexameric BMC-domain proteins (PduA of Pdu MCP, CcmK (1–4) of carboxysome and EutM of Eut MCP) contain small pores at the center of the hexamers (Figure 10.4); whereas, gated pores (in PduB, EutL, and CcmO) and [Fe-S] clusters (in PduT) are usually present at the middle of pseudohexameric trimers. The small pores in hexamers act as conduits for the transport of small molecules (Crowley et al., 2010; Chowdhury et al., 2014, 2015), while the gated pores are hypothesized to allow the movement of larger enzymatic cofactors across the shell while maintaining confinement of pathway intermediates (Pang et al., 2012; Thompson et al., 2014). In contrast, the [4Fe-4S] cluster at the place of the central pore region (Figure 10.4) is speculated to transfer electrons or involve in the transport of intact metal clusters across the shell (Chowdhury et al., 2014; Lee et al., 2019). MCP shells also contain several divergent types of BMC-domain proteins. Hexameric PduU and EutS typically contain an extension at the *N*-terminus that adopts a β-barrel structure at the central pore region. This may occlude the pore; presumably control the movement of the metabolites (Crowley et al., 2010; Tanaka et al., 2010). Alternatively, some BMC domain proteins like PduK and EutK have variable *C*-terminal extensions. The *C*-terminus of PduK has [Fe-S] cluster binding motif, whereas; *C*-terminus of EutK comprises a helix-turn-helix motif, a typical DNA binding domain (Crowley et al., 2010; Tanaka et al., 2010). However, their specific roles await establishments.

Although, paralogous shell proteins are structurally redundant in gross, however; minute variations confer specific function. Understanding MCP shell proteins would thus provide a new insight in modulating microenvironment for several biotechnology applications.

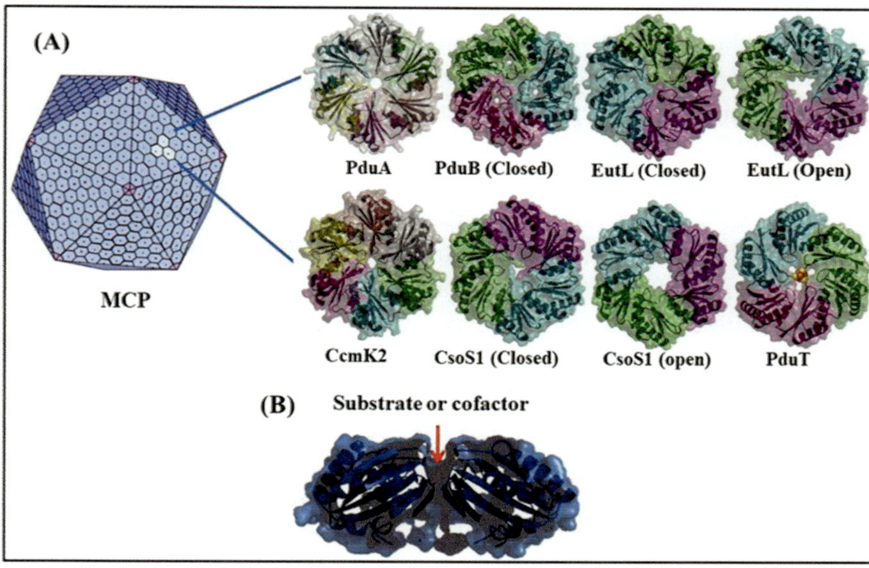

FIGURE 10.4 (A) Crystal structures of some of the MCP shell proteins; and (B) the central pores of the shell proteins are proposed to serve as conduit for substrates or cofactors.

10.4.1 SELECTIVE MOLECULAR TRANSPORT ACROSS MICROCOMPARTMENT (MCP) SHELL

Recent crystal structure reveals the presence of pores of different sizes down the center of hexamer (Crowley et al., 2010; Tanaka et al., 2010). It is proposed that the central pores in the hexameric or pseudohexameric shell proteins confer specific flux characteristics (Figure 10.4). Subsequent structural analyzes of different shell proteins also revealed that the pores are diverse in nature, consistent with the transport of various metabolites (Tanaka et al., 2009; Yeates et al., 2011, 2013; Chowdhury et al., 2014).

The canonical BMC hexamers like PduA, EutM, and CcmK (1–4) have small central pores (4–7 Å in diameter) that are presumed to control the diffusion of small substrate and metabolites across the shell. The narrowest point of these pores is formed by the convergence of six identical amino acid residues, one from each monomer of the BMC hexamer. The types of amino acid forming the constriction point presumably play a key role in transport selectivity, where primary factors influencing the permeability could be the size of the pore and its chemical properties (Kerfeld et al., 2005; Yeates et al., 2013; Chowdhury et al., 2014, 2015; Glasgow et al., 2015). In the case of carboxysomal shell proteins like CcmK1, CcmK2, and Ccmk4, the residues lined up the pore periphery are positively charged, which are speculated to aid the uptake of negatively charged bicarbonate, a substrate of carbonic anhydrase (Kerfeld et al., 2005; Chowdhury et al., 2014) while restricting uncharged molecular oxygen to enter the shell. This is how cyanobacteria evolve naturally to bypass wasteful photorespiration of RuBisCo (Kerfeld et al., 2005).

Lately, Chowdhury and co-worker have first time demonstrated the experimental evidence on how pore structure and chemical properties modulate permeability of an MCP shell (Chowdhury et al., 2015). In-depth studies including site-directed mutagenesis, assessment of growth phenotype, biochemical studies, and crystal structure determination have revealed that the key pore-forming amino acid, Ser 40 of PduA hexamer selectively allows entry of 1,2-PD (the substrate) into the MCP while inhibiting the outward diffusion of propionaldehyde (a toxic pathway intermediate) (Chowdhury et al., 2015). A mutation of Ser 40 with long-chain amino acids reduces the pore size, thereby impeding the inward diffusion of 1,2-PD (Chowdhury et al., 2015). Conversely, mutation of Ser 40 to Ala, does not change pore diameter but alters polarity of the pore, which thereby enhances the egress of toxic propionaldehyde intermediate out of MCP (Chowdhury et al., 2015). This study is no doubt a stepping stone to engineer pore architecture to achieve desired permeability in modulating substrate entry or product release.

Meanwhile, two other pseudohexameric trimers, the CsoS1D of α-carboxysomes and the EutL of the EutM, are reported to be present in two different conformations—pore open and closed. The pore open form has a large central pore (~10–12 Å), whereas in closed conformation, the pore seems to be completely occluded (Klein et al., 2009; Takenoya et al., 2010; Tanaka et al., 2010; Lee et al., 2019). It is hypothesized that the pores in these trimeric shell proteins are gated, which are proposed to modulate the entry and exit of the molecules. Recently, EutL has been shown to bind with ethanolamine in such a way that suggests allosteric regulation of the opening and closure of the pore (Bobik et al., 2015; Thompson et al., 2015). Later, crystallographic studies have proposed that PduB of the Pdu MCP might have a gated pore that admits the entry of bulkier cofactors in a similar fashion (Klein et al., 2009; Sagermann et al., 2009; Tanaka et al., 2010; Cai et al., 2013; Thompson et al., 2015). Genetic modification of trimeric BMC-domain proteins may thus offer an approach to engineering the movement of larger molecules across MCP shells.

Another divergent class of BMC trimer (PduT) is reported where the central pore region is occupied by a [4Fe-4S] cluster (Crowley et al., 2010; Pang et al., 2011). It is suggested that [4Fe-4S] may have a role in the transport of electrons across the shell or may involve in reconstructing Fe-S centers within the MCP lumen following oxidative damage. Electron transfer through central [4Fe-4S] cluster is hypothesized to help internal iron-sulfur enzyme(s) to perform redox reaction (Chowdhury et al., 2014). Hitherto, the exact role of PduT is not clear (Chowdhury et al., 2014); a detailed understanding of PduT function would possibly help develop synthetic system to carry out desired redox reaction inside its lumen.

While engineering microenvironment across the pore region seems to be a challenging endeavor, selectivity seen in natural pores suggests this would be feasible.

10.4.2 LOCALIZATION OF ENZYMES IN THE MICROCOMPARTMENT (MCP) INTERIOR

The most interesting feature of bacterial MCPs is that MCPs create passage to small molecules through its porous shell. But how the metabolic enzymes get encapsulated is still a mystery. The initial hint on the underlying mechanism of enzyme encapsulation inside MCP lumen was first described back in 2010. It has been found that some of the lumen enzymes of Pdu MCP of *Salmonella* contain *N*-terminal appendages (known as targeting peptides of about 18 amino acids long) that allow the enzymes to be localized inside the MCP lumen (Fan et al., 2010, 2012; Fan and Bobik, 2011), these extensions are not necessarily required for enzymatic activity though. It has also been demonstrated that this *N*-terminal targeting peptide can encapsulate a number of heterologous proteins, such as maltose-binding protein (MBP), green fluorescent protein (GFP), and glutathione *S*-transferase (GST) inside MCPs (Fan et al., 2010). Later studies showed that targeting peptides is widely spread among diverse MCP enzymes. A similar targeting peptide was reported in EutMs, which efficiently encapsulated heterologous proteins to recombinant, synthetic MCPs (Choudhary et al., 2012). In the case of β-carboxysome, a short *C*-terminal sequence is presumed to function similarly to the targeting peptide found in Pdu and EutM during encapsulation and assembly (Kinney et al., 2012; Chowdhury et al., 2014). Further studies on the mechanism of encapsulation revealed that the targeting peptide forms amphipathic α-helices that interact with *C*-terminal helices of shell proteins resulting in the successful recruitment of the enzyme inside MCP lumen (Fan et al., 2012; Chowdhury et al., 2014; Aussignargues et al., 2015). It is noteworthy to mention that several enzyme cargos of divergent MCPs do not contain obvious targeting sequences. It is proposed that encapsulation of those enzymes occurs by either protein-protein interactions or some other unknown mechanisms (Chowdhury et al., 2014; Aussignargues et al., 2015).

It is hypothesized that encapsulation of enzymatic cargo may influence on assembly, size, and shape of the MCP. Earlier studies revealed that empty MCP forms smaller and hollow shells (Mayer et al., 2016). A recent computational model demonstrated that at equilibrium, the shell size is influenced by the stoichiometry between cargo enzymes and the MCP shell proteins (Mohajerani and Hagan, 2018). However, a complete understanding of MCP assembly is necessary to repurpose MCP or to create synthetic mimic for several biotechnology applications.

10.5 ENGINEERING MICROCOMPARTMENTS (MCPS) TO REPURPOSE FOR INDUSTRIAL USAGE

Engineering efforts toward tailoring MCPs are being carried out extensively for their utilization in biotech-based industries. Two major domains where MCP engineering is aimed at heterologous cargo packaging inside MCP lumen and metabolic control by modulating shell permeabilities. Recent studies demonstrated the targeting of heterologous enzyme by fusing targeting peptide (Fan et al., 2010) at the *N*-terminus of the desired enzymes (Chowdhury et al., 2014) (Figure 10.5). Several proof-of-concept for successful encapsulation of desired cargo proteins have been reported lately (Lawrence et al., 2014; Jakobson et al., 2015; Quin et al., 2016).

On the other hand, engineering permeability of shell protein is required in order to enhance metabolic flux or limit intermediate diffusion across the shell to subsequently achieve ideal pathways with increased productivity in protein compartments. There are few instances where pore architecture has been altered for controlling metabolite transport (Chowdhury et al., 2015; Aussignargues et al., 2016; Lee et al., 2017; Slininger et al., 2017). However, more work is needed in this field to complete understanding the system in order to develop novel enclosures for several synthetic biology applications.

Additionally, synthetic MCP with a minimal number of shell proteins is being synthesized by several researchers (Parsons et al., 2010; Choudhary et al., 2012). Designing MCP as synthetic nanobioreactor for biofuel production was first demonstrated in 2014 (Lawrence et al., 2014). This work provides a foundation that led to the development of purpose-built MCPs.

10.5.1 ENGINEERING PORE PERMEABILITY

An engineering approach on changing the pore architecture of PduA shell protein was first initiated by Chowdhury et al. in 2015. A change of pore-lining Ser 40 to His does not change the pore diameter; rather it alters the substrate specificity (Chowdhury et al., 2015). It has been shown that Ser 40 in PduA is important for the transport of 1, 2-PD, a natural substrate of Pdu MCP. A substitution of Serine 40 to Histidine modifies the permeability across the shell and allows both 1,2-PD and glycerol (Chowdhury et al., 2015).

In another instance, the central pore of a trimeric BMC domain protein has been tailored to create a [4Fe-4S] cluster binding site in a synthetic shell (Aussignargues et al., 2016). The modified shells are able to perform electron transfer to internal redox reaction (Aussignargues et al., 2016). It has been shown that the central [4Fe-4S] cluster is stabilized by hydrogen-bonding network with the main chain of the synthetic shell (Aussignargues et al., 2016).

Construction of chimeric shell is an alternative strategy to mutation of single pore-lining residue in order to alter permeability property of a shell. In 2015, Cai et al. reported chimeric shell construction where β-carboxysomal CcmK2 hexamer has been replaced by α-carboxysomal CsoS1 (Cai et al., 2015). In a second instance, a chimeric Pdu MCP has been constructed by replacing the PduA protein with a paralogous BMC-hexamer from EutM (Lee et al., 2017; Slininger et al., 2017). A Δ*pduA*::*eutM* mutant exhibits increased substrate (1,2-PD) influx without having any abnormalities in MCP morphology. Intriguingly, MCPs of Δ*pduA*::*eutM* mutant seems to be better in sequestering propionaldehyde, 1-propanol and propionate as compared to wild type (Lee et al., 2017; Slininger et al., 2017).

Very recently, a site directed mutagenesis approach was employed on the central [Fe-S] cluster of PduT of *Salmonella* to modify permeability across MCP shells. A substitution of Cysteine 38 of PduT to Serine allows movement of larger cofactor like NAD/H across MCP shells, without egressing of aldehyde intermediate in the media to a toxic level (Chowdhury and Bobik, 2019). This study provides foundation of engineering trimeric BMC domain protein to control the movement of larger molecules with effective sequestration of the intermediate carbon simultaneously. Additionally, a change of Cystein 38 to either Leucine or Tryptophan suggests a method of blocking trimeric BMC domain proteins (Chowdhury and Bobik, 2019).

Hitherto, not much detailed studies were performed with other shell proteins. Engineering shell proteins to modify permeability property would help design customized shell with full control of metabolite transport or efficient product release for several biotechnology applications.

10.5.2 DESIGNING SYNTHETIC MICROCOMPARTMENT (MCP) SHELL

Studies towards developing synthetic MCPs for biotechnology applications are being conducted by several groups. The transfer of the entire *pdu* operon from *Citrobacter freundii* and its expression in *Escherichia coli* was the first demonstration of recombinant expression of MCP in heterologous host (Parsons et al., 2008). Similarly, the operon of α-carboxysome from *Halothiobacillus neapolitanus* was expressed in both Gram-positive *Corynebacterium glutamicum* (Baumgart et al., 2017) and Gram-negative *Escherichia coli* (Bonacci et al., 2012), separately. It was observed that the resulting strains were able to fix carbon dioxide easily (Bonacci et al., 2012; Baumgart et al., 2017). Later, attempts were made to express a minimal number of Pdu shell proteins from *Citrobacter freundii* into *Corynebacterium glutamicum* for the synthesis of value-added products (Huber et al., 2017). Very recently, 38 kb region from the *Salmonella enterica* serovar Typhimurium genome containing *pdu* and *cob/cbi* genes have been expressed in a number of hosts, namely *Salmonella bongori*, *E. coli*, *K. pneumoniae*, *Serratia marcescens*, *Cronobacter sakazakii* and different *Pseudomonas* species. These recombinant bacterial strains were able to express *pdu* operon and utilize 1,2-PD (Graf et al., 2018).

Transplantation of MCP operon has also been carried out between inter-kingdom taxonomic organisms. Genes of both α-carboxysome (Long et al., 2018) and β-carboxysome (Lin et al., 2014) were expressed in the chloroplast of *Nicotiana tabacum* and *Nicotiana benthamiana*, respectively. The transgenic plants were able to synthesize carboxysomes and exhibited photoautotrophic growth efficiently. Intriguingly, MCPs expressed in the transgenic plants were able to minimize the wasteful photorespiration caused by indigenous RuBisCo. This strategy of MCP bioengineering is a stepping stone towards enhancement of photosynthesis and crop yield (Long et al., 2018).

On the other hand, the production of a synthetic empty shell with a minimal number of shell proteins was first demonstrated by Parsons et al. in 2010, where genes encoding shell proteins of *pdu* operon of *Citrobacter freundii* was expressed in *E. coli* (Parsons et al., 2010). Later, minimal number of Pdu (Sargent et al., 2013) and Eut (Choudhary et al., 2012) shell proteins of *Salmonella enterica* were expressed in *E. coli* chassis to generate empty shell. Lately, an MCP cluster of unknown function from halophilic *Haliangium ochraceum* was also expressed in *E. coli*, which contains only one hexamer, two trimers and a pentamer (Lassila et al., 2014). In all the empty MCP shells heterologous GFP has been recruited successfully by fusion to a predicted targeting sequence (Lassila et al., 2014; Parsons et al., 2010; Choudhary et al., 2012; Sargent et al., 2013). An effective design of synthetic nanobioreactor was first depicted by Lawrence et al. where empty Pdu shell from *Citrobacter freundii* was transformed into ethanol bioreactor (Lawrence et al., 2014). Recently, a synthetic Pdu MCP has been repurposed to a 1,2-PD synthesizing bioreactor (Lee et al., 2016).

10.5.3 EXPORT OF HETEROLOGOUS PROTEINS INSIDE MICROCOMPARTMENT (MCP) LUMEN

Considerable progresses were made on engineering MCPs to contain a desired protein cargo. Empty MCP shells have been produced in heterologous host like *Escherichia coli* (Parsons et al., 2010; Lassila et al., 2014; Mayer et al., 2016). Desired proteins were heterologously expressed and encapsulated within empty MCP shells using short targeting peptides fused to their *N*-termini (Figure 10.5) (Fan et al., 2010; Choudhary et al., 2012; Kinney et al., 2012; Lawrence et al., 2014; Quin et al., 2016).

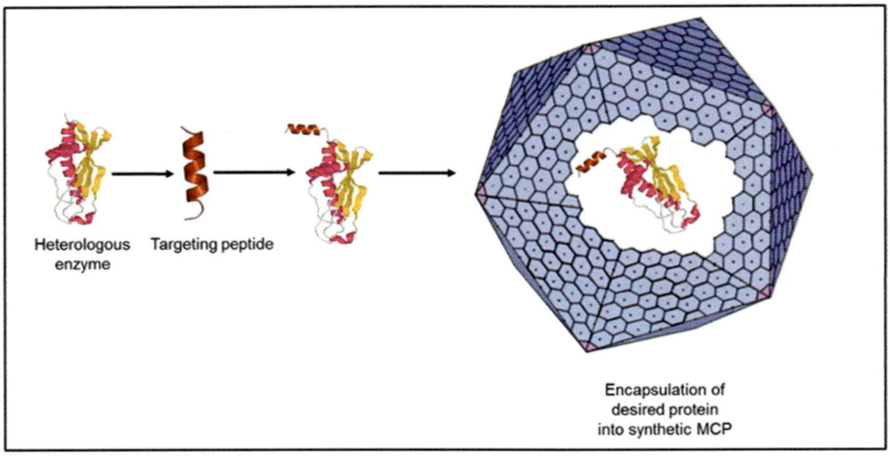

FIGURE 10.5 Schematic presentation of targeting protein of interest into a synthetic microcompartment by fusing targeting peptide at the *N*-terminus of the desired enzyme (Chowdhury et al., 2014).

Later, pyruvate decarboxylase and alcohol dehydrogenase were targeted to an empty MCP shell resulting successful repurpose of a Pdu MCP to an ethanol nanobioreactor (Lawrence et al., 2014). Subsequently, a number of MCP targeting sequences have been identified which may facilitate the encapsulation of multiple enzymes at desired stoichiometries (Jakobson et al., 2015; Quin et al., 2016). Lately, Lee et al. demonstrated encapsulation of as many as four heterologous enzymes namely dihydroxyacetone kinase, glycerol dehydrogenase, methylglyoxal synthase and 1,2-PD oxidoreductase by fusion of targeting peptide individually at their *N*-terminus (Lee et al., 2016). But studies also revealed that fusion of *N*-terminal targeting peptide results aggregation of the proteins thereby lowering of the activity (Lee et al., 2016), which may tailback the progress of customized nanobioreactor design. In 2017, Jakobson and coworkers have made considerable success to this issue. They have demonstrated *de novo* design of targeting peptide, where a canonical leucine-zipper motif was shown to recruit heterologous cargo efficiently inside a Pdu MCP (Jakobson et al., 2017). Although, there was minor polar aggregation of the fusion protein as observed under transmission electron microscope, however, the issue was resolved by the addition of 1,2-PD in the media (Jakobson et al., 2017).

In some cases, the process of encapsulation by fusing targeting peptide at the *N*-terminus of the enzyme may not be successful. The presence of targeting peptide near to active site at the *N*-terminus may likely interfere with protein's secondary structure and destroy the active site. In 2017, Huber and her coworkers have shown that targeting peptide fused to the *C*-terminus can recruit the eYFP protein successfully to the MCP lumen (Huber et al., 2017). Intriguingly, the targeting peptide at *C*-terminus is also capable of interacting with shell proteins during the process of encapsulation, like *N*-terminal targeting peptide (Huber et al., 2017).

Nevertheless, a control on cargo loading to MCP lumen is needed, which could be achieved by an engineering approach tuning the helix-helix interactions between targeting peptide and shell proteins. Further investigation will be required to set the criteria of effective helix-helix packing leading to efficient encapsulations without altering enzyme solubility and activity. A new strategy of precise delivery of heterologous cargo to the MCP lumen by fusing coiled-coil modules to both circularly permuted BMC hexamer and to desired heterologous cargo has been proposed elsewhere (Kirst and Kerfeld, 2019). However, all these interactions (between two helices and between two coiled-coil structures) occur on hydrophobic interfaces. Alternatively, electrostatic interaction may play in major role on successful encapsulation. The residues in the luminal surface of a hexameric BMC domain protein could be engineered to positively charge to promote compartmentalization of negatively charged cargo (Kirst and Kerfeld, 2019).

10.6 FUTURE PROSPECT OF MICROCOMPARTMENT (MCP) TECHNOLOGY

MCPs have provided great promise in designing subcellular nanobioreactor for the high-level synthesis of useful biochemicals, biorenewables, and pharmaceuticals. MCPs have indigenous properties to increase the yield of product by bringing reacting molecule in a close proximity and limit the loss of carbons for downstream reactions (Kerfeld et al., 2005; Cheng et al., 2008; Chowdhury et al., 2014). Moreover, MCPs have properties to trap toxic intermediate by isolating them from central metabolism. With these attributes MCPs thus offer great prospect in industrial applications to make unfavorable reaction feasible.

MCPs also have remarkable feat for its highly stable and robust shell proteins. It is reported that MCP shells can retain their structure in a range of temperature and pH (Kim et al., 2014). Therefore, the reactions sensitive to high temperature and pH can easily be performed in harsh industrial operating condition by exporting them inside MCPs. Recent analyzes of dynamic changes of MCP shell protein assembly also indicated the robustness of the shell (Faulkner et al., 2019), which is suggested to stabilize lumen enzymes from harsh environmental conditions. Interestingly, some of the MCP contains chaperon like protein (EutJ). It is hypothesized that EutJ might have a role in stabilizing aldehyde dehydrogenase and alcohol dehydrogenase inside EutM (Kaval and Garsin, 2018). However, function of EutJ is still under investigation.

Some of the MCPs house oxygen labile enzymes such is RuBisCo (in carboxysome), PduQ (in Pdu MCP) (Kerfeld et al., 2005; Cheng et al., 2012). It is presumed that MCP shell restricts the entry of molecular oxygen while allowing own substrates and cofactors (Kerfeld et al., 2005). So, it is envisioned that some oxygen sensitive reactions can be carried out easily by encapsulating sensitive enzymes inside synthetic MCPs. Recently, scientists develop hydrogen nanoreactor by compartmentalizing oxygen sensitive hydrogenase inside engineered virus capsid (Jordan et al., 2016). Proteinaceous MCPs are suggested to have great prospect in this regard.

At present, several engineering approaches are being exploited to modify MCP or generate synthetic MCPs for high level production of important chemicals or pharmaceuticals. Modifying pore architecture demands more prospects in controlling microenvironments for efficient substrate entry and/or product release (Crowley et al., 2010; Chowdhury et al., 2015; Chowdhury and Bobik, 2019). It is suggested that engineering pore of the canonical BMC hexamer would provide a great platform to alter substrate specificity and retention of intermediate. On the contrary, the trimeric shells are target for modifying transport of large cofactors, electron transfer and/or product release.

Only a limited number of heterologous proteins has been successfully exported so far into a MCP lumen (Lawrence et al., 2014; Lee et al., 2016; Quin et al., 2016), which is a bottle neck on developing effective metabolosome with multienzyme cascade. However, understanding the underlying principles of encapsulation of the proteins lacking targeting sequence in native MCPs would pave the way in making molecular chamber of multienzyme complex.

Besides developing intracellular nanobioreactor, MCPs can also be designed to serve as molecular delivery of different cargo molecules. Such cargo molecule could be either cytotoxic drugs and metal nanoparticles or cytotoxic proteins for targeted delivery. This strategy would have potential applications in several biomedical and nanomedicine industries. It is proposed that metal nanoparticle probes surrounded by self-assembled MCP protein shell can be used in magnetic imaging for tumor detections (Tsai and Yeates, 2011). Similarly, synthetic MCP shells can be used to surround cytotoxic drugs by attaching the drugs chemically on the surface of the MCP interior. In this way, cytotoxic drugs can be delivered to the target without imparting any side effects to normal cells (Tsai and Yeates, 2011; Chowdhury et al., 2014).

Despite several engineering effort to ameliorate MCP technologies toward broad biotechnology applications, the purification procedure of polyhedral body yet suffers several technical challenges. The existing procedure is laborious, time consuming and results in poor yield (Sinha et al., 2012; Chowdhury et al., 2015). A time saving small scale purification procedure was demonstrated earlier (Graf et al., 2018). However, there are numbers of MCP organelles being discovered, where purification procedure is still unknown. Polyhedral shells are being generated by synthetic method too. But lack of easy and simplified purification procedure limits the characterization process of this organelle and hinders the idea of making a cell-free system. Recently, a synthetic MCPs shell of *Haliangium ochraceum* have been isolated efficiently by affinity purification after addition of Strep-II tag to the pentameric vertex proteins (Hagen et al., 2018b). This procedure would enable researchers to develop a facile method of purification and subsequent analysis of the polyhedral shell by eliminating the need for technically demanding procedure.

Nevertheless, the use of MCP technology has become fertile ground for continuing scientific exploration. The proteinaceous compartment finds myriad potential uses in the broad areas of synthetic and systems biology, metabolic engineering, targeted molecular delivery, and other areas of nanotechnology as well as in biomedical industries. The on-going research on MCPs has provided new insights into their structures and functions, however many questions still remain unanswered. A complete understanding of the overall organization will help design MCP as a tool in the advance field of biotechnology research.

ACKNOWLEDGMENT

Chiranjit Chowdhury acknowledges Science and Engineering Research Board (SERB), Department of Science and Technology, Government of India for prestigious Ramanujan Fellowship vide sanction no. SB/S2/RJN-128/2017. The authors also thank Mr. Ritikraj Tillhari for his help in making some of the figures.

KEYWORDS

- **1,2-propanediol**
- **bacterial microcompartment**
- **biotechnology**
- **carboxysome**
- **ethanolamine**
- **Eut**
- **metabolic engineering**
- **nanobioreactor**
- **Pdu**
- **synthetic biology**

REFERENCES

Abdul-Rahman, F., Petit, E., & Blanchard, J. L., (2013). The distribution of polyhedral bacterial microcompartments suggests frequent horizontal transfer and operon reassembly. *J. Phylogen. Evolution Biol., 1*, 1–7.

Aussignargues, C., Paasch, B. C., Gonzalez-Esquer, R., Erbilgin, O., & Kerfeld, C. A., (2015). Bacterial microcompartment assembly: The key role of encapsulation peptides. *Commun. Integr. Biol., 8*, e1039755.

Aussignargues, C., Pandelia, M. E., Sutter, M., Plegaria, J. S., Zarzycki, J., Turmo, A., Huang, J., et al., (2016). Structure and function of a bacterial microcompartment shell protein engineered to bind a [4Fe-4S] cluster. *J. Am. Chem. Soc., 138*, 5262–5270.

Axen, S. D., Erbilgin, O., & Kerfeld, C. A., (2014). A taxonomy of bacterial microcompartment loci constructed by a novel scoring method. *PLoS Comput. Biol., 10*, e1003898.

Baumgart, M., Huber, I., Abdollahzadeh, I., Gensch, T., & Frunzke, J., (2017). Heterologous expression of the *Halothiobacillus* neapolitanus carboxysomal gene cluster in *Corynebacterium glutamicum*. *J. Biotechnol., 258*, 126–135.

Bobik, T. A., (2006). Polyhedral organelles compartmenting bacterial metabolic processes. *Appl. Microbiol. Biotechnol., 70*, 517–525.

Bobik, T. A., Havemann, G. D., Busch, R. J., Williams, D. S., & Aldrich, H. C., (1999). The propanediol utilization (PDU) operon of *Salmonella enterica* serovar Typhimurium LT2 includes genes necessary for formation of polyhedral organelles involved in coenzyme B(12)-dependent 1, 2-propanediol degradation. *J. Bacteriol., 181*, 5967–5975.

Bobik, T. A., Lehman, B. P., & Yeates, T. O., (2015). Bacterial microcompartments: Widespread prokaryotic organelles for isolation and optimization of metabolic pathways. *Mol. Microbiol., 98*, 193–207.

Bobik, T. A., Xu, Y., Jeter, R. M., Otto, K. E., & Roth, J. R., (1997). Propanediol utilization genes (PDU) of Salmonella typhimurium: three genes for the propanediol dehydratase. *J. Bacteriol., 179*, 6633–6639.

Bonacci, W., Teng, P. K., Afonso, B., Niederholtmeyer, H., Grob, P., Silver, P. A., & Savage, D. F., (2012). Modularity of a carbon-fixing protein organelle. *Proc. Natl. Acad. Sci. U. S. A., 109*, 478–483.

Brinsmade, S. R., Paldon, T., & Escalante-Semerena, J. C., (2005). Minimal functions and physiological conditions required for growth of *Salmonella enterica* on ethanolamine in the absence of the metabolosome. *J. Bacteriol., 187*, 8039–8046.

Cai, F., Sutter, M., Bernstein, S. L., Kinney, J. N., & Kerfeld, C. A., (2015). Engineering bacterial microcompartment shells: Chimeric shell proteins and chimeric carboxysome shells. *ACS Synth. Biol., 4*, 444–453.

Cai, F., Sutter, M., Cameron, J. C., Stanley, D. N., Kinney, J. N., & Kerfeld, C. A., (2013). The structure of CcmP, a tandem bacterial microcompartment domain protein from the beta-carboxysome, forms a subcompartment within a microcompartment. *J. Biol. Chem., 288*, 16055–16063.

Cheng, S., Fan, C., Sinha, S., & Bobik, T. A., (2012). The PduQ enzyme is an alcohol dehydrogenase used to recycle NAD+ internally within the Pdu microcompartment of *Salmonella enterica. PLoS One, 7*, e47144.

Cheng, S., Liu, Y., Crowley, C. S., Yeates, T. O., & Bobik, T. A., (2008). Bacterial microcompartments: Their properties and paradoxes. *Bioessays, 30*, 1084–1095.

Choudhary, S., Quin, M. B., Sanders, M. A., Johnson, E. T., & Schmidt-Dannert, C., (2012). Engineered protein nano-compartments for targeted enzyme localization. *PLoS One, 7*, e33342.

Chowdhury, C., & Bobik, T. A., (2019). Engineering the PduT shell protein to modify the permeability of the 1,2-propanediol microcompartment of *Salmonella. Microbiology.* doi: 10.1099/mic.0.000872.

Chowdhury, C., Chun, S., Pang, A., Sawaya, M. R., Sinha, S., Yeates, T. O., & Bobik, T. A., (2015). Selective molecular transport through the protein shell of a bacterial microcompartment organelle. *Proc. Natl. Acad. Sci. U. S. A., 112*, 2990–2995.

Chowdhury, C., Sinha, S., Chun, S., Yeates, T. O., & Bobik, T. A., (2014). Diverse bacterial microcompartment organelles. *Microbiol. Mol. Biol. Rev., 78*, 438–468.

Conrado, R. J., Varner, J. D., & DeLisa, M. P., (2008). Engineering the spatial organization of metabolic enzymes: Mimicking nature's synergy. *Curr. Opin. Biotechnol., 19*, 492–499.

Craciun, S., & Balskus, E. P., (2012). Microbial conversion of choline to trimethylamine requires a glycyl radical enzyme. *Proc. Nat.l Acad. Sci. U. S. A., 109*, 21307–21312.

Crowley, C. S., Cascio, D., Sawaya, M. R., Kopstein, J. S., Bobik, T. A., & Yeates, T. O., (2010). Structural insight into the mechanisms of transport across the *Salmonella enterica* Pdu microcompartment shell. *J. Biol. Chem., 285*, 37838–37846.

Dueber, J. E., Wu, G. C., Malmirchegini, G. R., Moon, T. S., Petzold, C. J., Ullal, A. V., Prather, K. L., & Keasling, J. D., (2009). Synthetic protein scaffolds provide modular control over metabolic flux. *Nat. Biotechnol., 27*, 753–759.

Erbilgin, O., McDonald, K. L., & Kerfeld, C. A., (2014). Characterization of a planctomycetal organelle: A novel bacterial microcompartment for the aerobic degradation of plant saccharides. *Appl. Environ. Microbiol., 80*, 2193–2205.

Fan, C., & Bobik, T. A., (2011). The N-terminal region of the medium subunit (PduD) packages adenosylcobalamin-dependent diol dehydratase (PduCDE) into the Pdu microcompartment. *J. Bacteriol., 193*, 5623–5628.

Fan, C., Cheng, S., Liu, Y., Escobar, C. M., Crowley, C. S., Jefferson, R. E., Yeates, T. O., & Bobik, T. A., (2010). Short N-terminal sequences package proteins into bacterial microcompartments. *Proc. Natl. Acad. Sci. U. S. A., 107*, 7509–7514.

Fan, C., Cheng, S., Sinha, S., & Bobik, T. A., (2012). Interactions between the termini of lumen enzymes and shell proteins mediate enzyme encapsulation into bacterial microcompartments. *Proc. Natl. Acad. Sci. U. S. A., 109*, 14995–15000.

Faulkner, M., Zhao, L. S., Barrett, S., & Liu, L. N., (2019). Self-assembly stability and variability of bacterial microcompartment shell proteins in response to the environmental change. *Nanoscale. Res. Lett., 14*, 54.

Frank, S., Lawrence, A. D., Prentice, M. B., & Warren, M. J., (2013). Bacterial microcompartments moving into a synthetic biological world. *J. Biotechnol., 163*, 273–279.

Garsin, D. A., (2010). Ethanolamine utilization in bacterial pathogens: Roles and regulation. *Nat. Rev. Microbiol., 8*, 290–295.

Giessen, T. W., & Silver, P. A., (2016). Encapsulation as a strategy for the design of biological compartmentalization. *J. Mol. Biol., 428*, 916–927.

Glasgow, J. E., Asensio, M. A., Jakobson, C. M., Francis, M. B., & Tullman-Ercek, D., (2015). Influence of electrostatics on small molecule flux through a protein nanoreactor. *ACS Synth. Biol., 4*, 1011–1019.

Graf, L., Wu, K., & Wilson, J. W., (2018). Transfer and analysis of *Salmonella* Pdu genes in a range of gram-negative bacteria demonstrate exogenous microcompartment expression across a variety of species. *Microb Biotechnol., 11*, 199–210.

Hagen, A. R., Plegaria, J. S., Sloan, N., Ferlez, B., Aussignargues, C., Burton, R., & Kerfeld, C. A., (2018a). *In Vitro* assembly of diverse bacterial microcompartment shell architectures. *Nano Lett., 18*, 7030–7037.

Hagen, A., Sutter, M., Sloan, N., & Kerfeld, C. A., (2018b). Programmed loading and rapid purification of engineered bacterial microcompartment shells. *Nat. Commun., 9*, 2881.

Havemann, G. D., Sampson, E. M., & Bobik, T. A., (2002). PduA is a shell protein of polyhedral organelles involved in coenzyme B(12)-dependent degradation of 1,2-propanediol in *Salmonella enterica* serovar Typhimurium LT2. *J. Bacteriol., 184*, 1253–1261.

Held, M., Quin, M. B., & Schmidt-Dannert, C., (2013). Eut bacterial microcompartments: Insights into their function, structure, and bioengineering applications. *J. Mol. Microbiol. Biotechnol., 23*, 308–320.

Horswill, A. R., & Escalante-Semerena, J. C., (1999). Salmonella typhimurium LT2 catabolizes propionate via the 2-methylcitric acid cycle. *J. Bacteriol., 181*, 5615–5623.

Huber, I., Palmer, D. J., Ludwig, K. N., Brown, I. R., Warren, M. J., & Frunzke, J., (2017). Construction of recombinant Pdu metabolosome shells for small molecule production in *Corynebacterium glutamicum*. *ACS. Synth. Biol., 6*, 2145–2156.

Hurtley, S., (2009). Spatial cell biology. Location, location, location. Introduction. *Science, 326*, 1205.

Iancu, C. V., Ding, H. J., Morris, D. M., Dias, D. P., Gonzales, A. D., Martino, A., & Jensen, G. J., (2007). The structure of isolated Synechococcus strain WH8102 carboxysomes as revealed by electron cryotomography. *J. Mol. Biol., 372*, 764–773.

Jakobson, C. M., Kim, E. Y., Slininger, M. F., Chien, A., & Tullman-Ercek, D., (2015). Localization of proteins to the 1,2-propanediol utilization microcompartment by non-native signal sequences is mediated by a common hydrophobic motif. *J. Biol. Chem., 290*, 24519–24533.

Jakobson, C. M., Slininger, L. M. F., & Tullman-Ercek, D., (2017). De novo design of signal sequences to localize cargo to the 1,2-propanediol utilization microcompartment. *Protein Sci., 26*, 1086–1092.

Jeter, R. M., (1990). Cobalamin-dependent 1,2-propanediol utilization by *Salmonella typhimurium*. *J. Gen. Microbiol., 136*, 887–896.

Jorda, J., Lopez, D., Wheatley, N. M., & Yeates, T. O., (2013). Using comparative genomics to uncover new kinds of protein-based metabolic organelles in bacteria. *Protein Sci., 22*, 179–195.

Jordan, P. C., Patterson, D. P., Saboda, K. N., Edwards, E. J., Miettinen, H. M., Basu, G., Thielges, M. C., & Douglas, T., (2016). Self-assembling biomolecular catalysts for hydrogen production. *Nat. Chem., 8*, 179–185.

Kaval, K. G., & Garsin, D. A., (2018). Ethanolamine utilization in bacteria. *MBio., 9*, pii: e00066–18.

Kerfeld, C. A., Aussignargues, C., Zarzycki, J., Cai, F., & Sutter, M., (2018). Bacterial microcompartments. *Nat. Rev. Microbiol., 16*, 277–290.

Kerfeld, C. A., Heinhorst, S., & Cannon, G. C., (2010). Bacterial microcompartments. *Annu. Rev. Microbiol., 64*, 391–408.

Kerfeld, C. A., Sawaya, M. R., Tanaka, S., Nguyen, C. V., Phillips, M., Beeby, M., & Yeates, T. O., (2005). Protein structures forming the shell of primitive bacterial organelles. *Science, 309*, 936–938.

Kim, E. Y., & Tullman-Ercek, D., (2014). A rapid flow cytometry assay for the relative quantification of protein encapsulation into bacterial microcompartments. *Biotechnol. J., 9*, 348–354.

Kim, E. Y., Slininger, M. F., & Tullman-Ercek, D., (2014). The effects of time, temperature, and pH on the stability of PDU bacterial microcompartments. *Protein Sci., 23*, 1434–1441.

Kinney, J. N., Salmeen, A., Cai, F., & Kerfeld, C. A., (2012). Elucidating essential role of conserved carboxysomal protein CcmN reveals common feature of bacterial microcompartment assembly. *J. Biol. Chem., 287*, 17729–17736.

Kirst, H., & Kerfeld, C. A., (2019). Bacterial microcompartments: Catalysis-enhancing metabolic modules for next-generation metabolic and biomedical engineering. *BMC Biol., 17*, 79.

Klein, M. G., Zwart, P., Bagby, S. C., Cai, F., Chisholm, S. W., Heinhorst, S., Cannon, G. C., & Kerfeld, C. A., (2009). Identification and structural analysis of a novel carboxysome shell protein with implications for metabolite transport. *J. Mol. Biol., 392*, 319–333.

Kofoid, E., Rappleye, C., Stojiljkovic, I., & Roth, J., (1999). The 17-gene ethanolamine (eut) operon of *Salmonella typhimurium* encodes five homologues of carboxysome shell proteins. *J. Bacteriol., 181*, 5317–5329.

Lassila, J. K., Bernstein, S. L., Kinney, J. N., Axen, S. D., & Kerfeld, C. A., (2014). Assembly of robust bacterial microcompartment shells using building blocks from an organelle of unknown function. *J. Mol. Biol., 426*, 2217–2228.

Lawrence, A. D., Frank, S., Newnham, S., Lee, M. J., Brown, I. R., Xue, W. F., Rowe, M. L., et al., (2014). Solution structure of a bacterial microcompartment targeting peptide and its application in the construction of an ethanol bioreactor. *ACS Synth. Biol., 3*, 454–465.

Lee, H., DeLoache, W. C., & Dueber, J. E., (2012). Spatial organization of enzymes for metabolic engineering. *Metab. Eng., 14*, 242–251.

Lee, M. F. S., Jakobson, C. M., & Tullman-Ercek, D., (2017). Engineered shell proteins confer improved encapsulated pathway behavior in a bacterial microcompartment. *BioRxiv., 106716*.

Lee, M. J., Brown, I. R., Juodeikis, R., Frank, S., & Warren, M. J., (2016). Employing bacterial microcompartment technology to engineer a shell-free enzyme-aggregate for enhanced 1,2-propanediol production in *Escherichia coli*. *Metab. Eng., 36*, 48–56.

Lee, M. J., Palmer, D. J., & Warren, M. J., (2019). Biotechnological advances in bacterial microcompartment technology. *Trends Biotechnol., 37*, 325–336.

Lehman, B. P., Chowdhury, C., & Bobik, T. A., (2017). The N terminus of the PduB Protein binds the protein shell of the Pdu microcompartment to its enzymatic core. *J. Bacteriol., 199*, pii: e00785–16.

Li, H., Zheng, G., & Zhu, S., (2018). Construction of an organelle-like nanodevice via supramolecular self-assembly for robust biocatalysts. *Microb. Cell Fact., 17*, 26.

Lin, M. T., Occhialini, A., Andralojc, P. J., Devonshire, J., Hines, K. M., Parry, M. A. J., & Hanson, M. R., (2014). β-Carboxysomal proteins assemble into highly organized structures in *Nicotiana* chloroplasts. *Plant J., 79*, 1–12.

Littlechild, J. A., (2015). Enzymes from extreme environments and their industrial applications. *Front. Bioeng. Biotechnol., 3*, 161.

Long, B. M., Hee, W. Y., Sharwood, R. E., Rae, B. D., Kaines, S., Lim, Y. L., Nguyen, N. D., et al., (2018). Carboxysome encapsulation of the CO_2-fixing enzyme rubisco in tobacco chloroplasts. *Nat. Commun., 9*, 3570.

Lopez-Gallego, F., & Schmidt-Dannert, C., (2010). Multi-enzymatic synthesis. *Curr. Opin. Chem. Biol., 14,* 174–183.

Mayer, M. J., Juodeikis, R., Brown, I. R., Frank, S., Palmer, D. J., Deery, E., Beal, D. M., et al., (2016). Effect of bio-engineering on size, shape, composition and rigidity of bacterial microcompartments. *Sci. Rep., 6,* 36899.

Mohajerani, F., & Hagan, M. F., (2018). The role of the encapsulated cargo in microcompartment assembly. *PLoS Comput. Biol., 14*, e1006351.

Moore, T. C., & Escalante-Semerena, J. C., (2016). The EutQ and EutP proteins are novel acetate kinases involved in ethanolamine catabolism: Physiological implications for the function of the ethanolamine metabolosome in *Salmonella enterica*. *Mol. Microbiol., 99*, 497–511.

Obradors, N., Badia, J., Baldoma, L., & Aguilar, J., (1988). Anaerobic metabolism of the L-rhamnose fermentation product 1,2-propanediol in Salmonella typhimurium. *J. Bacteriol., 170*, 2159–2162.

Pang, A., Liang, M., Prentice, M. B., & Pickersgill, R. W., (2012). Substrate channels revealed in the trimeric *Lactobacillus reuteri* bacterial microcompartment shell protein PduB. *Acta Crystallogr. D Biol. Crystallogr., 68*, 1642–1652.

Pang, A., Warren, M. J., & Pickersgill, R. W., (2011). Structure of PduT, a trimeric bacterial microcompartment protein with a 4Fe-4S cluster-binding site. *Acta Crystallogr. D Biol. Crystallogr., 67*, 91–96.

Parsons, J. B., Dinesh, S. D., Deery, E., Leech, H. K., Brindley, A. A., Heldt, D., Frank, S., et al., (2008). Biochemical and structural insights into bacterial organelle form and biogenesis. *J. Biol. Chem., 283*, 14366–14375.

Parsons, J. B., Frank, S., Bhella, D., Liang, M., Prentice, M. B., Mulvihill, D. P., & Warren, M. J., (2010). Synthesis of empty bacterial microcompartments, directed organelle protein incorporation, and evidence of filament-associated organelle movement. *Mol. Cell, 38*, 305–315.

Penrod, J. T., & Roth, J. R., (2006). Conserving a volatile metabolite: A role for carboxysome-like organelles in Salmonella enterica. *J. Bacteriol., 188*, 2865–2874.

Petit, E., LaTouf, W. G., Coppi, M. V., Warnick, T. A., Currie, D., Romashko, I., Deshpande, S., et al., (2013). Involvement of a bacterial microcompartment in the metabolism of fucose and rhamnose by *Clostridium phytofermentans*. *PLoS One, 8*, e54337.

Price, G. D., Badger, M. R., Woodger, F. J., & Long, B. M., (2008). Advances in understanding the cyanobacterial CO_2-concentrating-mechanism (CCM): Functional components, Ci transporters, diversity, genetic regulation and prospects for engineering into plants. *J. Exp. Bot., 59*, 1441–1461.

Price-Carter, M., Tingey, J., Bobik, T. A., & Roth, J. R., (2001). The alternative electron acceptor tetrathionate supports B12-dependent anaerobic growth of *Salmonella enterica* serovar Typhimurium on ethanolamine or 1,2-propanediol. *J. Bacteriol., 183*, 2463–2475.

Quin, M. B., Perdue, S. A., Hsu, S. Y., & Schmidt-Dannert, C., (2016). Encapsulation of multiple cargo proteins within recombinant Eut nanocompartments. *Appl Microbiol Biotechnol., 100*, 9187–9200.

Rae, B. D., Long, B. M., Badger, M. R., & Price, G. D., (*2013*). Functions, compositions, and evolution of the two types of carboxysomes: Polyhedral microcompartments that facilitate CO_2 fixation in cyanobacteria and some proteobacteria. *Microbiol. Mol. Biol. Rev., 77*, 357–379.

Rondon, M. R., Horswill, A. R., & Escalante-Semerena, J. C., (1995). DNA polymerase I function is required for the utilization of ethanolamine, 1,2-propanediol, and propionate by *Salmonella typhimurium* LT2. *J. Bacteriol., 177*, 7119–7124.

Sagermann, M., Ohtaki, A., & Nikolakakis, K., (2009). Crystal structure of the EutL shell protein of the ethanolamine ammonia-lyase microcompartment. *Proc. Natl. Acad. Sci. U. S. A., 106*, 8883–8887.

Sampson, E. M., & Bobik, T. A., (2008). Microcompartments for B12-dependent 1,2-propanediol degradation provide protection from DNA and cellular damage by a reactive metabolic intermediate. *J. Bacteriol., 190*, 2966–2971.

Sargent, F., Davidson, F. A., Kelly, C. L., Binny, R., Christodoulides, N., Gibson, D., Johansson, E., et al., (2013). A synthetic system for expression of components of a bacterial microcompartment. *Microbiology, 159*, 2427–2436.

Savage, D. F., Afonso, B., Chen, A. H., & Silver, P. A., (2010). Spatially ordered dynamics of the bacterial carbon fixation machinery. *Science, 327*, 1258–1261.

Schmid, M. F., Paredes, A. M., Khant, H. A., Soyer, F., Aldrich, H. C., Chiu, W., & Shively, J. M., (2006). Structure of *Halothiobacillus neapolitanus* carboxysomes by cryo-electron tomography. *J. Mol. Biol., 364*, 526–535.

Shively, J. M., Ball, F., Brown, D. H., & Saunders, R. E., (1973). Functional organelles in prokaryotes: Polyhedral inclusions (carboxysomes) of *Thiobacillus neapolitanus*. *Science, 182*, 584–586.

Sinha, S., Cheng, S., Fan, C., & Bobik, T. A., (2012). The PduM protein is a structural component of the microcompartments involved in coenzyme B(12)-dependent 1,2-propanediol degradation by *Salmonella enterica*. *J. Bacteriol., 194*, 1912–1918.

Slininger, L. M. F., Jakobson, C. M., & Tullman-Ercek, D., (2017). Evidence for improved encapsulated pathway behavior in a bacterial microcompartment through shell protein engineering. *ACS Synth. Biol., 6*, 1880–1891.

Sutter, M., Faulkner, M., Aussignargues, C., Paasch, B. C., Barrett, S., Kerfeld, C. A., & Liu, L. N., (2016). Visualization of bacterial microcompartment facet assembly using high-speed atomic force microscopy. *Nano Lett., 16*, 1590–1595.

Sutter, M., Greber, B., Aussignargues, C., & Kerfeld, C. A., (2017). Assembly principles and structure of a 6.5-MDa bacterial microcompartment shell. *Science, 356*, 1293–1297.

Takenoya, M., Nikolakakis, K., & Sagermann, M., (2010). Crystallographic insights into the pore structures and mechanisms of the EutL and EutM shell proteins of the ethanolamine-utilizing microcompartment of Escherichia coli. *J. Bacteriol., 192*, 6056–6063.

Talarico, T. L., Axelsson, L. T., Novotny, J., Fiuzat, M., & Dobrogosz, W. J., (1990). Utilization of glycerol as a hydrogen acceptor by *Lactobacillus reuteri*: Purification of 1,3-propanediol: NAD oxidoreductase. *Appl. Environ. Microbiol., 56*, 943–948.

Tanaka, S., Kerfeld, C. A., Sawaya, M. R., Cai, F., Heinhorst, S., Cannon, G. C., & Yeates, T. O., (2008). Atomic-level models of the bacterial carboxysome shell. *Science, 319*, 1083–1086.

Tanaka, S., Sawaya, M. R., & Yeates, T. O., (2010). Structure and mechanisms of a protein-based organelle in *Escherichia coli*. *Science, 327*, 81–84.

Tanaka, S., Sawaya, M. R., Phillips, M., & Yeates, T. O., (2009). Insights from multiple structures of the shell proteins from the beta-carboxysome. *Protein Sci., 18*, 108–120.

Thompson, M. C., Cascio, D., Leibly, D. J., & Yeates, T. O., (2015). An allosteric model for control of pore opening by substrate binding in the EutL microcompartment shell protein. *Protein Sci., 24*, 956–975.

Thompson, M. C., Crowley, C. S., Kopstein, J., Bobik, T. A., & Yeates, T. O., (2014). Structure of a bacterial microcompartment shell protein bound to a cobalamin cofactor. *Acta. Crystallogr. F Struct. Biol. Commun., 70*, 1584–1590.

Tsai, S. J., & Yeates, T. O., (2011). Bacterial microcompartments insights into the structure, mechanism, and engineering applications. *Prog. Mol. Biol. Transl. Sci., 103*, 1–20.

Yeates, T. O., Jorda, J., & Bobik, T. A., (2013). The shells of BMC-type microcompartment organelles in bacteria. *J. Mol. Microbiol. Biotechnol., 23*, 290–299.

Yeates, T. O., Kerfeld, C. A., Heinhorst, S., Cannon, G. C., & Shively, J. M., (2008). Protein-based organelles in bacteria: Carboxysomes and related microcompartments. *Nat. Rev. Microbiol., 6*, 681–691.

Yeates, T. O., Thompson, M. C., & Bobik, T. A., (2011). The protein shells of bacterial microcompartment organelles. *Curr. Opin. Struct. Biol., 21*, 223–231.

Yung, M. C., Bourguet, F. A., Carpenter, T. S., & Coleman, M. A., (2017). Re-directing bacterial microcompartment systems to enhance recombinant expression of lysis protein E from bacteriophage varphiX174 in Escherichia coli. *Microb. Cell. Fact., 16*, 71.

Zarzycki, J., Erbilgin, O., & Kerfeld, C. A., (2015). Bioinformatic characterization of glycyl radical enzyme-associated bacterial microcompartments. *Appl. Environ. Microbiol., 81*, 8315–8329.

CHAPTER 11

Management of Microbiological Hazards in the Food Processing Industries

TEJPAL DHEWA* and ANU KUMAR

**Corresponding author. E-mails: tejpal_dhewa07@rediffmail.com; tejpaldhewa@cuh.ac.in*

ABSTRACT

The occurrence of microbiological hazards in food industries is considered a serious and continuing problem at the global level, including primary production, manufacturers, retailers, and foodservice. The microbiological hazards not only causing illnesses to consumers but also adversely affecting the business of food industries. In recent times, food safety has been the prime objective of food manufactures, and several control measures are available with manufacturers. The selection of an appropriate and effective control method is an emergent requirement due to avoidance of the antimicrobial resistance problem as well as reduction of chemical burden in the environment. The proposed chapter will be focused on the technological advancement for managing potent microbiological hazards in various food processing industries. This chapter will also address both success stories in food safety and problems remaining to be dealt with. Moreover, the roles of different stakeholders and public distribution systems how food safety assurance can be achieved, which problems are to be prevented, and in particular, what type of knowledge is required for managing the production of safe foods for consumers.

11.1 INTRODUCTION

This chapter is dedicated to highlighting the challenges of food products because of contamination of foodstuff microbiologically and chemically

(Table 11.1). Food contamination is generally defined as foods that are spoiled by microbes such as bacteria, fungi, or parasites. Contamination is also due to some toxic substances that can make food unsafe for human consumption. In the meantime, it is emerging as a major problem throughout the world. If we talk about food contaminant that can be biological contaminant, chemical or physical contaminant (Figure 11.1) but former is regarded as the more common one. There are many routes through which such contaminants can enter into the food chain and convert food into a form that is unsafe for consumption. Here are some examples of common bacterial hazards depicted in Table 11.1. Another type of food contaminants such as Chemical food contaminants, e.g., pesticides, and heavy metals can also enter the food chain and convert foodstuff unfit and unsafe for mankind.

TABLE 11.1 Contamination of Foods Through Microbiological and Chemical Hazards

Type of Hazards	Examples of Different Class of Hazards
Microbiological Hazards	
Bacteria	*Bacillus cereus, Campylobacter jejuni, Clostridium botulinum, C. perfringens, Cronobacter sakazakii, Escherichia coli* 0104:H4, *E. coli* 0157:H7, *Listeria monocytogenes, Salmonella* spp., *Shigella* spp., *Staphylococcus aureus, Vibrio cholerae, V. parahaemolyticus, V. vulnificus, Yersinia enterocolitica*, etc.
Viruses	Bacteriophage, enteric virus (other than hepatitis A and noroviruses), hepatitis A virus, norovirus, Norwalk virus, Rotavirus, etc.
Parasites	*Cryptosporidium parvum, Entamoeba coli, E. histolytica, Trichinella spiralis*, etc.
Chemical Hazards	
Mycotoxins	Aflatoxin, deoxynivalenol (vomitoxin), fumonisin patulin and ochratoxin A, etc.
Natural toxins	Glycoalkaloids and other natural toxins *viz.* amanitins, cyanogenic furocoumarins, glycoside, goitrogens, gyromitrin, hypoglycin, lectin, muscarine, oxalic acid, phallotoxins,
Marine toxins	Decomposition and microscopic marine algae *viz.* brevetoxin, ciguatoxin, domoic acid, okadaic acid, saxitoxin, tetrodotoxin, etc.
Environmental contaminants	Arsenic, cadmium, lead, and mercury, etc.
Food additives	Bleaching agents, emulsifiers, food colors (natural and synthetic), food enzymes, gelling agents, glazing, and polishing agents, pH adjusting agents and preservatives, etc.

TABLE 11.1 *(Continued)*

Type of Hazards	Examples of Different Class of Hazards
Processing-induced chemicals	Acrylamide, ethyl carbamate (urethane) and furan, etc.
Pesticides/agricultural products	Azoxystrobin on peaches, Captan on cherries, clethodim on beans and thiocarbamate on apples, etc.
Veterinary drugs	–

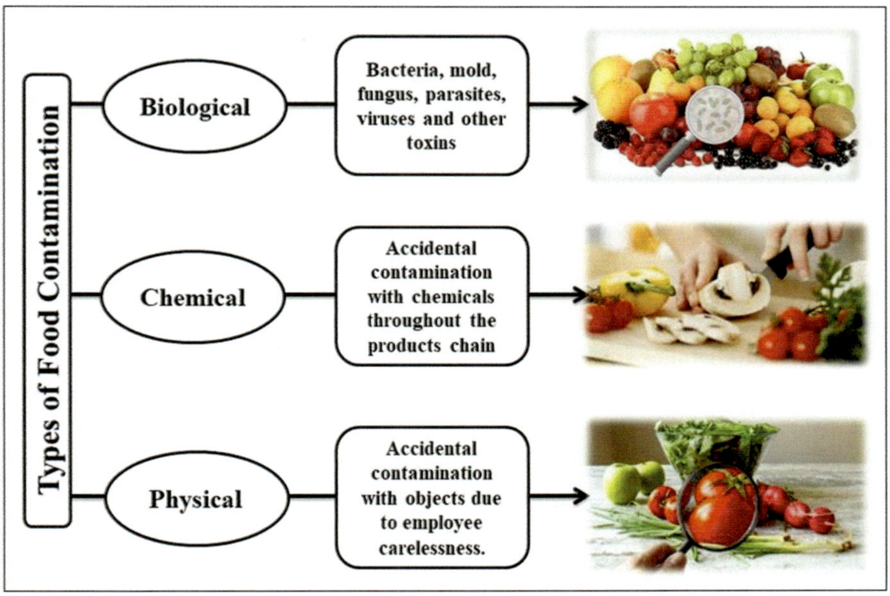

FIGURE 11.1 Food contamination and their types.

Contamination of food has been regarded as a worldwide challenge by World Health Organization (WHO) in several reports (Fukuda, 2015). Food contamination occurring in one place may adversely affect the normal life style and bad impact on the health of consumers residing on another side. In fact, several people are suffering from airborne diseases, food borne diseases or waterborne diseases. Due to this reason, a large number of people dying every year and as a result of consumption of toxic and contaminated food is becoming a leading cause of illness among millions of people all over the world. Presently food contamination is becoming a major threat throughout the world, and such challenges are enhancing day by day, and this list of food contamination challenges keeps growing.

The aim of the chapter entitled "management of microbiological hazards in food processing industries" to focus on possible sources of contamination, food handlers, and hygiene practices, approaches for management of microbiological hazards, International GMPs (good manufacturing practices), and HACCP (hazard analysis and critical control point), International food safety and quality standards, sanitary surveillance, safe food programs, and also to focus on the work, findings, and results of many researchers what they were confirmed in their experiments and what they were suggested for managing microbiological hazards in food industries.

11.2 POSSIBLE SOURCES OF CONTAMINATION

Literature indicates that the muscle tissue of a healthy living animal is free of microorganisms and the under the skin of animal carcass becomes sterile immediately after slaughter (McEvoy et al., 2000). Hence, contamination of raw meat maybe because of the slaughtering of stressed animals, as well as contact with external surfaces such as hair, gastrointestinal (GI), and respiratory tracts and/or other ambient environmental hazards. In the abattoir, contamination occurs with the microorganisms' introduction to direct meat contact with surfaces in operations performed during offloading, weighing, processing, cutting, and storage, as well as at the points of sale and distribution (Nørrung and Buncic, 2008; Sofos, 2008; Ali et al., 2010).

Further, Bas et al. (2006) reported that pathogens are transmitted passively from a source of contamination such as raw poultry to cooked food that is prepared for later consumption as cold foods. Typical microorganisms that are usually prevalent in raw meat include *Listeria monocytogenes*, *Salmonella*, *Staphylococcus aureus*, *Campylobacter (*on poultry), *Escherichia coli*, and *Escherichia coli* 0157: H7, amongst others (Ateba et al., 2008).

The unhygienic practices such as handling of raw materials in unhygienically, insufficient, and improper washing of the machines, knives, clothes, airborne microorganisms declared as the actual source of the contamination of the street foodstuff (Sousa, 2008).

11.3 FOOD HANDLERS AND RELATED HYGIENE PRACTICES

Meat cutting is important in meat processing as carcasses are deboned and cut into smaller and more desirable cuts, using hand tools and machines (Wang and Shanmugan, 2009). Hence, contamination of meat and risk behind this is becoming a serious issue and imposes a bad impact on the health of personnel handling foodstuff, their personal hygiene, and whatever their knowledge and awareness about the practice of food hygiene.

According to Norrung and Buncic (2008), the process of meat handling increases the possibility of microbial contamination because unhygienic practices during handling may lead to transmission of bacteria to the meat from working and other surfaces. Several studies have further indicated that foodborne illnesses occur due to poor handling of food (Van Tonder, 2004; Griffith, 2006).

Staphylococcus-related food poisoning has been linked to food handlers who are known to be carriers of this bacterium (in their skin, infected cuts, nose, throat, etc.), in meat establishments (Van Tonder, 2004). Besides, (Kusumaningrum et al., 2002) reported that many bacteria such as *Staphylococcus aureus*, *Escherichia coli*, and *Salmonella* spp., resides on hands and surfaces for hours or even days after initial contact. Besides, 97% of food consumers' illnesses in the USA were linked with irrelevant food handlers' practices in the foodservice industry. Food handlers may on some occasions serve as sources of contamination, especially as a result of some having GI illnesses or being in the convalescence phase when their symptoms have disappeared.

11.4 APPROACHES FOR MANAGEMENT OF MICROBIOLOGICAL HAZARDS

11.4.1 GOOD MANUFACTURING PRACTICES (GMPS), AND HAZARD ANALYSIS AND CRITICAL CONTROL POINT (HACCP)

In food industries, hazards management systems dealt with a number of relevant activities in the quality control, quality assurance department, and also evaluating the performance of the implemented actions. However, there are many relevant quality and food safety management systems are available based on GMPs and HACCP, adopted successfully in the food industry in Brazil (Tondo et al., 2015).

In order to prevent contamination of food items Good Agricultural and GMPs are essential. *Salmonella* spp. basically involved in contamination of foodstuff (CDCP, 2015). The emergence of drug resistance is also a major concern and the cross-contamination and antibiotic resistance must be taken as a serious concern in the food production chain, as studied by Capalonga et al. (2014) during a report of *S. enteritidis* known for spreadness of foodborne outbreaks in southern Brazil.

HACCP is relevant in order to ensure the safety of foodstuff. Management commitment is a key factor for a successful program and teamwork approach with full of qualified workers should be utilized for execution of a program work. HACCP is a popular science-based system for maintenance of cost-effective ongoing food safety program. All the stages which are involved in production of food and preparation can be evaluated properly and stages which are essential for safety of food products must be rectified systematically. By implementing HACCP, every stage which is involved in the production of food and preparation can be assessed systematically and all stages are crucial in terms of products safety are rectified systematically (Gaze, 2009). HACCP is primarily implemented for monitoring of food safety and give focus on identification and management of risks in food chains (the "farm-to-fork," or "plow-to-plate" approach) (Mortimore, 2000; Humphrey and Memedovic, 2006). Thus, identification, and reduction of potential hazards is the actual objective of the risk management.

Product should be traced or detected initially from manufacturer up to the consumer during the entire manufacturing process the product. That is why; HACCP is termed as a structured and a preplanned preventive system for identification of hazards during each step of the process. The important components of the regulations in the food industry to treating potential hazards are categorized in three main stages (Figure 11.2). The overall assessment process is as follows: (1) harvest cultivation, (2) collection, (3) transportation, (4) manufacturing processing, (5) packaging, (6) storage, and (7) supply to customers.

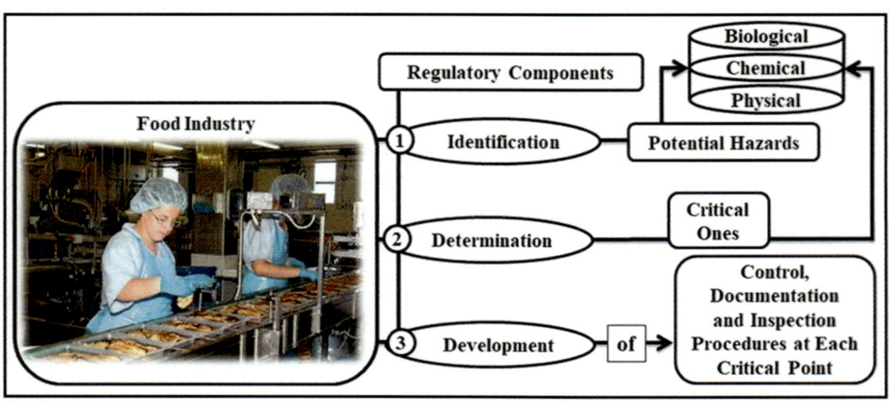

FIGURE 11.2 Important components of regulation for the food industry.

Basically, the process is usually too lengthy and tedious for any detailed evaluation. So, a better approach would be from cultivating to transporting products or in other terms we can say that from starting point to the product stage when product is at marketing distribution level. A very first and identification step is to draw a diagram of all stages of the process. Hazards identification should be applied at each and every stage to have their unique identification number. Hazards may be biological, chemical, or physical in nature and might be having harmful potential. Applications of HACCP principles in food industry are described in Figure 11.3.

Important Principles of HACCP		
Principles	**Objectives**	**Remarks**
Principle-1	Conduct a Hazard Analysis	Involves listing the steps in the process and identifying where significant hazards are likely to Occur
Principle-2	Identify the Critical Control Points	Point, step or procedure at which control can be applied and a food safety hazard can be prevented, eliminated or reduced to acceptable levels.
Principle-3	Establish Critical Limits	Maximum and/or minimum value to which a biological, chemical, or physical parameter must be controlled at a CCP to prevent, eliminate, or reduce to an acceptable level the occurrence of a food safety hazard.
Principle-4	Monitor CCP	HACCP team will describe monitoring procedures for the measurement of the critical limit at each critical control point.
Principle-5	Establish Corrective Action	The procedures that are followed when a deviation in a critical limit occurs.
Principle-6	Verification	Other than monitoring, that determine the validity of the HACCP plan and that the system is operating according to the plan.
Principle-7	Recordkeeping	A key component of the HACCP plan is recording information that can be used to prove that the a food was produced safely.

FIGURE 11.3 Applications of seven important principles of HACCP in food industry (UNL, 2019).

Source: For more details, readers must refer to the UNL (University of Nebraska-Lincoln) (2019). Principles of HACCP. HACCP documents and links. URL: https://food.unl.edu/seven-principles-haccp.

The most relevant motive of HACCP system is to control risks related to human health, as well as reduction of foodstuff adulteration along with implementing risk control measures during every production step which might be hazardous to human health most probably. HACCP may be applied for plant and animal cultivation through primary production for successful processing and manufacturing of foodstuff by the food industry. The followed by finishing of the final product and consumption by the consumer. But it has been recommended to establish GMPs and its Standard Operating Procedures before the implementing HACCP (Oliveira et al., 2016).

11.4.2 INTERNATIONAL FOOD SAFETY AND QUALITY STANDARDS

Emergence of novel technologies have helped to improve the quality of foodstuff worldwide Also, import, export, and foodborne outbreaks have helped in this regard. Maintenance of quality of food items is generally a vital requirement and international food safety and quality standards must be used as a reference for maintaining the quality of food, such as WHO, the Food and Agriculture Organization of the United Nations (FAO), and the World Trade Organization (WTO) (Zach et al., 2012). The WHO is indulged in monitoring foodborne disease outbreaks and run and sponsors different health care programs to prevent foodborne diseases (Zach et al., 2012). As far as a concern with the health of mankind Food safety is termed as an essential public health function by the WHO (WHO, 2016a). In 1963, Joint FAO/WHO Food safety Program was established, has the Codex Alimentarius Commission that is responsible for establishing international food standards as a reference for ensuring food quality and safety, and ultimately improving the trade relation between countries (FAOUN, 2016c).

11.4.2.1 AUDITING PROCESS

The word "audit" has been taken from the Latin word "auditus," which means hearing. The audit procedure usually includes reviewing the documents and conducting checks and interviews at regular intervals to confirm compliance with a standard. Although traditional procedures of audits are available for evaluation of the structure and construction of the sites. Such procedures are also meant for visual assessment cleanliness condition, but such procedures may not always be the best indicators. The inspection of the whole manufacturing process from a raw material through preparation to services can be achieved successfully under auditing process (Griffith, 2005). Audits are really very much essential and transparent tools in order to maintain food safety standards, any relevant certification, assuring that the standards are maintained. This transparency can enhance the opportunities regarding collaboration of stakeholders within the supply chain, and it will automatically improve safety and efficiency during the promotion of the continual improvement (GFSR, 2016).

11.4.2.2 THE ROLE OF SAFETY AND QUALITY CONTROL SYSTEMS

Food crisis is the main reason behind the popularity of food quality and safety in the media. In most of the time, terms food quality and food safety are not clearly discriminated. Leading food industries involved in the development of quality management systems for the better improvement of food safety and monitoring processes (Rohr et al., 2005). According to Holleran et al. (1999), "quality assurances systems are designed for assurance of customers that production processes are consistently delivered.

HACCP assessment is a valuable tool for verifying manufacturing of safe products and their distribution. Two types of assessments under HACCP used by manufacturers, one is in house assessment and other by using external experts. The paper of Poksinska et al. (2006) analyzed the study, i.e., survey report related to auditing process which was conducted in 269 ISO 9001:2000-certified organizations in Sweden. Findings depicted the claim of studied organizations about certification audits and such auditing procedure assists them to improve their QMSs as well as in increasing motivation for quality work.

11.4.3 SANITARY SURVEILLANCE

The local population is lacking in knowledge regarding good practices for the processing of foodstuff and this must be an urgent priority to control contamination. Various competent authorities are quite indulged in this regard and have been still working hard on a check system known as sanitary surveillance to prevent contamination in the food production chain and to control contamination by *T. cruzi* (Fiocruz, 2013). Sanitary surveillance is basically indulged in promoting health and safety concerns (Silva ACP et al., 2012). The actions of which have been carried out since the 19th century, changing to adapt themselves to the changes in the economic models and with social development (Lucchese et al., 2008).

In 2003, surveillance on world-level agenda regarding Influenza was conducted and control measures constructed for minimizing morbidity and mortality secondary to the annual influenza epidemics (Ibiapina et al., 2005). A poultry surveillance system was formulated in evidence ensuring food safety by diagnosis of the influenza A virus at a very early stage that is much critical (Ping et al., 2015). Information gained through surveillance system under national intervention plan indicated that H3N2 and H1N2, rather than H5N1, circulated in Brazil in 2003 (Ibiapina et al., 2005).

11.4.4 FOOD SERVICES AND SAFE FOOD PROGRAM

Hygienic practices must be adopted in concerns with food served to young ones in primary care centers. Care must be taken in concerns with meals served in schools at child care centers and nursery schools. Young ones are more susceptible to food borne pathogens. Keeping this in view, maintenance of microbiological quality of foodstuff to ensure their safety is of major concern which can reduce instances of emerging foodborne diseases (FDA, 2016a).

Many Food Safety Management Systems are there in the food industry investigating the performance of the implemented actions. GMPs and HACCP are regarded as well-known Food Safety Management Systems and are available as a relevant tool for quality management in the food processing industry in Brazil (Tondo et al., 2015). The main objectives of this program are circulating food safety tools among various food industries (Tondo et al., 2015).

Various international organizations such as WHO, the FAO of the United Nations, and the WTO are responsible for establishing and maintaining food quality and safety standards (Zach et al., 2012). WHO denoted safety of food as a vital and relevant public health function (WHO, 2016a).

Safety of foods is ensured by local agencies, federal In the USA as these are taking care for safety. At present food safety programs and regulatory affairs has been recently reconstructed in European Union (EU), and is fully regulated by EU legislation which is applicable to EU member countries (Chen et al., 2015).

11.5 SUMMARY AND CONCLUSION

The establishment of various strategies must be needed for maintenance of food quality and this is also important for ensuring the safety of this foodstuff. We are just concluding here a solution to overcome challenges regarding food safety by the adoption of good practices, SOPs, and HACCP in this country. Literature survey also depicted the growing need to investigate personnel hygiene practices and food handlers and to monitoring the methods of conducting their daily routine. Focus should be there on eradicating contamination rather than spreading it. Legal requirements must adapt for specific production systems from the lab scale to pilot scale industries, producers. Effective training of producers is very much necessary by competent authorities so that regulations can be effectively executed. However, adopting food safety programs and quality management many places of the world is a challenging concern due to some factors, e.g., unawareness about food safety regulations and their importance, lack of their training in safe food program and also irregular inspection by competent authorities. Effective awareness regarding food safety among population is much needed and persons should be educated about the knowledge of food services and safe food program. Various food safety and management programs adopted by training local leaders in GMP and HACCP practices. This can overcome this issue.

KEYWORDS

- **antimicrobial resistance**
- **food adulteration**
- **food contamination**
- **food hygiene**
- **food monitoring**
- **food poisoning**
- **food processing**
- **food safety**
- **hazard analysis**
- **microbiological hazards**

REFERENCES

Ali, N. H., Farooqui, A., Khan, A., Khan, Y. K., & Kazmi, S. H., (2010). Microbial contamination of raw meat and its environment in retail shops in Karachi, Pakistan. *Journal of Infection in Developing Countries, 4*(6), 382–388.

Ateba, C. N., Mbewe, M., & Bezuidenhout, C. C., (2008). Prevalence of *Escherichia Coli* O157:H7 strains in cattle, pigs and humans in North West Province, South Africa. *South African Journal of Science, 104*(1, 2), 7–8.

Bas, M., Ersun, A. S., & Kivanc, G., (2006). The evaluation of food hygiene knowledge, attitudes, and practices of food handlers' in food businesses in Turkey. *Food Control, 17*(4), 317–322.

Capalonga, R., Ramos, R. C., Both, J. M. C., Soeiro, M. L. T., Longaray, S. M., Haas, S., & Tondo, E. C., (2014). Salmonella serotypes, resistance patterns, and food vehicles of salmonellosis in southern Brazil between 2007 and 2012. *Journal of Infection in Developing Countries, 8*, 811–817.

CDCP (Center for Disease Control and Prevention), (2015). *Salmonella: Prevention.* URL: https://www.cdc.gov/salmonella/general/prevention.html (accessed on 27 September 2021).

Chen, K., Wang, X. X., & Song, H. Y., (2015). Food safety regulatory systems in Europe and China: A study of how co-regulation can improve regulatory effectiveness. *Journal of Integrative Agriculture, 14*, 2203–2217.

FAOUN (Food and Agriculture Organization of the United Nations). (2016c). *Codex Alimentarius International Food Standards.* URL: https://www.fao.org/fao-who-codexalimentarius/en/ (accessed on 27 September 2021).

FDA (Food and Drug Administration). (2016a). *Food Safety: Importance for At-Risk Groups.* URL: https://www.fda.gov/food/consumers/people-risk-foodborne-illness (accessed on 27 September 2021).

FIOCRUZ (Fundação Osvaldo Cruz). (2013). Chagas Disease Doença de Chagas. Agência Fiocruz de Notícias. URL: https://agencia.fiocruz.br/doen%C3%A7a-de-chagas (accessed on 27 September 2021).

Fukuda, K., (2015). Food safety in a globalized world. *Bull. WHO, 93*, 212.

Gaze, R., (2009). *HACCP: A Practical Guide* (4th edn. pp. 1–88). Chipping Campden: Campden BRI.

GFSR, (2016). *Food Safety Audits.* URL: http://globalfoodsafetyresource.com/food-safety/food-safety-audits# (accessed on 27 September 2021).

Griffith, C. G., (2005). Are we making the most of food safety inspections? A glimpse into the future. *Br. Food J., 107*, 132–139.

Griffith, C. J., (2006). Food safety: Where from and where to? *British Food Journal, 108*(1), 6–15.

Holleran, E., Bredahl, M. E., & Zaibet, L., (1999). Private incentives for adopting food safety and quality assurance. *Food Policy, 24*, 669–683.

Humphrey, J., & Memedovic, O., (2006) *Global Value Chains in the Agrifood Sector* (pp. 1–60). Vienna, Austria: United Nations Industrial Development Organization.

Ibiapina, C. C., Costa, G. A., & Faria, A. C., (2005). Avian influenza A (H5N1) – the bird flu. *Jornal Brasileiro de Pneumologia, 31*, 436–444.

Kusumaningrum, H. D., Van, P. M. M., Rombouts, F. M., & Beumer, R. R., (2002). Effects of antibacterial dishwashing liquid on food borne pathogens and competitive microorganisms in kitchen sponges. *Journal of Food Protection, 65*, 61–65.

Lucchese, G., (2008). Health surveillance in the single health system. In: De Seta, M. H., Pepe, V. E., & Oliveira, G. O., (eds.), Organizers (pp. 33–47). Health Management and Surveillance: Ways current ways of thinking and doing. Rio de Janeiro: Fiocruz.

McEvoy, J. M., Doherty, A. M., Finnerty, M., Sheridan, J. J., McGuier, L., Blair, I. S., McDowell, D. A., & Harrington, D., (2000). The relationship between hide cleanliness and bacterial numbers on beef carcasses at a commercial abattoir. *Letters in Applied Microbiology, 30*(5), 390–395.

Mortimore, S., (2000). "An example of some procedures used to assess HACCP systems within the food manufacturing industry. *Food Control, 11*, 403–413.

Nørrung, B., & Buncic, S., (2008). Microbial safety of meat in the European Union. *Meat Science, 78*, 14–24.

Oliveira, C. A. F. D., Cruz, A. G. D., Tavolaro, P., & Corassin, C. H., (2016). Food safety: Good manufacturing practices (GMP), sanitation standard operating procedures (SSOP), hazard analysis and critical control point (HACCP). In: *Antimicrobial Food Packaging* (pp. 129–139).

Ping, S., Geng, S., Ting-ting, L., Yu-shui, L., Ting, F., & Hua-nan, W., (2015). Methods to detect avian influenza virus for food safety surveillance. *Journal of Integrative Agriculture, 14*, 2296–2308.

Poksinska, B., Dahlgaard, J. J., & Eklund, J. A. E., (2006). From compliance to value-added auditing—experiences from Swedish ISO 9001:2000 certified organizations. *Total Qual. Manage Bus., 17*, 879–892.

Rohr, A., Luddecke, K., Drusch, S., Muller, M. J., & Alvensleben, R. V., (2005). Food quality and safety—-consumer perception and public health concern. *Food Control, 16*, 649–655.

SENAI (National Service for Industrial Apprenticeship). (2012). *SENAI Actions for Sustainable Development*. Brasília: SENAI/DN.

Silva, A. C. P., & Pepe, V. L. E., (2012). Health surveillance: Field of health promotion and protection. In: Giovanella, L., Escorel, S., Lobato, L. V. C., Noronha, J. C., & Carvalho, A. I., (eds.), Organizers (pp. 709–738). Health policies and system in Brazil. Rio de Janeiro: Fiocruz.

Sofos, J., (2008). Challenges to meat safety in the 21st century. *Meat Science, 78*, 3–13.

Sousa, C. P., (2008). The impact of food manufacturing practices on food borne diseases. *Brazilian Archives of Biology and Technology, 51*, 815–823.

Tondo, E. C., Casarin, L. S., Oliveira, A. B., Martello, L., Silva, Jr. E. A., & Gelli, D., (2015). Advances in food safety in Brazil (Avanços da segurança de alimentos no Brasil). *Vigilância Sanitária em Debate, 3*, 122–130.

UNL (University of Nebraska-Lincoln), (2019). *Principles of HACCP*. HACCP documents and links. URL: https://food.unl.edu/seven-principles-haccp (accessed on 27 September 2021).

Van, T. I., (2008). *A Survey of Process Hygiene and Associated Food Handler Practices in a Retail Group in the Western Cape*, South Africa. PhD thesis, South Africa: School of Agriculture and Environmental Sciences, Central University of Technology, Free State.

Wang, J., & Shanmugan, D. K., (2009). Cutting meat with bone using an ultrahigh-pressure abrasive water jet. *Meat Science, 81*(4), 671–677.

WHO (World Health Organization), (2016a). *About WHO*. http://www.who.int/about/en (accessed on 27 September 2021).

Zach, L., Doyle, M. E., Bier, V., & Czuprynski, C., (2012). Systems and governance in food import safety: A U.S. perspective. *Food Control, 27*, 153–162.

Index

B

D

F

G

H

M

O

P

Q

R

T

U

V

W

Y

Z

For Product Safety Concerns and Information please contact our
EU representative GPSR@taylorandfrancis.com Taylor & Francis
Verlag GmbH, Kaufingerstraße 24, 80331 München, Germany